Mechanical Design of
Structural Materials in Animals

Mechanical Design of
Structural Materials in Animals

JOHN M. GOSLINE

PRINCETON UNIVERSITY PRESS
Princeton and Oxford

Published by Princeton University Press, 41 William Street,
Princeton, New Jersey 08540
In the United Kingdom: Princeton University Press,
6 Oxford Street, Woodstock, Oxfordshire OX20 1TR

press.princeton.edu

Cover image courtesy of Shutterstock

ISBN 978-0-691-17687-1

Library of Congress Control Number: 2017945569

British Library Cataloging-in-Publication Data is available

This book has been composed in Minion Pro and ITC Avant
Garde Gothic

Printed on acid-free paper. ∞

Printed in the United States of America

10 9 8 7 6 5 4 3 2 1

Contents

Preface   ix

SECTION I: BACKGROUND

Chapter 1   **Introduction to Materials Engineering**   3
    1. Nature Builds with Polymers   4
    2. The Vast Majority of Natural Materials Are
        Fiber-Reinforced Composites   5
    3. Biomaterials Exhibit Hierarchical Complexity
        of Structure   5
    4. Biomaterials Are Remarkably Diverse   5
    5. The Quality of Mechanical Design in Animals   8

Chapter 2   **Principles of Materials Engineering**
    and Mechanical Testing   10
    1. Solids– Reversible Deformation and Ideal Elasticity   10
    2. Stress-Strain Curves   12
    3. Ultimate Properties   13
    4. Poisson's Ratio and the Relationship between
        Elastic Moduli   18
    5. Fluids, Flow, and Viscosity   20

Chapter 3   **Viscoelasticity**   22
    1. Hysteresis and Resilience   22
    2. Creep and Stress Relaxation   24
    3. Viscoelastic Models   25
    4. Time-Temperature Superposition   31
    5. Dynamic Mechanical Testing   33

SECTION II: THE STRUCTURAL BASIS FOR MATERIAL PROPERTIES

Chapter 4   **The Structural Origin of Elasticity and Strength**   43
    1. Bond Energy Elasticity   43
    2. The Theoretical Strength of Materials   46

Chapter 5   Fracture Mechanics   50
1. Stress Concentrations   50
2. The Work of Fracture   50
3. The Realized Strength of Materials   53
4. Fracture Toughness   56

Chapter 6   The Molecular Origins of Soft Elasticity   59
1. Flexible Linear Polymers   59
2. The Thermodynamics of Random-Coiled Molecules   61
3. Entropy Elasticity   62
4. The Effects of Cross-Links   66
5. Experimental Measurements   70

Chapter 7   The Molecular Origins of Viscoelasticity   74
1. Diffusion and Entanglement   74
2. Viscosity and Chain Length   76
3. The Glass Transition   79
4. An Example: Elastin   81

Chapter 8   The Design of Composite Materials   84
1. Fiber and Matrix   84
2. The Effects of Fiber Angle   86
3. Reinforcement Efficiency   91
4. The Strength of Composite Materials   93

SECTION III: THE MECHANICAL DESIGN OF TENSILE MATERIALS

**Chapter 9   The Structural Design of Collagen:
Tendons and Ligaments**   103
1. Crystalline Polymers and Tensile Fibers   103
2. The Evolution of Collagen   107
3. Tropocollagen, the Collagen Molecule   107
4. The Assembly of Collagen Fibrils   112
5. The Structural Organization of Collagen Fibers
in Tendons and Ligaments   118
6. Mechanical Properties: Stiffness, Strength, Resilience,
and Toughness   125
7. The Structural Design of Tendons and Their
Fatigue Lifetime   133
8. The Nanomechanics of Tendons and Ligaments   138
9. Echinoderm Ligaments and Mutable
Connective Tissues   144

Chapter 10   The Structural Design of Spider Silks   152
1. The Functional Diversity of Spider Silks   152
2. The Mechanical Properties of Spider Silks   164

3. The Network Structure of Major Ampullate Silks   170
4. Silk Formation in the Gland/Spinneret Complex   174
5. The Functional Design of Spider Draglines   180

SECTION IV: THE MECHANICAL DESIGN OF RIGID MATERIALS

Chapter 11 The Structural Design of Bone   191
1. The Structural Hierarchy of Bone   192
2. Bone Cells   200
3. The Composite Structure of Bone Material   201
4. Nanoscale Composite Models for Bone   206
5. The Mechanical Properties of Bone   214
6. The Adaptations of Bone   225

Chapter 12 The Structural Design of Insect Cuticle   231
1. The Evolution of Insect Cuticle   231
2. The Crystal Structure of Chitin and Cellulose   233
3. The Structure of Chitin Microfibrils   236
4. The Stiffness and Strength of Chitin Microfibrils   238
5. The Organization of Chitin Microfibrils in the
Cuticle Composite   239
6. The Protein Matrix and Its Sclerotization   243
7. The Mechanical Properties of Rigid Cuticle   249
8. Hardness Testing and Nanoindentation Studies   260
9. The Structural Design of Soft Cuticles   263
10. The Functional Consequences of Short-Fiber Composites
in Insect Cuticle   271

SECTION V: THE MECHANICAL DESIGN OF PLIANT BIOMATERIALS

Chapter 13 The Evolutionary Origins of Pliant Biomaterials   277
1. The Evolution of Collagen Fibrils   277
2. Sea Anemone Anatomy and Function   280
3. Mesoglea: Viscoelastic Properties   281
4. Mesoglea: A Composite Model   284
5. Matrix Chemistry   287

Chapter 14 Rubberlike Proteins   290
1. Microfibrillar Elastomers   290
2. The Amino-Acid Composition of Protein Rubbers   298
3. The Amino-Acid Sequence and Cross-Linking
of Protein Rubbers   301
4. The Network Structure and Mechanical Properties
of Protein Rubbers   304

5. The Dynamic Mechanical Properties and Fatigue
of Rubberlike Proteins   308

**Chapter 15 Pliant Matrix Materials and the Design of Pliant
Composites**   314
1. The Mechanical Design of Mucus   314
2. The Mechanical Design of Cartilage   319
3. The Mechanical Design of Vertebrate Skin   327
4. The Mechanical Design of Vertebrate Arteries   337

SECTION VI: CONCLUDING COMMENTS

Chapter 16 Final Thoughts   353

List of Symbols   359

Bibliography   363

Index   375

Preface

"We believe that the study of mechanical design in organisms using the approach of the mechanical engineer and the materials scientist can promote an understanding of organisms at all levels of organization from molecules to ecosystems." Thus wrote Steve Wainwright, Bill Biggs, John Currey, and John Gosline in 1976 in the preface to their classic text *Mechanical Design in Organisms*. Now, 42 years later, Gosline again takes up the challenge of understanding and explaining the way animals are constructed to survive in nature.

This is a book about principles, in particular the design principles that underlie how molecules interact with each other to produce the functional attributes of biological materials: strength, stiffness, the ability to absorb and perhaps store and release energy, and the ability to resist the fatigue of a lifetime of physical insults. Instead of being a review of the literature, the text focuses on those aspects of the literature that elucidate the pertinent theory that explains how molecules are arranged into macromolecular structures and how those structures are then built up into and used by whole organisms. In particular, Gosline develops the theory of discontinuous fiber-reinforced composites, a theory that he innovatively uses to explain the properties of everything from the body wall of sea anemones, to spider silks, insect cuticle, tendons, ligaments, and bones. Although the theories discussed here are treated in depth, Gosline's elegant explanations make them accessible to a broad audience.

Sadly, John Gosline lost his extended battle with cancer before his completed manuscript could be sent to press. To see it through, we have made the minor revisions necessary to accommodate the wishes of two reviewers, and have edited the text to put it into the format required by the press. These changes involved splitting one large chapter into what is now chapters 1–8, standardizing the symbols, and distributing the panels of some composite figures to bring them closer to their discussion in the text. However, none of our revisions in any way change the intent, tenor, or style of Gosline's prose.

Mark Denny
Robert Shadwick

SECTION I

BACKGROUND

Chapter 1

Introduction to Materials Engineering

The science of biology differs from other basic sciences in one important manner. Whereas all fields ask the fundamental questions: *what*, *where*, *when*, and *how*, only in biology do we also ask *why*. This is because the things studied in chemistry, physics, geology, etc., have complex structures that lead to interesting and often important properties, but these things do not have inherent functions in the sense that their properties are organized to achieve specific goals. For example, the physical interaction of photons with a crystal of quartz creates interesting optical effects, but it is difficult to ascribe any purpose to this interaction. In biology, however, we recognize that all living things are organized around a complex set of goals that lead to the growth and reproduction of organisms and to the survival and evolution of species. Further, our observations clearly indicate that the complex structures in organisms have properties that are strongly associated with these goals; that is, they have important functions. Once we accept this notion that function is the essence of biology, we can understand the importance of *design*, which is simply the relationship between the structure and the function of the various components that exist in living systems.

Applied science or engineering provides a strong parallel to biology in that the objects produced in engineering are all inherently functional. That is, synthetic structures have been designed for a specific purpose. Indeed, engineers have spent many years studying how to create good designs, and the *principles of design* derived for synthetic structures can provide us with interesting insights into the way that organisms have solved similar problems. Here we must recognize that the word "design" can be used in two ways. Firstly, design can represent the process by which a functional structure is created, and in this sense engineers are designers. In biology, we recognize that evolution is the designing process, and evolution works in a very different manner from engineering design. Most importantly, for engineers the process of design is a series of conscious decisions about the method of approaching some predetermined goal, and engineering design principles provide a guide for attaining these goals. For evolution as designer there are no predetermined goals, with the exception of survival. Thus, selection of various, possibly random, alterations in the structure of an organism may lead to a change in the organism's function, which then presumably has an effect on the fitness or survival of that organism. The second definition of design may simply be, as stated above, *the relationship between structure and function of a machine or an organism.* This is an important distinction, and this second definition of design is the one that will be used here because we are not concerned with the process of evolution, rather we are interested in understanding how organisms function. In this context, we should recognize that design is a rather neutral term in that there can be good designs and bad designs. Hopefully the design process will lead to the creation of good designs, and a major goal of biomechanics is to use engineering design principles to analyze biological systems to evaluate the quality of biological designs. For example, materials science provides a conceptual basis for understanding the design of structural materials, which can be applied to an analysis of the design of materials that have evolved in plants and animals. This is the focus of the current chapter.

Consideration of the effective use of energy by living organisms is an important unifying principle that can be used to link biomechanics to animal performance, behavior, ecology, and evolution. We pay considerable attention to locomotor energetics in this book. This linkage is generally expressed with the concept of a balanced energy budget for an organism, which states that there must be a balance of energy input and utilization for an animal to survive, reproduce, and prosper. The key to energy budgets is that net energy input (energy intake as food less metabolic waste) must be partitioned between a variety of conflicting uses, including basal metabolic costs, activity costs, somatic growth, and reproductive growth. Somatic growth is necessary for animals to reach sexual maturity, and reproductive growth is necessary to produce offspring for the next generation, and in general animal fitness is directly linked to success in producing offspring. Thus, any reduction in activity costs leaves more energy available for growth and reproduction. Locomotion in most animals is an energy-demanding form of activity, but because feeding usually involves locomotion, it is also rather directly linked to food intake. Thus, biomechanical designs that reduce the energetic cost of movement could play a crucial role in the overall energy balance for an animal by making more energy available for reproduction, hence improving the fitness of the animal.

Biomechanics can also reveal insights into the design of animal skeletons, an approach that treats animals as machines and allows us to develop an understanding of the internal mechanisms that underlie structural support and locomotion. Thus, we can investigate the materials that skeletons are constructed from and the shapes and interactions of the skeletal elements that together form the functional whole of the animal machine. Materials scientists have developed an extensive set of design principles that relate the molecular and microscale structure of materials to functionally important mechanical properties, such as strength, stiffness, and toughness. This is the subject of chapters 2–8. Before we delve into the details of biomaterials design, it will be useful to consider some of the general features of the structural materials that have evolved in living systems.

1. NATURE BUILDS WITH POLYMERS. This statement probably best represents the nature of biomaterials, as there are few, if any, biomaterials that are completely devoid of organic polymers. The polymers are typically protein or polysaccharide in general chemical structure, but the details of the macromolecular structure in any particular material are both complex and elegantly tailored to the specific function of the material in a living organism. One of the dominant themes we will follow is the relationship between biopolymer sequence and three-dimensional polymer conformation.

In the case of structural proteins, which dominate in animal materials, the amino-acid sequence to protein conformation relationship provides the clearest link between genetics and a biomaterial's structure and function. Protein polymers in structural biomaterials are exceptionally complex and diverse, and this diversity of microstructure provides amazing control over material properties. Polysaccharide polymers are dominant in plants but also common in animal materials, although their structural diversity is less than is found in proteins. Again, polymer sequence controls microstructure, and hence determines material properties.

2. THE VAST MAJORITY OF NATURAL MATERIALS ARE FIBER-REINFORCED COMPOSITES. Many of the structural biopolymers mentioned above are assembled into fibrous structures that play dominant roles in the materials where they are found. For example, the protein polymer collagen self-assembles into nanoscale filaments that organize into parallel arrays. In the composite we call bone, collagen filaments are infiltrated by inorganic (calcium phosphate) crystals to form a stiff and strong composite material. Similarly, cellulose molecules self-assemble into nanoscale filaments that associate with other, less structured, polysaccharides or with lignin to form a completely organic fiber-matrix composite, such as that found in plant cell walls. These two examples illustrate a broad diversity of rigid composite materials in biology, but other materials combine protein or polysaccharide filaments with softer matrices that create soft composites, with very different properties.

3. BIOMATERIALS EXHIBIT HIERARCHICAL COMPLEXITY OF STRUCTURE. That is, nanoscale structures, which arise from the self-assembly of individual protein or polysaccharide molecules within or in association with individual cells, assemble into several levels of structural organization in the formation of macroscale materials and structures, such as a tendon or a bone or a hoof. There are at least two possible roles for hierarchical complexity. First, it may provide structural mechanisms for fine-tuning mechanical properties and for controlling anisotropy of properties. Second, the hierarchical organization of structural materials may arise from mechanisms that allow the cellular synthesis of macromolecules to be organized into macroscopic material systems that are orders of magnitude larger than the cells and the molecular building blocks that they synthesize. At present we can recognize hierarchical organization in many structural biomaterials, but we are only beginning to understand the roles that the various levels in the structural hierarchy play in the formation and the function of the hierarchical materials.

4. BIOMATERIALS ARE REMARKABLY DIVERSE. However, they can be grouped into three structural-design motifs. The majority of artificial structures that we encounter every day are essentially rigid. Stationary structures, like classic Greek temples and the vast array of buildings and bridges we see in the modern world, are supported by large columns or beams that are made from rigid materials, such as stone, steel, concrete, or wood. These columns and beams function by resisting compression and bending forces, created largely by gravitational loading, with minimal deformation. Most artificial mobile structures are also rigid and are built largely from rigid materials, but mobile structures, such as airplanes, are frequently built from lightweight,

high-performance materials such as aircraft aluminum and, more recently, carbon-fiber composites. The key feature of rigid-material designs is that they have high stiffness and hence resist deformation under all loading regimes, which means they strongly resist deformation under compression, tension, and shear loading. The horse shown in fig. 1.1A may seem to be supported by rigid columns of bone, but this is a gross over-simplification because animals are highly mobile, and they achieve mobility by having structural systems that are deformable. Individual bones are indeed rigid structures, and are made from a stonelike mineralized organic composite. However, in order to achieve mobility of support, animals employ a wealth of highly deformable components that are made of two very different alternate material-design motifs (fig. 1.1B). These motifs will be called tensile materials and pliant materials, to clearly distinguish them from the rigid materials that are highlighted in fig. 1.1A.

Because most biomaterials can be described as fiber-reinforced composites, we will start by thinking about the design of tensile materials because they are generally constructed as parallel arrays of essentially pure polymeric fibers, so their design is relatively simple. In addition, these tensile materials provide the fiber phase for many of the rigid and pliant composites we see in nature. Tensile elements play a unique set of roles in both artificial and natural structures. Their function, generally, is to be sufficiently stiff and strong to effectively transmit large forces in tension with little deformation and hence little energy storage. At the same time, they must be flexible (i.e., have little resistance to bending) so that the tensile forces can be transmitted, through suitable devices, around corners. The common artificial devices that meet this requirement are called ropes or cables, and sailboats and sailing ships employ numerous tensile structures that play key roles in the boats' propulsive systems. Masts and spars were traditionally made of wood, a rigid biological composite; although in modern vessels they are made from aluminum or carbon-fiber composites. But masts must be supported by "stays," spars must be controlled by "sheets," and sails must be raised by "halyards." In sailboats the ropes that form the sheets and halyards pass through pulleys so that their tensile forces can be transmitted around corners. Pliant composites are not often found in artificial structures, but they are abundant in animals.

If we look inside the horse's forelimb in greater detail (fig. 1.1B) we see a combination of rigid, tensile, and pliant structural materials that form its skeletal support system. The rigid bones do indeed form structural-support columns, but the bones alone do not form a stable column. The individual bones are arranged at angles, and they would simply collapse under their own weight if not stabilized by other structural elements. A dynamically stable vertical column is created by structural components that control the rotation and translation of the angled bones. Tensile structures called ligaments hold the bones together in structures we call joints, so that compressive forces can be transferred from bone to bone. Ligaments prevent bones from sliding apart (i.e., dislocating) because they are rigid in tension, and at the same time the ligaments allow the bones freedom of rotation because they are flexible. Tendons, another tensile structure, attach muscles to bones, and this allows active muscular contractions to stabilize the limb column in static loading and to move the limb in locomotion. Ligaments and tendons are made from the tensile material collagen, in the form of rope-like bundles of essentially parallel collagen fibers. In addition to collagen, silk and silklike protein fibers are also abundant in tensile systems, although collagen is the primary fiber-forming biopolymer in animals.

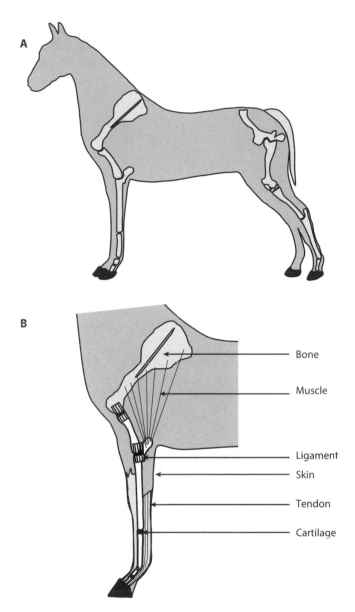

Figure 1.1. Skeletal materials. The bony skeleton of the horse when viewed on its own might lead one to believe that the animal is supported by stone columns, rather like the marble columns of Greek temples. In reality, support is achieved by the combination of rigid elements called bones that are interconnected with and attached to muscles through tensile elements called ligaments and tendons. The whole animal is surrounded by the pliant composite skin, and the contact surfaces of bone are lined with another pliant composite, cartilage.

Bone, the rigid composite material that forms the major structural elements in the vertebrate skeleton, is a composite of collagen fibers infiltrated by mineral crystals of calcium phosphate, but there is a broad diversity of rigid composite materials in biology. The fiber-forming polysaccharide chitin is the reinforcing fiber in the rigid composite that forms arthropod cuticle, and cellulose is the ubiquitous fiber-forming polymer in the plant world, where it forms the reinforcing fiber in plant cell walls. In addition to tensile elements like tendons and ligaments and rigid elements made from bone, the horse employs a broad range of pliant materials for multiple functions, two of which are illustrated here. First, note that the animal is covered with a robust, stretchy skin, a pliant composite made from collagen fibers, the rubberlike protein elastin, and a variety of highly hydrated "matrix" molecules that fill the spaces between these two structural fiber systems. In addition, the contact surfaces where bones meet are lined with another pliant composite, cartilage, which is formed from a three-dimensional mesh of collagen fibers embedded in a proteoglycan gel. Cartilage forms a soft, energy-absorbing surface that transfers compressive forces between bones without allowing the rigid bony surfaces to impact each other. The cartilage also plays an important role in lubricating the sliding of joint surfaces during limb rotation.

5. THE QUALITY OF MECHANICAL DESIGN IN ANIMALS. One notion that is very strong amongst biologists is that organisms have evolved to exhibit outstanding design, perhaps even optimal design. One important goal of the next chapter will be to consider this proposition and evaluate the quality of design in structural biomaterials. In carrying out this evaluation we will need to make quantitative estimates of design quality, and to do this we need to be able to quantify functional attributes of biomaterials and material systems. These functional attributes are called properties and are usually mechanical properties like the stiffness or the strength of a structural material. Engineering theory provides us with a number of properties that we can use to quantify the design quality of biomaterials. The next chapter introduces you to some important engineering properties. In addition, however, we will need to develop other measures based on these properties that will allow us to assess the quality of design in a system. That is, the properties alone are usually not sufficient to quantify design quality.

First, there are many properties that can be defined and measured, but all are not equally important in determining the quality of a design in achieving its function. Thus, it is essential that the functional goals be understood so that appropriate properties are measured. In addition, however, good properties alone do not necessarily provide the answer to the question of design quality because in the construction of a complex machine, like a bird or an airplane, there are many compromises required between complex and often conflicting requirements. As a consequence, it is frequently very useful to analyze design quality in terms of cost-benefit ratios, in which a benefit may be some measurable mechanical output (i.e., an important mechanical property), and a cost may be some important consequence of this output (i.e., a measurable biological, metabolic, or energetic input required to achieve the output).

As you will see, there are many types of cost-benefit ratios, but because the energetic consequences of skeletal design and locomotor performance are so crucially important in biomechanics, we will use *efficiency* as one of the cost-benefit ratios. Efficiency, however, is used much too frequently to describe all aspects of design. We will use efficiency only to define the quality of an energy-exchange mechanism, and efficiency will always be defined as the ratio of energy recovered from an energy-transformation process divided by the energy input to that system. Other measures of design quality may have a similar appearance, as a ratio of some system output (e.g., stiffness, strength, speed, etc.) divided by a system input (e.g., weight, cost of synthesis, metabolic cost, etc.), but you must resist calling these ratios efficiencies. Their units will vary depending on the nature of the inputs and outputs (note that efficiency has no units; it is the ratio of two energies), and we will call these *effectiveness ratios*. As you proceed through this book you will discover quite a number of ratios that can be used to quantify different aspects of the quality of mechanical design in skeletal systems.

Chapter 2

Principles of Materials Engineering and Mechanical Testing

In this chapter, we define the properties of materials that are important in the design of organisms, and explore the tests that allow us to measure these properties.

1. SOLIDS—REVERSIBLE DEFORMATION AND IDEAL ELASTICITY. When an external force is applied to an object made from a solid material (that is, when the object is loaded), it will deform, and, as shown in fig. 2.1, if this material is elastic the extent of the deformation will be in proportion to the magnitude of the applied force, F. Note on the time-deformation graph (fig. 2.1A) that an upward-pointing arrow indicates the application of force, and a downward-pointing arrow indicates the removal of the force. Note also that deformation appears to be instantaneous and that it does not vary with time as long as the force is constant. This time-independent behavior is observed for many rigid objects as long as the time scale for loading and unloading is long relative to the resonant response of the spring-mass system made from this material. Elastic recoil appears to occur instantaneously when the force is removed, and initial dimensions are recovered in full. As a consequence, the energy stored in the deformation can be recovered during elastic recoil. The linear increase in deformation, d, with increasing force (fig. 2.1B) shows the relationship called Hooke's Law, where the spring constant, K, a measure of the stiffness of the object, is indicated by the slope of the force-deformation relationship. That is,

$$F = Kd. \tag{2.1}$$

In order to determine the elastic properties of the material from which an object is constructed, it is necessary to account for the dimensions of that object. This is illustrated in fig. 2.1C, where a rectangular sample is loaded in tension (i.e., stretched) to increase its length. In analyzing this situation we create two terms: stress, σ, which is the force, F, divided by the cross-sectional area, A, and strain, ε, which is the stretch, ΔL, divided by the initial length of the sample, L_0, as follows:

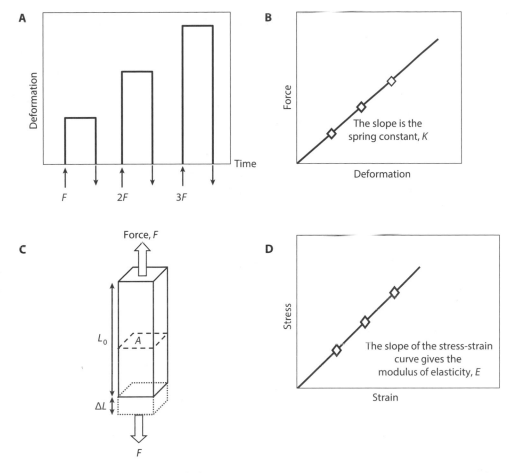

Figure 2.1. Ideal elasticity and stress-strain curves. **Panel A.** Deformation of an elastic object under load. Deformation under a fixed load (*F*; load applied at upward arrows) remains constant with time and recovers completely when load is removed (downward arrows). Deformation is proportional to the magnitude of the load. **Panel B.** The slope of the linear relationship between load and deformation gives the spring constant, *K*, of the object. **Panel C.** In order to quantify the properties of the materials from which the object is constructed, we define terms that quantify the intensity of the force and relative magnitude of the deformation. Stress, which is force (*F*) per unit area (*A* = cross-sectional area), and strain, which is deformation (Δ*L* = stretch) relative to the initial size of the test sample (*L*$_0$ = initial length), are used to create a stress-strain curve, shown in **panel D.** The slope of the stress-strain curve provides the modulus of elasticity, or the stiffness of the material that the object is made from.

$$\sigma = \frac{F}{A}, \qquad \varepsilon = \frac{\Delta L}{L_0}. \tag{2.2}$$

For "ideal elasticity" there is a linear relationship between stress and strain, and the ratio of stress divided by strain provides a measure of the stiffness of the material. That is, when stress is plotted against strain (fig. 2.1D), the slope of the stress-strain curve indicates the stiffness of the material, called the modulus of elasticity, *E*:

$$E = \frac{\sigma}{\varepsilon}. \tag{2.3}$$

The test sample in fig. 2.1C is loaded in tension; thus the modulus would be called its tensile modulus of elasticity. If the arrows indicating the force had been directed inward to compress the object, the stiffness would represent the compressive modulus.

It is important that we understand the units that are used to quantify stress, strain, and stiffness. According to Newton's second law, force, F, is equal to Ma, where M is mass and a is acceleration. In the SI system the unit of force is the newton, and one newton is equal to one kg m s^{-2}. Since the acceleration of gravity is $g = 9.8$ m s^{-2}, one newton is approximately equal to the force that gravity exerts on a 100 g mass. Stress is force per unit area, and therefore stress will have the units of newtons per square meter (N m^{-2}). One newton per square meter is also equal to one pascal (Pa), which is the SI unit for pressure, but can also be used for stress. Therefore, N m^{-2} and Pa are equivalent. The stresses that are relevant for materials testing are usually very large, so it is common to express them in GPa, where 10^9 Pa = 1 GPa. Strain is the ratio of two lengths, and therefore has no units. A strain of $\varepsilon = 0.25$ means a 25% extension. The modulus of elasticity, which is the ratio of stress/strain (i.e., the slope of the stress-strain curve), will have the same units as stress (N m^{-2}). In fact, the magnitude of the modulus can be interpreted as the stress required to achieve a strain $\varepsilon = 1.0$, or to extend the test sample by 100%. However, few materials can be deformed by 100% without breaking.

2. STRESS-STRAIN CURVES. As shown in fig. 2.2, the slope of the stress-strain curve varies with the stiffness of any particular material. Stiff materials, like glass, have

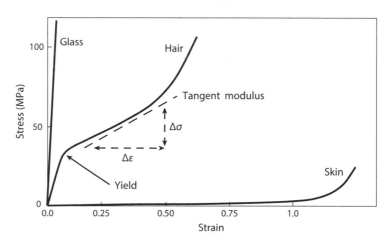

Figure 2.2. Stress-strain characteristics vary among biological materials. Stress-strain curves of real materials can be more complex than the linear stress-strain curve in fig. 2.1D. Hair, made of the protein composite keratin, shows high initial stiffness, followed by a yield where stiffness falls, but, with continued strain, stiffness rises again. These more complex properties arise from major shifts in the structural organization of proteins in the keratin. Skin initially has very low stiffness because it contains a rubberlike protein called elastin, as well as wavy collagen fibers that straighten with extension and cause stiffness to rise at large strains. $\Delta\varepsilon$, change in strain; $\Delta\sigma$, change in stress.

a high modulus and thus have very steep stress-strain curves; less stiff materials, like rubber, have much flatter curves. Note that the stress-strain curve for glass is straight, indicating linear elasticity, whereas the stress-strain curves for two familiar biological materials are distinctly nonlinear. Such nonlinear elasticity is quite common in biological materials.

To quantify the stiffness of nonlinearly elastic materials it is necessary to specify the slope of the stress-strain curve at different extensions, and one way to do this is to use a tangential modulus of elasticity, E_{tan}, which is simply the slope of a line tangent to the stress-strain curve, as shown by the dashed line next to the stress-strain curve for hair in fig. 2.2.

$$E_{tan} = \frac{\Delta\sigma}{\Delta\varepsilon} \tag{2.4}$$

Thus, for some materials, there is not a single value of modulus to define its stiffness, but stiffness varies with the extent of deformation. This is generally the case for biomaterials. In addition, it is common for biomaterials to exhibit rapid changes in stiffness with deformation, such as the yield point indicated for hair. In engineering materials, yield behavior frequently indicates the start of a failure process that leads to fracture. In biomaterials, however, yield often indicates a shift into a different functional regime for the material, not the start of failure processes. In the case of hair, unfolding of helical protein molecules provides an energy-absorbing mechanism that dramatically toughens the keratin from which the hair is constructed.

The stress-strain curve for skin is typical of the behavior of a broad range of pliant composite materials, otherwise known as soft tissues. In this case the very low initial stiffness and high extensibility are due to the presence of a rubberlike protein called elastin. The rise in stiffness at high strain is due to the presence of diffuse and wavy collagen fibers in this composite. At high extension collagen fibers become aligned with the applied load, and stiffness rises because of the tensile stiffness of collagen fibers.

3. ULTIMATE PROPERTIES. In addition to being able to measure the way a material resists the application of external forces, it is also important to determine the conditions that cause a material to fail. Failure can mean many things, depending on the specific function required of a material in a structure, so it is perhaps convenient to consider first the conditions under which materials will break. If we do this, we can define several additional material properties: strength, extensibility, and toughness.

Strength is defined as the maximal stress achieved when a sample of material is loaded until it breaks (indicated by an X in fig. 2.3). If the sample is loaded in tension, then the strength is called its tensile strength, and will be denoted as σ_{max}. Extensibility is defined as the maximal extension (or strain) achieved before a sample breaks, and it will be denoted as ε_{max}. These properties are illustrated graphically in fig. 2.3A.

Toughness is a rather more complex concept that relates to the ease or difficulty of breaking a material. It has a great deal to do with the amount of energy required to cause a material to fracture, but it is difficult to quantify. One way of quantifying toughness is to calculate the area under the stress-strain curve (fig. 2.3B). The area under a force-extension curve will have the units of energy (force, N, times distance, m, = work, expressed in joules, J). Since stress and strain are normalized force and normalized

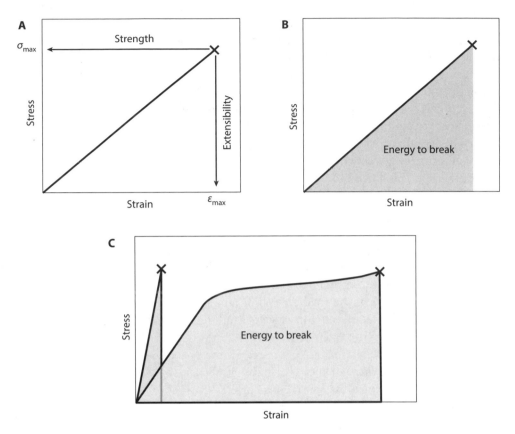

Figure 2.3. Strength, extensibility, and energy to break. **Panel A.** Strength is defined as the maximal stress achieved when a material fails (σ_{max}), and extensibility is the strain achieved at failure (ε_{max}). **Panel B.** The energy to break is determined as the area under the stress-strain curve. **Panel C.** This graph illustrates the properties of two materials that are equally strong but have very different toughness. The material that fails at higher strain absorbs more energy before it breaks and is therefore tougher than the other material.

deformation, the area under a stress-strain curve has the units of energy per unit volume ($J\ m^{-3}$). Because this material has a linear stress-strain relationship, the area under the stress-strain curve is energy per volume to break, ϕ_{break}:

$$\phi_{break} = \frac{\sigma_{max} \times \varepsilon_{max}}{2}. \tag{2.5}$$

For materials with nonlinear stress-strain relationships (fig. 2.3C), the energy to break must be determined by integration. Fig. 2.3C shows that it is possible to have two materials with the same strength, but very different energy to break. This difference in energy to break implies that one of these materials is easy to break, that is it is "brittle," and the other is more difficult to break, that is it is "tough." This graph actually presents a relatively reasonable comparison between the tensile properties of glass, a brittle material, and bone, a reasonably tough biological composite made from protein fibers stiffened by inorganic crystals. There are other properties that are used to define the toughness of materials that have to do with the energy to create new fracture surfaces when a test sample breaks. We will consider these comparisons later when we analyze the process of fracture in greater detail.

It is also common to express the energy to break in terms of the energy per unit mass of material (i.e., J kg^{-1}) as opposed to energy per unit volume (J m^{-3}). This is achieved by dividing energy per unit volume by density (kg m^{-3}). Similar properties based on stiffness or strength that take density into account can provide insights into materials design in mobile structures, where the costs associated with movement may be an important design criterion. This is certainly the case in flying machines, such as airplanes and birds, because greater stiffness or strength per unit mass should directly affect the cost of movement. However, stiffness- or strength-per-density ratios will also be important in assessing the design of static structures, as self-loading in large structures, such as skyscrapers or trees, will put high stress on the structural components at their base, and lower density materials with equal strength or stiffness will be more useful in these structures.

An alternate form of properties that takes density into account is the use of specific properties, where a material property, such as strength, is divided by the specific gravity of the material (specific gravity is defined as the density of the material divided by the density of water). Since specific gravity has no units, this ratio will leave the original units of the material property unchanged, and specific properties provide a convenient way to make comparisons between materials that take account of material densities. Indeed, specific properties provide important effectiveness ratios for assessing the quality of a material where the cost of moving that material may be important for the overall structural design of the object or an organism.

Table 2.1 provides some typical values for the stiffness and strength of several materials that you should be familiar with. This table is provided because the units for modulus and strength can be difficult to understand. For example, it is difficult to interpret what a tensile stiffness $E = 2 \times 10^{10}$ N m^{-2} for bone means, because it is not easy to visualize what 20 billion newtons would do to a piece of bone that had a cross-sectional area of one square meter. Note, however, that the stiffness of tendon collagen is about 1000 times that of a rubber band. Note also that bone and wood are both about 20 times stiffer than tendon. These kinds of comparisons should provide an intuitive feeling for the relative stiffness. The stiffness values are not exact, and the values for strength are even more approximate. More accurate values for relevant biomaterials will be introduced where appropriate in the sections that follow.

In addition, the table provides approximate values for the materials' specific gravity, and the final two columns on the right indicate the specific modulus and specific strength for these materials. The most interesting comparisons here will be between low-density organic biomaterials, such as spider silk or wood, and high-density materials such as steel. Note that the elastic modulus of silk and wood are about 20-fold less than that of steel, but they are within a factor of 2–3 when expressed as specific modulus. The comparison of specific strengths is even more impressive, with dry wood having a specific strength roughly equivalent to that of high-tensile steel, and spider silk having a specific strength that is about 6 times greater than that of the steel.

The stiffness and strength values in this table have been determined using the definitions for stress and strain given in eq. (2.2), in which cross-sectional area and strain

Table 2.1. Mechanical Properties for Selected Structural Materials

Material	Modulus ($N\,m^{-2}$)	Strength ($N\,m^{-2}$)	Specific Gravity	Specific Modulus ($N\,m^{-2}$)	Specific Strength ($N\,m^{-2}$)
Jello	$\approx 10^3$	$\approx 10^2$	1.0	$\approx 10^3$	$\approx 10^2$
Rubber	10^6	10^7	1.3	$\approx 10^6$	$\approx 10^7$
Collagen (tendon)	10^9	10^8	1.3	8×10^8	8×10^7
Spider dragline silk	10^{10}	10^9	1.3	8×10^9	8×10^8
Wood (dry)	10^{10}	10^8	0.5	1.4×10^{10}	2×10^8
Bone	2×10^{10}	2×10^8	2.3	10^{10}	9×10^7
Glass	2×10^{10}	$\approx 10^7$	2.6	8×10^9	$\approx 4 \times 10^6$
Steel	2×10^{11}	$\approx 10^9$	7.8	2.5×10^{10}	$\approx 1.3 \times 10^8$
Diamond	1×10^{12}	$\approx 10^9$	3.5	3×10^{11}	$\approx 3 \times 10^8$

are based on initial dimensions. These definitions are commonly called engineering stress and engineering strain, but it is important to note that there are other definitions for stress and strain that may prove useful for describing the properties of materials that have large failure strains, as sample dimensions can change dramatically at large extension. As a sample is strained in tension to high strain, its cross-sectional area will decrease, and we can define true stress as:

$$\sigma_t = \frac{F}{A},\qquad(2.6)$$

where A is the actual cross-sectional area at each extension. If we define a new term, the extension ratio, $\lambda = L/L_0$ (i.e., $\lambda = \varepsilon + 1$), then if the material deforms at constant volume, the true stress will be:

$$\sigma_t = \frac{\lambda F}{A_0}.\qquad(2.7)$$

Thus, at 100% extension ($\lambda = 2$), the cross-sectional area will be just half of A_0, and the true stress will be twice the engineering stress calculated for the same force. If materials change volume with deformation, which is the case for most rigid materials, the true stress values will have to take into account the magnitude of the volume change.

Similarly, if each increment of extension is referenced to the actual sample length at the start of that increment, then we can integrate all increments to provide true strain, ε_t, as:

$$\varepsilon_t = \int_{L_0}^{L_{final}} \frac{dL}{L} = \ln\frac{L}{L_0}.\qquad(2.8)$$

Fig. 2.4 shows the potential impact of using true stress-strain as opposed to engineering stress-strain data. The stress-strain curves are for a natural rubber band, loaded to failure. The engineering stress-strain curve indicates that the rubber band stretched

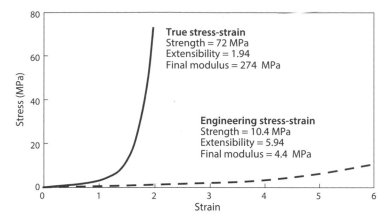

Figure 2.4. Material properties of natural rubber. These material properties are based on engineering stress-strain and true stress-strain curves. Note that true stress-strain increases apparent stiffness and reduces apparent extensibility relative to that calculated with engineering stress-strain values. This results in very large differences in the calculated values for strength and modulus.

about sevenfold before breaking, making the engineering failure strain $\varepsilon_{max} \approx 6$. With conversion to true stress and strain, the properties are dramatically elevated. In this case, strength is increased sevenfold, and the final stiffness is about 60-fold higher. So, for stretchy biomaterials, like skin, artery wall, and spiders' viscid silk, the differences can be significant. Generally, using true stress and strain for materials that stretch by more than about 20% will introduce significant differences in properties, and this includes most pliant materials in biology. Interestingly, calculating the energy to break ($J\ m^{-3}$) from the area under a stress-strain curve is not affected by switching to true stress-strain data because the increase of true stress is balanced by the decrease of true strain. So, the areas under the two curves in fig. 2.4 are identical.

The question arises of which kind of stress-strain values should be used, and the answer is that it depends on the way that the values are to be used. Clearly, for exceptionally rigid materials like steel or concrete, and for that matter for bone and other mineralized biomaterials, that fail at small strains, engineering stress-strain is appropriate for essentially all analyses. Pliant materials, which are abundant in biology, provide numerous opportunities for the use of true stress-strain values. However, high extensibility does not necessarily mean that true stress-strain must always be used. True stress-strain analyses are most useful in assessing the quality of molecular design in structural materials. Building on the example in fig. 2.4 for the properties of rubber, if we were to compare the tensile strength of the rubberlike viscid silk that forms the catching spiral in a spider's web, which fails at strains of about 4, with the nylon-like radial threads in the web, which fail at a strain of about 0.3, comparisons of strength based on engineering stress would indicate that the dragline silk was about five to six times stronger than the viscid silk. Comparison made on the basis of true stress would indicate that the two materials have very similar strengths. This makes the viscid silk the more remarkable of the two materials, because synthetic polymeric fibers such as

Kevlar can be much stronger than spider dragline silk, but there are no natural or synthetic rubbery materials that are as strong as viscid silk. In addition, true stress-strain analyses provide more useful information about the tangential modulus of extensible materials, as we will see when we consider the design of elastic arteries.

However, if we were to use tensile strength to predict the force required to break a structure made from a particular material, such as a spider's dragline or viscid silk thread, engineering strength would provide the most direct route to this answer. Multiplying the initial cross-sectional area of the silk strand by the silk's tensile strength expressed as engineering stress would yield a prediction of the breaking force in newtons directly. Multiplying the resting cross-sectional area by the strength expressed in true stress would greatly exaggerate the breaking force for the viscid silk thread.

4. POISSON'S RATIO AND THE RELATIONSHIP BETWEEN ELASTIC MODULI. The lateral contraction of materials deformed in tension, mentioned above in the discussion of true stress, need not occur at a rate that maintains constant volume, as we assumed for the rubber band example in fig. 2.4. The change in volume with deformation in materials is quantified by the material's Poisson's ratios, which are defined as:

$$v_{xy} = -\frac{\varepsilon_y}{\varepsilon_x}, \quad v_{xz} = -\frac{\varepsilon_z}{\varepsilon_x}, \tag{2.9}$$

where ε_x is the longitudinal strain imposed on the material, and ε_y and ε_z are the lateral strains in the y and z directions. For isotropic materials, such as steel, glass, concrete, etc., where stiffness is equal in all directions, the Poisson's ratios in the y and the z directions are equal, and values typically are in the range from 0.35 to 0.2. This means that these materials increase in volume to varying degrees when they are stretched and decrease in volume when they are compressed. Rubber has a Poisson's ratio of very close to 0.5, which means that its volume does not change significantly as it is stretched or compressed, and thus the rubber band analyzed in fig. 2.4 does actually deform at constant volume. Interestingly, cork, a foamlike biomaterial, has a Poisson's ratio of approximately 0, which means that it does not contract laterally when it is stretched nor expand laterally when compressed.

The type of strain used in eq. (2.9) is usually engineering strain, as this is the strain typically used for common engineering materials. However, it is important to note that the assumption that a Poisson's ratio of 0.5 indicates deformation at constant volume applies only at small strains if engineering strain is used. For higher strains it is necessary to use true strain (Denny, 1988, 182, 194).

Up to this point we have considered mechanical tests that place a sample in tension; that is by the application of a pair of opposing forces that act directly in line with each other to elongate the sample. Alternatively, one can apply in-line forces that push inward onto the sample to load it in a compression test. In this type of test, we apply a compressive stress, which will result in a compressive strain, each defined as above for tensile testing. The stress-strain relationship will yield an initial slope, which will give the compression modulus of elasticity for the material. The compression and tension moduli for many engineering materials are quite similar. In biomaterials, however, this is not necessarily the case. The compression stress-strain curve may remain linear to

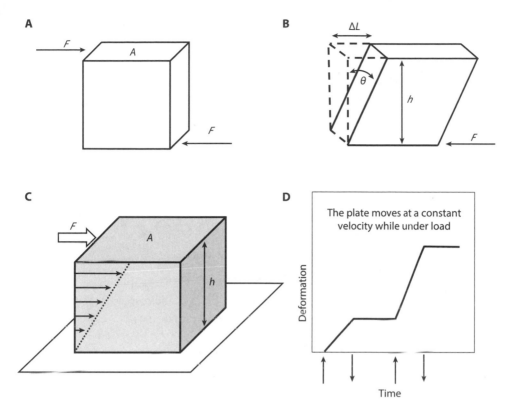

Figure 2.5. Shear deformation of an elastic cube and a fluid cube. **Panel A.** Shearing forces, *F*, applied to the top and bottom of a block made from a solid, elastic material will cause the cube to be deformed in shear, as shown in **panel B**. *A*, area of top and bottom surfaces; *h*, sample height; Δ*L*, lateral shift of top surface; *θ*, angle created by lateral shift. In **panel C** we see a block of fluid resting on a solid surface and capped with a solid plate, *A*. If a shearing force is applied to the plate, the fluid below the plate will be dragged to the right by the moving plate, and this movement will cause a linear velocity gradient (indicated by arrows) to develop within the fluid. **Panel D**. A constant force, *F*, will cause the top plate to move at a constant velocity that is proportional to the magnitude of the force. There is no elastic recoil in this system because the shearing of fluid layers will dissipate the mechanical energy applied to move the top plate into heat through frictional processes in the fluid's flow. *Upward arrows*, load applied; *downward arrows*, load removed.

failure or may show more complex behaviors, and compression strength and failure strain may be quite different from the values observed for the same material in tension.

The tension and compression tests described above are modes of deformation in which a pair of opposing forces act directly in line with each other. However, if the two opposing forces are arranged so that they do not act directly in line, but one is offset relative to the other, then the material under load will experience deformation in shear. This situation is illustrated in fig. 2.5A and B. The off-axis forces (*F*) acting on top and bottom surfaces of the block of material will create a shear stress, *τ*:

$$\tau = \frac{F}{A}, \tag{2.10}$$

where A is the area of the top and bottom surfaces. This shear stress will shift the top surface laterally relative to the bottom surface by a distance ΔL, and will cause a shear strain, γ:

$$\gamma = \frac{\Delta L}{h}. \tag{2.11}$$

That is, the shear strain is the distance that the top surface moves relative to the bottom surface, normalized by the sample height (h), which is also the tangent of the angle (θ) created. For low strains, $\gamma \approx \theta$ when θ is expressed in radians. If the material tested is elastic, then it should be possible to create a stress-strain plot from this test, and the slope of the stress-strain curve will be the shear modulus of elasticity, G:

$$G = \frac{\tau}{\gamma}. \tag{2.12}$$

Although we will not be dealing extensively with the shear deformation of solid structures in animal skeletons, shear deformation does occur in solid mechanics during torsional loading.

For isotropic materials, as described above, the tensile (and compressive) modulus is related to the shear modulus through the Poisson's ratio, as follows:

$$G = \frac{E}{2(1+\nu)}. \tag{2.13}$$

Thus, for rubber ($\nu \approx 0.5$) $E = 3G$, and for steel ($\nu \approx 0.3$) $E = 2.6G$. Biological materials are rarely isotropic because of their complex, fiber-reinforced composite structure, and as a consequence Poisson's ratios vary in different directions depending on the organization of reinforcing fibers in these composites. Bone, for example, is considered to be an orthotropic material, which means that it has different elastic moduli in three principal axes relative to a typical bone's tubular structure. This means that six different Poisson's ratios are needed to describe bone's material behavior. The situation becomes even more complex in pliant composites, which typically deform at constant volume because the tissue volume is essentially determined by the water constrained within the polymer and fiber mesh that makes up the composite. However, Poisson's ratios can vary dramatically from 0.5 in different directions because of the complexity introduced by the orientation of reinforcing fibers and because fibers can reorient with strain. Thus, multiple strain-varying Poisson's ratios are needed to document the changes in material properties observed as pliant composite materials are deformed.

5. FLUIDS, FLOW, AND VISCOSITY. If a "block" of fluid is placed between two flat parallel solid surfaces, and shearing forces, F, are applied to the top and bottom surfaces (fig. 2.5C), the fluid will be sheared and will exhibit the phenomenon of viscous flow. That is, the top surface will be displaced relative to the fixed bottom surface, and the fluid within will be sheared. There is a large difference between the shear deformation of an elastic solid and that of a fluid. This can be seen by comparing the time-deformation plots in fig. 2.1A and fig. 2.5D. Again, the upward-pointing arrows

indicate the time when force is applied, and the downward-pointing arrows indicate the removal of that force. Fluids loaded in shear show several important characteristics that distinguish them from solids (fig. 2.5D). First, the deformation is not instantaneous, but it occurs continuously, at a constant rate or velocity (u) under a constant load. Second, the plate's velocity increases with an increase in the magnitude of the shearing force. Finally, there is no elastic recovery when the force is removed; deformation is permanent.

The movement of the plate creates a velocity gradient in the fluid, as indicated by the arrows on the front face of the fluid block in fig. 2.5C, and this creates a shear strain rate, $\dot{\gamma}$, that is equal to the velocity, u, of the top plate divided by the fluid thickness, h:

$$\dot{\gamma} = \frac{u}{h}. \tag{2.14}$$

Shear strain rate has the units of s^{-1}. In the process of shearing a fluid, mechanical work, W, input by the external force, F, is equal to Fx, where x is the horizontal movement of the top plate, and this work is transformed into heat by frictional processes between the layers of fluid in the velocity gradient that is formed. In most instances of fluid flow, as illustrated above, the shear strain rate is proportional to the shear stress, $\tau = F/A$. That is, there is a linear relationship between shear stress and the shear strain rate, the slope of which provides a new material property, viscosity, η:

$$\eta = \frac{\tau}{\dot{\gamma}}, \tag{2.15}$$

which is a measure of how strongly a fluid resists deformation under shear loads. This linear behavior is called Newtonian viscosity, and the units for viscosity are Pa s. For some fluids, notably those containing polymeric molecules, viscosity may vary with strain rate. The concept of viscosity is crucial for animal locomotion in fluids and will be discussed in detail in chapter 3. For our current discussion of materials design, the key features to keep in mind are that the force required to create fluid movement rises as velocity increases, and fluid movements show no elastic recoil because all energy required to deform the fluid is dissipated as heat in frictional interactions between the layers of fluid molecules. In many biomaterials, shearing forces are created when macromolecules change their shape or when fibers reorient. When this happens, energy is dissipated as heat through internal friction. That is, many solid materials, particularly those that are constructed from polymeric materials, dissipate mechanical energy as heat when they deform, and energy dissipated through friction is not available for elastic recoil. Materials that exhibit this kind of behavior are called viscoelastic, and viscoelastic materials are common in biology.

Chapter 3

Viscoelasticity

Most biological materials and some engineering materials exhibit a combination of elastic and viscous properties. This aspect of materials design is complex, but we will start with two relatively simple ideas: mechanical hysteresis and time-dependent elasticity.

1. HYSTERESIS AND RESILIENCE. Mechanical hysteresis, H, arises from energy dissipation by viscous processes, and it is a measure of the energy dissipated in a load-unload cycle of a solid material. Fig. 3.1A shows how the results of a stress-strain test can be used to determine hysteresis. In this case, the sample is strained (in tension, compression, or shear) at a constant strain rate, and stress is recorded continuously, as opposed to the discrete load increments illustrated in figs. 2.1 and 2.5. The three graphs indicate the energies associated with the extension and elastic recoil of a material sample. The shaded area in the first panel indicates the energy per unit volume required to extend the material, ϕ_{in}, and the shaded area in the second panel indicates the energy recovered in elastic recoil when the material is unloaded, ϕ_{out}. These two energies are not equal, and as a consequence energy is lost as heat in the cycle, as indicated in the third panel by the shaded loop, ϕ_{lost}. The energies per unit volume obtained from this test can be used to calculate the hysteresis of the material, H, a measure of the fraction of energy lost, as well as the material's resilience, R, a measure of the efficiency of elastic-energy storage, as follows.

$$H = \frac{\phi_{lost}}{\phi_{in}},$$

(3.1)

$$R = \frac{\phi_{out}}{\phi_{in}}.$$

(3.2)

Hysteresis and resilience are material properties that quantify the efficiency of energy transformations, and they can be used to assess the quality of materials that have evolved to function in energy storage and exchange systems. For example, collagen in the limb tendons of terrestrial vertebrates is known to function in the storage of elastic energy in running locomotion, and load-unload cycles on limb tendons indicate that collagen's resilience can be as high as 0.95 (i.e., 95% efficient). This is one of the features of collagen that makes it such an effective material in skeletal structures. Energy

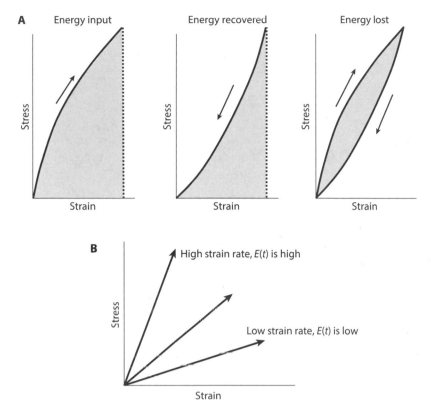

Figure 3.1. Viscoelasticity of solid materials. **Panel A**. The results of load-cycle tests on a sample of a solid material. The first graph shows the energy input during the extension of the sample as stress rises, and the shaded area under this curve represents the energy per volume, ϕ_{in}, required to strain the sample. The second graph shows the unload portion of the load cycle, and the shaded area under the curve shows the energy per volume recovered as the sample recoils to its initial dimensions. The third graph combines the load and unload curves to show the full load cycle, and the shaded area within the loop indicates the energy per volume dissipated by viscous processes during the load-unload cycle. Ratios of these energies provide the resilience and the hysteresis of the material being tested, as defined in eqs. (3.1) and (3.2). **Panel B** shows a different manifestation of viscoelasticity in materials. In this case a sample of material is deformed at constant strain rate, but the three lines indicate the behavior of the material in three different tests, at low, medium, and high strain rate. Note that stiffness is higher when strain rate is high. This rise in stiffness is a clear indication that viscous processes are occurring at the molecular level. $E(t)$, time-dependent modulus, or stiffness.

dissipation is clearly not good for springlike tendons, but other materials may function in mechanisms where energy dissipation is important. The silks found in spider orb webs, for example, are known to be very poor energy-storing materials; their resilience being ≈ 0.35, and thus their hysteresis is ≈ 0.65. This means that 65% of the energy required to strain the silks is lost as heat, leaving only 35% of the energy to power elastic recoil. Clearly, these silks would not work well as replacements for collagen-based tendons, so why do they work well in spider webs? Orb webs are used as aerial filters in the capture of flying insects, and their key mechanical function is to absorb the kinetic energy of the flying insects without breaking, so that the spider can rush out and immobilize its prey. High strength and extensibility, leading to an exceptional energy to break, are also key properties for web silks, but if the silks had high resilience, the energy absorbed by the web would be stored, and elastic recoil might catapult the prey back out of the web.

Another feature of viscoelasticity arises from the fact that viscous forces in fluids rise with the velocity of shearing, or more correctly as the shear strain rate, $\dot{\gamma}$, increases. Thus, if there are elastic and viscous components within a material, then the material's stiffness should vary with strain rate. This is illustrated in a very simple way in fig. 3.1B, which shows three stress-strain curves for a single material sample that is deformed at three different but constant strain rates. At high strain rate the material is stiffer than it is at lower strain rates, because internal viscous forces are larger at high strain rates. It is also possible that other properties, such as strength and extensibility, could be influenced by strain rate, but we will deal with those later. It is important to note that these kinds of time-dependent material properties are manifest in many different and complex ways. Strain rate is just one way that internal viscous forces are modulated. The viscosity of fluids is strongly influenced by temperature. That is, increased temperature lowers fluid viscosity, and decreased temperature raises viscosity. As a consequence, viscoelastic properties are also temperature dependent. Temperature dependence can also occur from phase changes that arise from the melting of polymeric crystals and of polymeric glass, such as Plexiglas, and both of these events are affected by changes in the solvent environment within the polymer. So the time-dependent properties illustrated in fig. 3.1B can arise from all of these parameters (strain rate, temperature, and solvents) in various combinations. Time-dependent elasticity is exceptionally complex and is also profoundly important in understanding the quality of design in structural biomaterials.

2. CREEP AND STRESS RELAXATION. There are a number of experimental protocols that can be used to document viscoelastic behavior in materials. All tests for viscoelasticity must quantify stress, strain, and "time" in some appropriate manner. The simplest approach is to design an experiment where either stress or strain is held constant, and the other variable is measured over time, commonly called a creep test. The experiment illustrated in fig. 3.2A is a creep experiment, in which a constant stress (σ_0) is applied to a sample, and strain is followed over time, $\varepsilon(t)$. The example shown here indicates that strain rises rapidly at the application of stress, but over time the rate of strain declines gradually, and strain eventually reaches a fixed value after some long period of time. If the stress is removed, the material recoils quickly at first, but it requires a long time to recoil completely to its initial length. This kind of behavior is characteristic of a viscoelastic solid. Energy is stored and elastic recovery is observed, but it is elasticity that is delayed by viscosity. We can use the data in fig. 3.1 to calculate the material's time-dependent modulus, $E(t)$, or stiffness. That is, we can calculate $E(t)$ from the known applied stress (σ_0) and the observed time-varying strain $\varepsilon(t)$, as follows:

$$E(t) = \frac{\sigma_0}{\varepsilon(t)}, \tag{3.3}$$

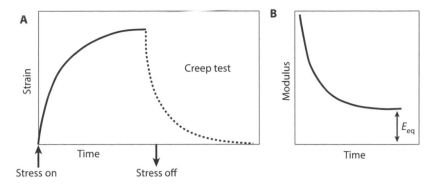

Figure 3.2. Viscoelastic materials can creep. This figure shows another manifestation of viscoelastic behavior in a solid material: viscoelastic materials can show changes in properties over time because viscous processes may allow sample strain to continue over time when external forces are constant. In this example, called a creep test, a constant stress is applied to a sample, and strain is plotted with time. **Panel A.** Initially, strain rises rapidly, but with time the strain rate falls, reaching a constant strain after some long period of time. When the stress is removed, there is a reverse of this process, with recoil that is rapid initially, but slowing as the sample recoils to its initial dimensions. **Panel B** shows how a time-dependent elastic modulus, $E(t)$ would appear for this material, as described in eq. (3.3). This curve only applies to the loading portion of panel A. At the start of the experiment the material appears to have high stiffness, $E(t)$, because viscous processes delay the deformation of the material. As time progresses, however, the strain rate falls and $E(t)$ reaches a constant level determined by the inherent elastic stiffness of the system. E_{eq}, equilibrium modulus.

and fig. 3.2B shows the form of the modulus-versus-time behavior we would obtain from these creep data. At short times the strain is low, but the strain rate is high (i.e., strain is rising rapidly, fig. 3.2A). This implies that the initial high stiffness is due to viscous processes that play a prominent role in this material. As time proceeds the strain rate falls, and the sample approaches a fixed strain at long times. This equilibrium strain indicates that the true elastic stiffness of the material has been reached, and that at this point in time the viscous processes no longer contribute to the material's behavior.

3. VISCOELASTIC MODELS. To more fully understand the origins of viscoelasticity we will consider simple computational models that describe some of the behaviors shown in fig. 3.1. Fig. 3.3 shows how we can assemble elastic models (fig. 3.3A: a spring with modulus E) and viscous elements (fig. 3.3B: a dashpot with viscosity η) in various ways to generate viscoelastic "mechanisms." The spring element can be thought of as a coil spring, and it represents ideal elastic behavior. The dashpot is a plunger in a fluid-filled tube that represents purely viscous behavior. These models can be assembled in various ways to illustrate different aspects of viscoelasticity. Fig. 3.3C shows a spring arranged in parallel with a dashpot, creating a Voigt model that can be used to describe the behavior of a viscoelastic solid in an experimental protocol, a creep test, as in fig. 3.2A. In this case, however, it is possible to demonstrate how the relative magnitudes of

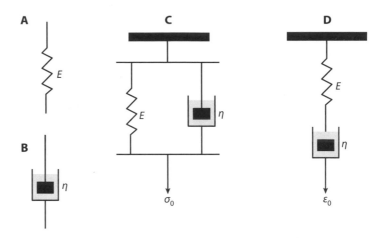

Figure 3.3. Viscoelastic models. **Panels A and B** show models for ideal elastic and ideal viscous behavior, where the elastic model is a spring, and the viscous model is a fluid-filled dashpot. E, modulus; η, viscosity. **Panel C.** The form of a Voigt model, which is a dashpot placed in parallel with a spring. This model illustrates the creep behavior of a viscoelastic solid. It is assumed that the applied stress, σ_0, will remain constant over time, and this will cause the spring and the dashpot to strain equally over time. Initially the strain will be small because the dashpot has not had time to extend, but as time proceeds the dashpot extends, as does the spring. At long times the spring carries all of the applied stress, and the system reaches its elastic equilibrium. **Panel D.** A Maxwell model, where the dashpot is in series with the spring. This model illustrates the stress-relaxation behavior of a viscoelastic fluid. It is assumed that the applied initial strain, ε_0, will be distributed between the spring and dashpot and will change over time, causing the stress to fall over time. At short times the spring will be maximally strained because the dashpot will not have had time to extend, and the stress will be high. As time progresses, however, the dashpot will extend and the spring will contract, causing the stress to fall, ultimately to zero as all the energy input in the initial stretch is dissipated by the dashpot.

elasticity (E) and viscosity (η) influence the behavior of the system. Recall that the units of elastic modulus are Pa, and the units of viscosity are Pa s. Thus the ratio of the viscosity divided by the elasticity has the units of s, and this ratio defines the viscoelastic time constant of the system, t_{ve}, in s as:

$$t_{ve} = \frac{\eta}{E}. \tag{3.4}$$

The time-dependent creep strain of this model is described by:

$$\varepsilon(t) = \frac{\sigma_0}{E}\left[1 - e^{-\frac{t}{t_{ve}}}\right]. \tag{3.5}$$

In this model a stress, σ_0, is applied at time = 0 and remains constant thereafter. With this model both the spring and the dashpot must strain together, and the form of the creep function is shown as the solid line in fig. 3.4A, plotted on a linear time scale.

At time zero all of the stress acts to extend the dashpot, and the model strains rapidly because the spring is initially unstretched and therefore cannot take up any of the load. As the system extends, however, a rising portion of the stress is transferred to the spring, and the strain rate falls exponentially, until at long times the spring carries the entire load. Thus, the equilibrium strain at long times is due entirely to the stiffness of the spring, and energy stored in the spring will allow the model to recoil to its initial dimensions, as illustrated in fig. 3.2A. Fig. 3.4B shows the creep response of the same Voigt element plotted on a $\log_{10}$ time scale. This figure demonstrates the utility of using logarithmic time scales in documenting viscoelastic properties. Very short times are greatly expanded, and long times are compressed, and as we will see later, viscoelastic material properties can vary across enormous time scales. It is therefore common to use logarithmic time scales to document these behaviors. The time-dependent modulus, $E(t)$, for the Voigt model can be calculated with eq. (3.3), and the form of this curve is shown in fig. 3.2B.

Fig. 3.3D shows a Maxwell model, which contains a spring and a dashpot in series, and this element can be used to model the experimental protocol called stress relaxation. The model is initially deformed to fixed strain level, ε_0, at time $t = 0$, creating an initial stress, σ_0, and the stress required to maintain ε_0 is followed as a function of time, $\sigma(t)$. The time-dependent relaxation of this model is described as:

$$\sigma(t) = \sigma_0\left(e^{-\frac{t}{t_{ve}}}\right). \tag{3.6}$$

Again, the time constant, t_{ve}, is the ratio of viscosity, η, divided by elasticity, E. The behavior of the Maxwell model in stress-relaxation is plotted with a dashed line in fig. 3.4A on a linear time axis, and in fig. 3.4B in on a logarithmic time axis. At the start of a stress-relaxation test, following the initial strain, all of the stress is carried by the stretching of the spring because at very short times the dashpot will not have time to extend. However, as time proceeds, the tension in the spring extends the dashpot, and

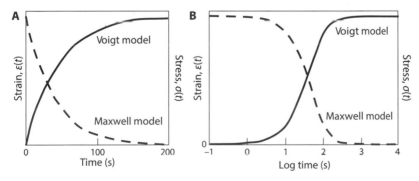

Figure 3.4. Comparing Voigt and Maxwell models. **Panel A.** The time responses of the Voigt and Maxwell models as a function of linear time, from time = 0 to 200 s. The *solid line* shows the strain, $\varepsilon(t)$, behavior for the Voigt model, and the *dashed line* shows the stress, $\sigma(t)$, behavior for the Maxwell model. **Panel B.** The Voigt and Maxwell model behaviors in panel A are plotted here on a $\log_{10}$ time scale, indicating the model's behaviors from 0.1 to 10,000 s.

the stress declines exponentially to zero because the extension of the dashpot allows the spring to recoil to its resting length. Because there is no long-term elastic behavior with this model, it is used to illustrate the properties of viscoelastic fluids. Now, with time-varying stress and constant strain, it is possible to quantify the time-dependent elastic modulus of this model, as:

$$E(t) = \frac{\sigma(t)}{\varepsilon_0}. \tag{3.7}$$

Eqs. (3.3) and (3.7) suggest that creep and stress-relaxation experiments produce identical time-dependent modulus values for viscoelastic solids and fluids. This is qualitatively correct, but in fact the creep and stress-relaxation protocols produce somewhat different values for viscoelastic moduli, and this can be demonstrated by the use of the standard linear solid (SLS) model shown in fig. 3.5A. The SLS has three elements, two springs and a dashpot, which in the case of the model shown here can be described as a Voigt model in series with a spring. The SLS can be used to model both creep and stress-relaxation, and fig. 3.5B shows the response of an SLS model in these two situations. In the model behavior illustrated here, the spring E_1 is stiffer than E_2. In a creep test the model shows a small instantaneous strain at short times, which arises from the stretching of spring E_1. The dashpot restricts strain on spring E_2 at short times, but as time progresses the dashpot extends, allowing spring E_2 to extend to its equilibrium length. In stress-relaxation, the model develops an initial high stress that arises from the instantaneous stretching of the stiff spring, E_1. With time the dashpot allows spring E_2 to extend slowly to an equilibrium strain. Now, when the strain data from the creep test and the stress data from the stress-relaxation test are used to calculate the time-dependent modulus, $E(t)$, using eqs. (3.3) and (3.7), we see in fig. 3.5C that the two tests do not produce identical $E(t)$ profiles. This is because the time constants in creep and in stress-relaxation are not identical, and the stress-relaxation modulus falls more rapidly than the creep modulus.

The solution to this problem is to define different material properties for these two test protocols. In stress-relaxation we can use time-dependent stress to measure time-dependent stiffness, $E(t)$, as:

$$\sigma(t) = E(t)\varepsilon_0. \tag{3.8}$$

Using a similar form of equation, we can use time-dependent strain to measure a new property called time-dependent compliance, $D(t)$, as:

$$\varepsilon(t) = D(t)\sigma_0, \tag{3.9}$$

where the units of compliance are the inverse of those for stiffness—i.e., Pa^{-1}. The theory of linear viscoelasticity provides methods for converting properties between stiffness and compliance, but these are beyond the scope of this overview. However, this distinction between two seemingly very similar experimental protocols provides an important insight for biologists who want to investigate the functional role of a structural tissue in an organism. Very likely the best test protocol is one that closely mimics the mode of deformation that occurs in the natural system, and we will see other experimental test

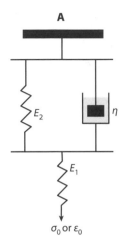

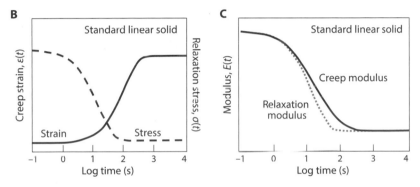

Figure 3.5. The standard linear solid. **Panel A.** A standard linear solid model, which essentially combines Voigt and Maxwell models, as the dashpot in this model is both in parallel and in series with springs. E_1, E_2, moduli; η, viscosity; ε_0, initial strain; σ_0, applied stress. **Panel B.** The creep and stress-relaxation behavior of the standard linear solid model, illustrated in panel A, where the *solid line* shows the strain of the model in a creep test, and the *dashed line* shows the stress of the model in a stress-relaxation test **Panel C.** Although the creep and stress-relaxation curves look like the inverse of one another, the viscoelastic moduli, $E(t)$, calculated from these two tests are not identical.

protocols that are more useful in structural systems that function under rapidly changing load/strain regimes. We will explore these other test methods in chapter 7; however, it will be useful to consider whether these simple models on their own can accurately model real materials.

The short answer is that the simple models are useful in illustrating concepts of viscoelasticity, but they should not be taken as exact models for real materials—although they are in some rare cases. To illustrate this, fig. 3.6A shows a comparison of creep data for a sample of bovine knee cartilage plotted together with a calculated creep response for a Voigt model. The stiffness and viscosity parameters for the Voigt model have been adjusted to center the model response on the time and strain scale of the experimental data. Note that the majority of the Voigt model's strain occurs over two logarithmic decades of time (between about 1 and 100 s), whereas the cartilage creep response occurs over more than four decades of log time. This is the case for most polymeric systems, and viscoelastic models that mimic real systems must be constructed from multiple response units. For example, fig. 3.6B shows a model constructed from four Maxwell models arranged in parallel. Each Maxwell model has its own stiffness and viscosity, providing a broad range of relaxation times, and the responses of all of the model elements are summed to provide a composite response. It is important to note that all elements in this composite model must strain together; that is, the two horizontal lines that the elements are attached to must remain parallel as they move apart. The calculated stress-relaxation profile for this model, $E(t)$, is plotted in fig. 3.6C. The stiffness of the springs and the viscosity of the dashpots have been adjusted to produce an $E(t)$ that matches reality.

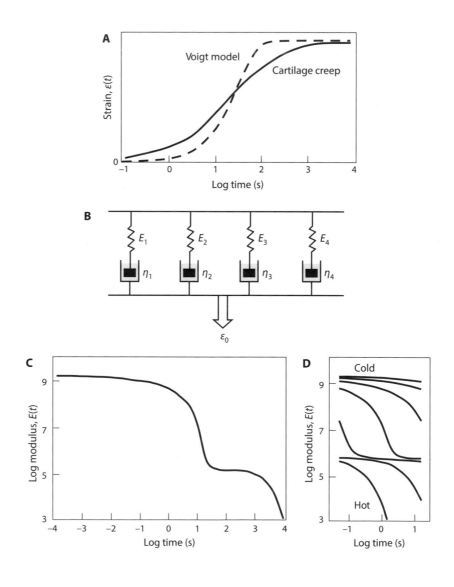

4. TIME-TEMPERATURE SUPERPOSITION. This brings us to another important aspect of viscoelastic properties and their measurement. It is rarely possible for a single mechanical test to quantify viscoelastic properties over the enormous time scale needed to document the full behavior of polymeric materials. One of the most useful ways of expanding the time scale of viscoelastic measurements is to use the "time-temperature superposition principle," which allows the use of temperature shifts to expand time scales in experiments to quantify the viscoelastic properties of materials. Again, calculations based on simple models will allow us to understand this process. The entire procedure arises from the observation that the viscosity of fluids is strongly influenced by temperature. That is, making fluids hotter reduces their viscosity, and this means that they will flow faster. The stiffness of solids, on the other hand, is relatively unaffected by changes in temperature, particularly if temperature is well below the melting point of the solid. Thus, in the simple Voigt and Maxwell viscoelastic models, which contain a spring to represent an elastic solid and a dashpot to represent a viscous fluid, the time constant for the deformation of the model is determined by the ratio of the dashpot's viscosity divided by the spring's elasticity, as shown in eq. (3.4). If we lower temperature, viscosity will increase, which means that the dashpot will extend more slowly. Thus, we will have to wait longer for the model to extend. Conversely, if we raise temperature and lower viscosity, the model will extend more quickly.

Now, we apply this principle to the four Maxwell elements in the model shown in fig. 3.6B. Fig. 3.6D shows calculated properties for the model in fig. 3.6B, but in this

Figure 3.6. (*left*) The viscoelastic response of real polymer systems. **Panel A** compares the creep-strain response of a Voigt model with the experimental results of a creep test carried out on a sample of bovine knee cartilage. Note that the time axis spans five decades of $\log_{10}$ time, ranging from 10^{-1} to 10^4 s. The time constant of the Voigt model has been adjusted so that it matches that of the cartilage material, and it is clear that the real material response covers a much wider time scale. Redrawn from Setton et al. (1993). **Panel B** shows a viscoelastic model that contains four Maxwell models arranged in parallel. If this model is subjected to a stress-relaxation protocol, it will exhibit a viscoelastic behavior that is determined by the full range of time constants assigned to the Maxwell models. E, elastic modulus; η, viscosity; ε_0, initial strain. **Panel C** shows the stress-relaxation response of the model in panel B with time constants that range from 0.002 to 200 s, and the model response is plotted on a graph with log-log axes. That is, both the modulus and the time are plotted on $\log_{10}$ axes because modulus values cover six decades of log stiffness and time spans eight decades of log time. This behavior is quite typical of many real polymer systems, and the broad range of time scales makes it difficult for experimenters to measure these kinds of behavior in a single experiment. **Panel D.** It is possible to infer the mechanical behavior of viscoelastic materials over a broad time range by combining experiments carried out over relatively narrow time scales with other experiments over the same time scale but at different temperatures by employing the time-temperature superposition principle, as described in the text.

case the properties are only calculated for a narrow time span from 0.1 to 10 s. The calculations, however, are repeated seven times with changes in the magnitude of time constants for each element. As suggested above, the time constants were changed by altering only the dashpot viscosity of each element, by the same factor. The spring stiffness for each element remains unchanged, and the shift in viscosity would have been created by changes in temperature. Fig. 3.6D shows seven curve segments, each generated with a different temperature-shifted time constant. If we assume that the data in fig. 3.6C were collected at room temperature, then the curve segments at the top of fig. 3.6D were calculated for very low temperatures that have dramatically increased the viscosity of all four dashpots. Thus, we are seeing high stiffness behaviors in the one-second time range in fig. 3.6D that were previously seen in the millisecond time range in fig. 3.6C. That is, the stiffness is high and relatively constant over the time range plotted in fig. 3.6D. Thus, lowering temperature and hence raising the viscosity of all elements shifts viscoelastic behaviors to longer times, and the system as a whole is relaxing more slowly from its initial high stiffness. Curve segments at the bottom of fig. 3.6D were calculated for elevated temperatures (decreased viscosity) and hence with short time constants. This shifts the behaviors seen at very long times in fig. 3.6C to the one-second time range in fig. 3.6D.

Thus, fig. 3.6D provides a dissection of the full stress-relaxation response of this material into short, overlapping segments obtained by carrying out seven tests that span a short and convenient time range over a broad range of temperatures. Now we can see the benefit of this dissection because it is possible to reverse this process and assemble the full stress-relaxation response by shifting the individual segments horizontally until they overlap with their adjacent neighbors. To do this we would shift the top curve segments to the left to short times, and the curve segments at the bottom to the right to long times. By doing this we will construct what is called a "master curve" for the polymer that spans a time range that is orders of magnitude broader than the limited time range of individual tests at each temperature. In the example here, a viscoelastic response spanning eight decades of log time and six decades of log modulus is constructed from experiments carried out over two decades of log time, but at a range of temperatures. This is the essence of the time-temperature superposition principle.

Other interesting aspects of the properties shown in fig. 3.6C are the major shifts in stiffness seen over this broad range of time. At very short times (10^{-4} s) stiffness is high, at about 3 GPa, which is the typical elastic modulus for glassy plastics like Plexiglas, or polystyrene. However, at longer times (or higher temperatures) stiffness falls dramatically by about three orders of magnitude in a process known as the glass transition. This phase of the viscoelastic response curve was modeled with three Maxwell elements that each had a 2 GPa spring, but they had dashpots with varying viscosities that together created the fall in stiffness associated with the glass transition. The next feature of the full curve is the plateau in stiffness starting at about 20 s, leading into a second fall in stiffness starting at about 1000 s. This plateau and fall was created by a fourth Maxwell element with a 0.5 MPa spring and a dashpot with a very high viscosity, creating a relaxation with a time constant of 2000 s. Because the model has only Maxwell elements, we know that this material is ultimately a viscoelastic liquid, and the stiffness will relax to zero rapidly beyond 10^4 s.

The midpoint of the glass transition falls at about 10 s on the time scale in fig. 3.6C, as this is the time where $E(t)$ has fallen about halfway through the full glass transition. If these properties were for Plexiglas or polystyrene, we could estimate the temperature at which this master curve was created, as follows. The glass-transition temperature for these polymers is about 100°C, and since the midpoint of the time scale of the graph is at 1 s, we might guess that this master curve was created for a temperature somewhat below the glass-transition temperature, perhaps at about 80°C. If we were to document the behavior of this polymer at normal room temperatures, where it behaves as a rigid elastic solid, we would have to shift this curve to the right by many decades of log time, shifting the fully elastic behavior seen at 10^{-4} s in the graph to time regimes of the order of 10^{3} s.

The two kinds of springs used in this model—stiff gigapascal springs and soft megapascal springs—arise from two fundamentally different molecular mechanisms. Elasticity in rigid structural materials arises from the deformation of chemical bonds between atoms. Elasticity in soft-polymeric materials arises from changes in the shape of large, flexible polymer molecules. We will deal with these elastic mechanisms when we consider the molecular origins of material properties in chapters 4 and 6.

5. DYNAMIC MECHANICAL TESTING. As we found in chapter 2, stress-relaxation and creep testing provide useful experimental methods for determining the behavior of material systems that function in long-term loading regimes, but they are not as useful in instances when structures are loaded cyclically, over short time periods. The skeletal structures associated with moving limbs in animals may experience periodic loading regimes as short as 10^{-3} s in insect flight, and as long as tens of seconds in terrestrial and aquatic locomotion. Cyclic-loading experiments provide an excellent way to characterize the quality of the materials used in such dynamic systems. There are two basic types of testing protocol for dynamic properties; one employs resonant oscillations of spring-mass systems and the other employs forced oscillations. Resonant systems are relatively simple to create, but their frequency span is limited to the natural resonance of the apparatus. Forced-oscillation methods, often called dynamic mechanical analysis (DMA), allow measurements across a broad range of frequencies, and for this reason they are most commonly used.

Fig. 3.7A illustrates the basic organization of a resonant system, which contains a material sample, in this case a vertically oriented beam, anchored at one end to a solid support and carrying a moving mass at the other. The mass is pushed up by a distance, y_0, and it is released to start its resonant oscillation. The system will oscillate at its resonant frequency, ω_n, expressed in rad s^{-1}, which is determined by the square root of the ratio of the beam's stiffness, k, and mass, M, of the moving object, as:

$$\omega_n = \sqrt{\frac{k}{M}}, \tag{3.10}$$

or, alternately, for frequency in cycles per second (f_n, Hz) as:

$$f_n = \frac{1}{2\pi}\sqrt{\frac{k}{M}}. \tag{3.11}$$

If the beam is made from a perfectly elastic material, it will oscillate at its natural frequency with no decline in amplitude over time. However, if the beam is made from

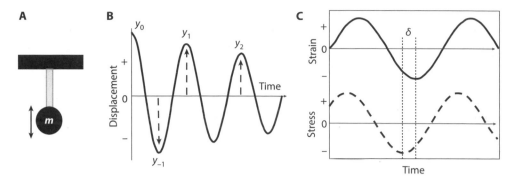

Figure 3.7. Dynamic mechanical testing of viscoelastic materials. **Panel A** shows a test sample for a damped free-oscillation test of a viscoelastic material. The gray sample is clamped to a rigid support at the top and to a moving mass (M) at the bottom. If this model is set into motion by pushing the mass upward and releasing, it will oscillate up and down at the resonant frequency of the spring-mass system. **Panel B** shows the decay in the amplitude of oscillation caused by the viscoelasticity of the sample material. Tracking the fall in the amplitude of this oscillation will provide information about the viscoelastic properties of the material at the oscillation frequency, as described in the text. **Panel C** shows the sinusoidal oscillations in stress and strain of a test sample in a forced-oscillation test system. Both stress and strain traces are constant-amplitude sine functions, but the stress trace is shifted in advance by a phase angle, δ, relative to the strain trace. The magnitude of the phase shift provides a direct indication of the viscous contribution to the properties of the material.

a viscoelastic material, the amplitude of oscillations will decline over time, and this decline allows us to calculate the resilience of the beam's material (fig. 3.7B). The logarithmic decrement, Δ, is used to quantify the decline in amplitude for one full cycle of the oscillation, and it is defined as:

$$\Delta = \ln\frac{y_1}{y_2}. \tag{3.12}$$

The equation that describes the decline of deflection, y, over time for a resonant viscoelastic beam is:

$$y = y_0 \cos(2\pi\omega_n) \exp(-\Delta\omega_n t). \tag{3.13}$$

The resilience, R, of the beam material can be calculated as the ratio of squared amplitudes of successive peaks, and it is often specified as:

$$R = \frac{y_2^2}{y_1^2}. \tag{3.14}$$

However, in this case the resilience is calculated over two load-unload events, and it is useful to use the ratio of amplitudes per half-cycle as this ratio quantifies the efficiency of a single unload-to-reload event. That is, the movement from y_{-1} to 0 involves elastic recoil of the loaded beam from a negative maximum to the midpoint, where all the elastic potential is converted either into heat through internal friction or into kinetic energy of the moving mass. Then as the mass moves through the midpoint to the opposite extreme at y_1, the kinetic energy is transformed into the elastic deformation of the beam and into heat through internal friction. For this single load-unload event, resilience is calculated as:

$$R = \frac{y_1^2}{y_{-1}^2}.$$ (3.15)

The log decrement, Δ, used for this example was set at 0.05, and the resilience per half-cycle is 0.82, or 82% elastic-energy recovery. The disadvantage of this kind of test is that it requires a resonant system, which is normally found at a single frequency, although it is possible to change the mass and/or stiffness of the test object to change the resonant frequency to some degree. However, it is much easier to apply forced oscillations created by an external driver, where oscillation frequency can be controlled directly.

Forced-oscillation methods typically use electromechanical devices to deform test samples with small-amplitude sinusoidal movements over a broad range of frequencies, and electronic transducers are used to record the sinusoidal deformation and loading force, as shown in fig. 3.7C. Data analysis involves measuring the ratio of amplitudes of the strain (ε_0) and stress (σ_0) waveforms, which provides a measure of dynamic stiffness. That is, the dynamic strain applied is:

$$\varepsilon = \varepsilon_0 \sin \omega t,$$ (3.16)

where ω is the frequency in radians per second. For ideal elasticity the stress developed with this sinusoidal strain would be perfectly in phase with the strain, but for viscoelastic materials there is a phase shift, δ, between stress and strain because the viscous component of stress is proportional to deformation velocity, which varies with $\cos \omega t$. The combination of sine and cosine waves at the same frequency is a phase-shifted wave of different amplitude at the same frequency, giving stress as:

$$\sigma = \sigma_0 \sin (\omega t + \delta).$$ (3.17)

The ratio of the stress and strain amplitudes defines a complex modulus, E^*, as:

$$\frac{\sigma_0}{\varepsilon_0} = E^* = E' + i E'',$$ (3.18)

and this complex modulus is composed of an in-phase elastic component and an out-of-phase viscous component, called respectively the storage modulus, E', and the loss modulus, E''. Storage and loss moduli can be calculated as:

$$E' = E^* \cos \delta, \quad E'' = E^* \sin \delta.$$ (3.19)

The ratio of the storage and loss moduli gives the loss tangent, or damping coefficient, of the material:

$$\frac{E''}{E'} = \tan \delta.$$ (3.20)

The graph in fig. 3.8 shows the viscoelastic behavior of the rubberlike protein elastin, which provides soft elasticity in vertebrate tissues such as arteries and skin (Gosline

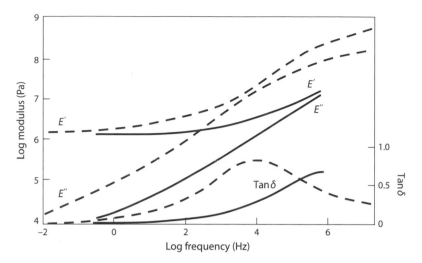

Figure 3.8. Results of dynamic tests on the elastic protein elastin. The *solid lines* show the behavior of elastin at the in vivo level of hydration; the *broken lines* show its behavior at reduced water content. These curves span many decades on the frequency scale, and they were created by using the temperature-shift procedures, as described in fig. 3.6. At full hydration elastin functions as a highly resilient (low tanδ) rubbery material, with a storage modulus of about 1 MPa. This behavior extends from the lowest frequencies plotted (ca. 0.1 Hz) to about 50 Hz, but at frequencies above this the material becomes stiffer and more viscous. At reduced hydration (*dashed lines*) the viscoelastic behavior is shifted about two decades of frequency to the left, and the full curve now spans nine decades of log frequency. *E'*, storage modulus; *E"*, loss modulus; tanδ, loss tangent.

and French, 1979). This particular elastin came from the elastic ligament that provides support in the neck of a cow. The graph presents two master curves, each created by the temperature-shifting process described in fig. 3.6C and D, and the data span nine decades of log time and five decades of log modulus. Both master curves were created by shifting data to a reference temperature of 36°C, close to the body temperature of the animal. So these data should be relevant to the function of elastin in the animal.

In this graph short-term phenomena are on the right because the "time" axis is actually a frequency axis. That is, at $\log_{10}$ of frequency equal to 6 we are looking at the behavior of elastin being vibrated at 10^6 cycles per second, or once every microsecond. Long-term behavior is on the left of the graph, where the $\log_{10}$ of frequency is −2, and a single sinusoidal oscillation takes 100 s. The key feature of this graph is the transition from a rigid glass-like state at high frequency to a soft rubberlike state at low frequency. The data plotted with solid lines show the behavior of fully hydrated elastin, as it would exist in the living animal. We see that the storage modulus (*E'*) is about 100 times greater than the loss modulus (*E"*) at the low-frequency end (ca. 1 Hz), where elastin functions in the animal. At frequencies above about 100 Hz, stiffness rises rapidly, and the loss modulus is nearly as large as the storage modulus at 10^5 Hz. This rise in stiffness indicates the entry into the glass transition, which is seen more clearly in the data set plotted with broken lines. These data were obtained from elastin samples that had been partially dehydrated by the removal of about 15% of their bound water. This water plays a crucial role in maintaining the molecular mobility that allows elastin to function as a rubberlike material, and the loss of the water enhances the internal viscosity of elastin. Now, we see that the span of the dynamic properties extends to about 10^7 Hz, and the entire curve appears to have been shifted to the left by about two log decades on the frequency scale. This is a very clear example of how both temperature and solvent content can affect viscoelastic properties. We will consider the molecular origins of these shifts in properties in the next chapter.

Elastin functions in the major arteries providing an elastic tube that expands with each beat of the heart, and this elasticity plays a key role in minimizing the work of the heart. The ability of elastin to function as a dynamic spring is documented by the $\tan\delta$ data, which are plotted at the bottom of fig. 3.8. Again, the solid line is for fully hydrated elastin and the dashed line is for partially dehydrated elastin. At frequencies of the order of cardiac frequencies (ca. 1 Hz) $\tan\delta$ is about 0.02, whereas for the partially dehydrated elastin at 1 Hz $\tan\delta$ is about 0.16. This suggests that there may be a large loss in the resilience of elastin if it dehydrates, and this is indeed the case.

In certain instances, $\tan\delta$ can be used to calculate the resilience of a viscoelastic material. Fig. 3.9 shows dynamic stress and dynamic strain plotted against each other to form a stress-strain loop, which is elliptical in shape but in other regards looks like the load-unload cycles that we used in fig. 3.1 to determine resilience in constant-strain-rate load cycles. In fig. 3.9, however, the load cycle has been adjusted by applying a prestrain 0.05 to the sample, so that it oscillates entirely in tension, between strains of 0 and 0.1. The stress oscillates around 0.05 MPa, and the sample is maintained in tension throughout the load cycle. The reason for using pretension is that soft tissues typically are made from filaments of collagen and elastin, which can be stretched in tension, but which collapse in compression. The inner, dashed, loop is calculated for $\tan\delta = 0.1$, a situation that roughly matches the conditions for fully hydrated elastin tested at about 100 Hz. The solid outer loop has been drawn for $\tan\delta = 0.5$, which is roughly the value for the partially dried elastin at 100 Hz.

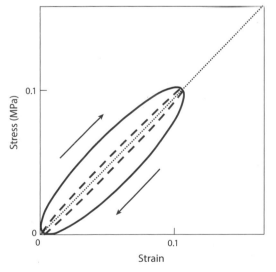

Figure 3.9. Quantifying hysteresis. A stress-strain plot of the elliptical hysteresis loop created when stress and strain are plotted against each other, creating a loop whose area is equal to the energy per unit volume dissipated per cycle. Dynamic test data like these can be used to calculate the resilience of the material, as described in the text.

By eye we can see that there are large differences in the size of the loops as $\tan\delta$ increases from 0.1 to 0.5, and these imply large differences in resilience. However, it is not intuitively clear how to measure the areas to calculate resilience at high values of $\tan\delta$ because the peak stress and peak strain do not occur at the same place on the loop. However, as long as $\tan\delta$ is less than about 0.1, it is possible to calculate resilience directly from it with reasonable accuracy. For samples that are loaded alternately in tension and compression around zero strain (as in fig. 3.7C), resilience per half-cycle can be calculated as:

$$R \approx \exp(-\pi \tan\delta). \tag{3.21}$$

Interestingly, if we consider the case illustrated in fig. 3.9, where the stress-strain loop is offset so that the sample is strained entirely in tension (or compression), we can calculate the resilience of this system as:

$$R \approx \exp\left(-\frac{1}{2}\pi \tan\delta\right). \tag{3.22}$$

Again, this equation applies as long as $\tan\delta$ is about 0.1 or less. It is possible to determine resilience from stress-strain cycles with $\tan\delta > 0.1$, but this requires numerical integration of appropriate areas within and under the load and the unload arms of the cycle (see Ker, 1981; Lazan, 1968).

Dynamic mechanical testing involves the measurement of stress and strain amplitudes, as well as the phase difference between strain and stress. Modern computer-based signal processing makes the analysis of these periodic waveforms quite easy, and therefore the measurement of dynamic properties is now a standard routine for the characterization of viscoelastic materials. However, there is one major limitation to this method: it really only works at small strains, because the properties defined (i.e., E', E'', $\tan\delta$) do not actually specify the magnitude of the applied strain; it simply disappears into the calculated moduli. The analysis presented assumes linear viscoelastic behavior over the strain ranges employed in dynamic tests, and the range of linear viscoelastic behavior may be much smaller than the full extensibility of the material in question. For extensible materials, like elastin and other rubberlike biomaterials that can be stretched by 100% or more, linear viscoelastic behavior is seen up to strains of 0.1 to 0.2. For rigid polymeric materials linear viscoelasticity is limited to strains of the order of 0.01.

What linear viscoelasticity means is that if a sample is loaded dynamically over a range of low strain amplitudes and/or prestrains, the dynamic properties will not change. However, at high strains many biomaterials show dramatic shifts in mechanical behaviors, as illustrated in fig. 3.10, which shows constant-strain-rate tests of mammalian elastin obtained over a broad range of strain rate, temperature, and water content. At full hydration, high temperature, and low strain rate, elastin is a soft rubber, but as hydration levels are reduced, temperatures are lowered, and strain rates are increased elastin is shifted through the glass transition into a rigid polymeric glass. Clearly, the shapes of these stress-strain curves vary dramatically. In particular, the curves seen in various stages of the glass transition show abrupt changes in slope that reflect major shifts in the molecular mechanisms involved. All of these tests involve strain to failure, and many of the changes in curve shape reflect differences in the mechanisms responsible for failure. The analysis of failure processes is, however, beyond the scope of

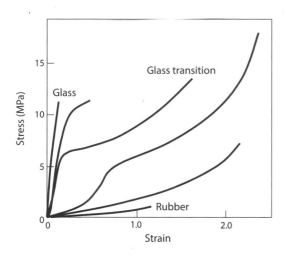

Figure 3.10. The spectrum of visco-elastic behavior. Results for constant-strain-rate tests to failure on porcine aortic elastin samples tested at a variety of strain rates, hydration levels, and temperatures to document the full range of viscoelastic behavior in this material. Stress-strain curves at the bottom right, labeled *rubber*, represent tests under in vivo conditions of high temperature, full hydration, and modest strain rate, and we see low stiffness and strength, with a failure strain of about 1. The other extreme, labeled *glass*, represents tests at low temperature, reduced hydration, and high strain rate, and we see greatly increased stiffness and strength, but greatly reduced extensibility. In between these extremes, in the area labeled *glass transition*, we see increased stiffness, strength, and extensibility, which together indicate that the material elastin becomes quite tough in the middle of its glass-transition zone.

linear-viscoelasticity theory. We will deal with these issues separately in a later chapter. Other changes in curve shape are not associated with failure processes, but rather arise from structural transitions in polymeric materials that expand the functional response of a biomaterial. We have already seen one example of this kind of transition in fig. 2.2, where we identified a yield zone where helical protein structures uncoiled to toughen keratin. We will see more when we look at the design of the three classes of biomaterials in later chapters.

SECTION II

THE STRUCTURAL BASIS
FOR MATERIAL PROPERTIES

Chapter 4

The Structural Origin of Elasticity and Strength

This section deals with the principles of design of structural biomaterials, and we decided at the beginning that the word "design" will be taken as the relationship between the structure and the function. Up to this point we have concentrated on identifying properties that can be measured for structural materials, with the assumption that the properties quantify the functional attributes of materials, and we have seen that there are many functional attributes that we can assign to materials. As we will see in later chapters, appropriately defined material properties can be used to assess the quality of design in structural biomaterials. Now we want to consider the structural origins of the material properties that we have defined. First, in this chapter we will consider the origins of elasticity and strength in rigid materials, where properties arise from the nature of the chemical bonds that hold atoms and molecules together in these materials. Then, in succeeding chapters we will consider the origins of elasticity and viscoelasticity in polymeric materials, where elasticity and viscosity are determined by the ability of organic polymer molecules to change their shape without necessarily breaking the covalent bonds between the monomer units that create the structural polymers.

1. BOND ENERGY ELASTICITY. We begin with a consideration of the structural origins of ideal elasticity of rigid materials; to be specific, we will explore bond-energy elasticity. The linear, reversible, ideal elasticity illustrated in fig. 4.1 is characteristic of a number of common engineering materials, such as glass and steel; however, this linear behavior is typically seen only at very small strains. The structural origins of elastic stiffness (i.e., modulus of elasticity) and material strength can be traced directly to the properties of the chemical bonds that hold individual atoms together in these materials. Fig. 4.1 shows the principles of how chemical bonding is related to material properties. Fig. 4.1A shows the basic features of bond energies and bond forces between a pair of ions that are held together by electrostatic attraction; that is, one ion has a positive charge and the other has a negative charge. At equilibrium, the two atoms are separated by a bond distance, r_0, which is at the bottom of the potential-energy curve created by the sum of the attractive energy due to the electrostatic attraction between the ions and the repulsive energy that develops when the two ions move closely together so that their electron

clouds overlap significantly. This bond-potential curve, $U_{net}(r)$, can be modeled with equations of the following form, where r is the separation distance of the two atoms:

$$U_{net}(r) = -\frac{A}{r} + \frac{B}{r^n}. \tag{4.1}$$

The first term on the right is the attractive energy, which in the case of electrostatic forces falls with r^{-1}. This term has a negative sign because as the ions move together their electrostatic attraction brings them to a lower energy state. The second term is the positive, repulsive energy that falls with r to some large negative power that is somewhere in the range of r^{-9}. The exponent of this term varies for each kind of bond interaction and with the nature of the individual atoms involved. The magnitude of this exponent can range from about 7 to 12, and it must be determined experimentally. The depth of the trough in this curve indicates the magnitude of the bond energy for this ion pair, and breaking the bond requires that the ions are moved apart by a sufficient distance that $U_{net}(r)$ rises to zero.

The application of this bond-energy relationship to mechanical behavior requires that we calculate the force required to displace one of the ions either closer to or farther away from its opposite, and this can be done by taking the first derivative of the bond-energy equation above, as:

$$F_{net}(r) = \frac{A}{r^2} - \frac{nB}{r^{n+1}}. \tag{4.2}$$

The force-displacement curves for this ion pair are shown in the lower graph in fig. 4.1A. The solid line indicates the net force, F_{net}, and the two dashed lines indicate the attractive (above) and repulsive (below) forces that determine the force-extension behavior of this bond. Note that the electrostatic force rises quite gradually as the ions move together, but it remains above zero at quite large separations. The repulsive force, on the other hand, rises rapidly as the ions move together, and it falls rapidly to zero as the ions move apart. The net force, F_{net}, shows us the elastic behavior of this bond, which we can translate into the elastic behavior of materials. Before we do this, however, it is important to note that although this analysis is for ionic bonds, a similar analysis can be made for covalent or metallic bonding processes and thus can be applied in general to materials of all types.

Fig. 4.1B shows the F_{net} versus bond length plot on a larger scale so that we can see more clearly the origins of the stiffness and strength of materials. First, note that at r_0 the force rises linearly in tension (r increasing) and decreases linearly in compression (r decreasing), as indicated by the dashed line. This is the origin of linear elasticity as described by Hooke's law. Note also that in tension the linear behavior does not extend very far; that is, quite soon the slope of the force curve in tension starts to decline. Elastic behavior in materials is quantified by stress and strain, and we can estimate the strain on the bond easily, as follows. The initial length, L_0, of this "material" is the equilibrium bond length, r_0, and with the aid of the dashed line we can see that the linear behavior extends for only a few percent of increase in r (ΔL) before the slope begins to drop. So, linearly elastic, rigid materials are only linear at small strains.

The stress requires that we determine the force per unit area, and it is obvious what the force per bond is: it is $F_{net}(r)$. The area is less obvious, but if we assume that this

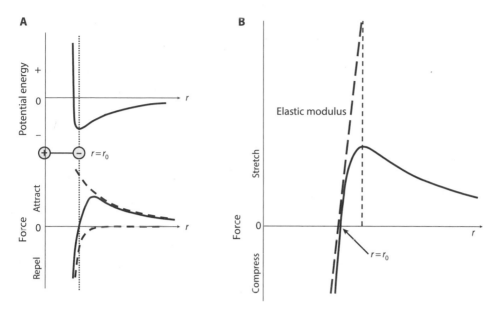

Figure 4.1. Bond energy elasticity and the strength of materials I. **Panel A** (*top graph*) shows the bond-energy potential between a pair of ions that are held together by electrostatic attraction. At equilibrium, the two ions are separated by a bond distance, r_0, which is at the bottom of the potential-energy curve created by the sum of the attractive energy due to the electrostatic attraction between the ions and the repulsive energy that develops when the two ions move closely together so that their electron clouds overlap. The *bottom graph* shows the form of the attractive force (*upper dashed line*) and the repulsive force (*lower dashed line*), which together form the bond-force curve (*solid line*). **Panel B** shows the bond-force curve in greater detail. Note that this curve is straight as it crosses the zero-force line, and it is this feature of chemical bonding that gives bond-energy-elastic materials their linear elastic properties in both tension and compression. However, as bond length increases significantly beyond the zero-force equilibrium position the slope of the force curve declines, and the force reaches a maximum at an apparent bond strain of about 0.25. This peak in bond force sets the maximal extension and strength of a bond-energy-elastic material.

ion pair is within a block of solid material with a crystalline structure, we can imagine the number of bonds, with known separations, r_0, that make up a given cross-sectional area of the material. So, taking F_{net} as a surrogate for stress, we can see that this curve provides a great deal of information about the behavior of a material made from these atoms. The important feature of the F_{net} curve is that at about 25% extension of the bond from its rest state r_0, the force reaches a peak, and this peak clearly indicates the point where a bond will fail, even though additional energy is required for the force to fall to zero. The logic here is that once the externally applied force reaches the peak of the force curve, no additional force is needed to continue the extension of this bond because the ion being pulled is now more bonded to the pulling object than it is to the other ion in the diagram.

2. THE THEORETICAL STRENGTH OF MATERIALS. Now, let's consider the strength of this material, and we will begin with the diagrams in fig. 4.2. Here we see a crystal lattice of atoms, held together by ionic bonds of length r_0 (dashed lines). This could be a crystal of sodium chloride, with bond length $r_0 = 0.28$ nm and bond energy of 756 kJ mole^{-1}. One approach to estimating the strength of this crystal is to estimate the energy required to split it in two along a single plane of atoms, as shown in fig. 4.2B. This represents the smallest number of bonds that need to be broken to break the crystal. We can estimate the strength of the crystal by calculating the mechanical energy input in a stress-strain test required to form the two new surfaces that are created when the crystal is broken, as follows. The area under a tensile-stress-strain curve equals the energy per unit volume, ϕ_{break}, required to break a test sample (fig. 2.3B and eq. [2.5]), as:

$$\phi_{break} = \frac{\sigma_{max} \times \varepsilon_{max}}{2}. \qquad (4.3)$$

We can also rearrange eq. (4.3) to express strain, ε, in terms of stress and modulus, as $\varepsilon = \sigma/E$, and then we can calculate ϕ_{break} in terms of stress and modulus, as:

$$\phi_{break} = \frac{\sigma_{max}^2}{2E}. \qquad (4.4)$$

Since ϕ_{break} is energy per unit volume, we can multiply both sides of the equation by the length of the test sample to give us energy per unit cross-sectional area, and if we choose the length to be the bond length, r_0, we can calculate the energy required to break all the bonds in one layer in the crystal, as shown in fig. 4.2B.

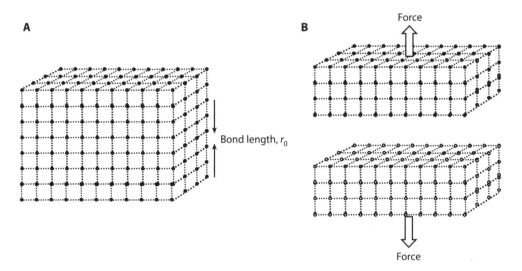

Figure 4.2. Bond energy elasticity and the strength of materials II. **Panel A** shows a diagram of a small portion of a crystalline lattice of atoms, which has been fractured in **panel B**. The newly created, atomically smooth fracture surface represents the smallest number of bonds that must be broken to fracture this material sample, as is used in the analysis of theoretical strength of bond-energy-elastic materials. In addition, the creation of one side of the fracture surface requires an energy, γ_{surf}, which is the surface energy of the material.

$$\frac{\text{Energy}}{\text{Area}} = \frac{\sigma_{max}^2 r_0}{2E}. \tag{4.5}$$

Energy per unit area is a measurable quantity that can be determined for solids when they are heated near their melting point. It is essentially the same as the surface tension determined for liquids, as it defines the energy required to create a unit of new surface area of a solid, γ_{surf}. In the case of fracture, two new surfaces are created, and thus we can we can write:

$$\frac{\text{Energy}}{\text{Area}} = 2\gamma_{surf} = \frac{\sigma_{max}^2 r_0}{2E}, \tag{4.6}$$

and we can solve for the stress, which will provide us with an estimate of the theoretical tensile strength of the material, $\sigma_{max,th}$:

$$\sigma_{max,th} = 2\sqrt{\frac{E\gamma}{r_0}}. \tag{4.7}$$

Table 4.1 shows us the results of calculations based on this equation for a variety of materials. Columns three and four give inputs of surface energy and modulus, and the fifth column shows the theoretical strength of each material. Interestingly, the values for the ratios of the modulus to the theoretical strength vary somewhat, but on average this analysis calculates the theoretical strength as about one-quarter of the tensile modulus, E. The two columns on the right-hand side provide values for measured strength and modulus divided by strength, and they tell a very different story. Each of the six materials has calculated values for bulk samples, but in the case of iron, glass, sodium chloride (NaCl), and zinc there are additional values for special cases of these

Table 4.1. Material Surface Energy and the Strength of Rigid Materials

Material	Form	Surface Energy ($J\,m^{-2}$)	Modulus, E (GPa)	Theoretical Strength, $\sigma_{max,th}$ (GPa)	$E/\sigma_{max,th}$	Measured Strength, σ_{max} (GPa)	E/σ_{max} (GPa)
Iron	Bulk, cast	2	210	46	4.6	0.2	1050
	Whisker					14	**15.0**
Glass	Bulk	0.54	80	14	5.0	0.07	1140
	Fiber					4	**20**
NaCl	Bulk	0.115	44	8	5.5	0.05	850
	Whisker					1.1	**40**
Copper	Bulk	1.65	120	31	3.9	0.14	860
	Whisker					3	**40**
Zinc	Bulk	0.75	90	18	5.0	0.07	1290
Aluminum	Bulk	0.9	73	18	4.1	0.06	1220

Data from Martin (1972) and Brenner (1956).
Bold indicates the special cases of glass fibers and "whiskers" of several materials, as discussed in the text.

materials. Looking at the modulus/strength ratios for the bulk samples, we see that they are of the order of 1000. This suggests that, on average, bulk samples of real materials break at stresses that are hundreds of times less than predicted by the theoretical analysis above. Why are these materials so weak? Now look at the special cases of glass fibers and "whiskers" of several materials. Here we see that the modulus/strength ratio falls dramatically, with the measured modulus/strength ratio approaching the theoretical ratio to within a factor of three in some cases.

Before we consider these special cases, it will be useful to consider whether the bond-energy information in fig. 4.1B can give us any additional insights into the theoretical strength of materials. This diagram has not been calculated for any particular ionic bond pair, but its general form can provide useful information about the strength of materials. First, note that the form of the force-extension profile is not at all similar to the linear stress-strain relationship assumed in eq. (4.3) for the analysis above. The dashed line labeled "elastic modulus" is the form of the stress-strain behavior assumed in the analysis of theoretical strength above, and this line clearly does not fit the $F_{net}(r)$ behavior for the bond. Rather, the $F_{net}(r)$ curve rounds off and reaches a maximum at a strain of about 0.25. That is, the peak of $F_{net}(r)$ occurs at $r \approx 1.25r_0$, and, as mentioned above, this force maximum clearly indicates the failure point for this bond. Interestingly, because the analysis above indicated that the theoretical strength of materials is about one-quarter of the elastic modulus, we can use the $F_{net}(r)$ graph to see how well the calculated values for theoretical strength fit with the bond-energy profiles. If a linear-elastic material fails at a stress that is about one-quarter of its modulus, it must fail at a strain of about 0.25. Thus, the dashed line in fig. 4.1B should be approximately equal to the stress-strain relationship for the theoretical strength calculation used above; however, in this case it is the linear force-extension behavior for a single bond within the material sample. Note that this line reaches a force that is more than twice the maximum force predicted in the $F_{net}(r)$ curve at $r \approx 1.25r_0$. Thus, the linear stress-strain analysis is likely overpredicting theoretical strength values, and we can more reasonably estimate that the theoretical strength should be at about one-tenth, rather than one-quarter, of the elastic modulus.

With this adjustment in hand, consider the strength of the special cases for iron and glass in Table 4.1. Bulk cast iron is very brittle and fails at quite low stress, but under appropriate conditions it is possible to make iron crystallize into very thin crystalline whiskers (i.e., short, thin fibers). When whiskers with diameters of the order of $2\,\mu$m are bent under a microscope, their strength can be calculated from the minimum radius of curvature of the whisker before it breaks. Strengths under these conditions can reach 14 GPa. Similarly, sodium chloride and copper whiskers and glass fibers made commercially for use in fiber-reinforced composites have tensile strengths in the gigapascal range. Of particular note are glass fibers, with diameter of about $9\,\mu$m and a tensile strength of >4 GPa, that are used commercially in fiber-reinforced composites. What is it about these thin filaments that allows them to be 50–100 times stronger than bulk objects made from the same molecular structures? The answer is that the failure "mechanism" shown in fig. 4.2B does not describe how macroscopic objects actually break. It is very unlikely that a failure will occur on a single plane of atoms, where all bonds will fail simultaneously. Rather, failure is initiated by flaws or defects in the structure of materials, and fracture processes occur through the sequential breaking of chemical bonds because stresses inside material samples are concentrated on small volumes of material at the surfaces of flaws. This process is the subject of the field of fracture mechanics, which plays a major role in materials design for artificial structures, and, as we will see in the next chapter, for understanding the quality of design in structural biomaterials as well.

Chapter 5

Fracture Mechanics

In this chapter we will explore the manner in which materials break. As we will see, the mechanics of fracture explain why real materials are weaker than we would predict solely from the strength of their bonds.

1. STRESS CONCENTRATIONS. Figs. 5.1A and 5.1B provide illustrations that explain how structural defects, or flaws, might affect the failure processes. Fig. 5.1A shows the pattern of stress transfer through two blocks of material, one without flaws and one with a large notch on one side. The flaw size is greatly exaggerated for this illustration, but in real materials flaws on the micrometer scale can initiate failure processes, as we will soon see. The block on the left is apparently flawless, and this is indicated by the uniform distribution of dashed lines running from the top to the bottom surface of this block. These lines are called stress trajectories, as they are meant to describe the direction that stress is transferred through the block. Note also the force vectors located on the top and bottom surfaces; their uniform distribution indicates that force is uniformly distributed across the surfaces, and as a consequence there is uniform stress, σ_0, running through the block. The notched block is loaded uniformly on its surfaces, but the notch prevents stress trajectories from running straight through the block. Since stress cannot be transferred through the open gap of the notch, the stress trajectories must divert around the notch. As a consequence, the stress trajectories bunch together as they go around the notch, and their density is highest right at the tip of the notch. This collection of stress trajectories indicates a stress concentration, where the local stress may be many times larger than σ_0. Stress concentrations at flaws are key to understanding why real materials are so weak.

2. THE WORK OF FRACTURE. The right block in fig. 5.1A is shaded in gray to the left of the notch to indicate areas of the block that transfer stress through the material. Because this portion of the block is stressed, it is also strained, and this means that it stores elastic strain energy. The unshaded portion to the right is not stressed, and hence does not contain stored strain energy. The failure of the notched block will occur because the stored strain energy is released when the flaw grows in length, and it is this release of strain energy that breaks the bonds in the material, causing fracture. The analysis presented below was created by A. A. Griffiths in 1921, in a paper entitled "The Phenomena of Rupture and Flow in Solids." It is illustrated in Figs. 5.1B–D.

Fig. 5.1B provides a view of the strain-energy release that will happen if the notch grows by a small increment in length, ΔL. The gray-shaded band indicates the volume of strained material that will become unstrained when the notch increases in length by ΔL. However, there are special conditions that are required for the catastrophic rupture of the sample. The energy balance analysis created by Griffiths first calculates the strain energy released with crack growth, and then compares the energy released to the energy required to create new fracture surfaces by breaking the chemical bonds at the tip of the crack. The analysis starts with the diagram in fig. 5.1C, which shows a plate of material, with unit thickness, that has a long, thin cut at its center, with a length = $2L$. This cut creates a double-ended crack with two crack tips, one at each end. The drawing to the left shows how stress trajectories would exist in this test sample before fracture; note the stress concentrations at the tips of the crack. The area bounded by the dashed circle that surrounds the flaw is devoid of stress trajectories and hence contains no stored strain energy. The area outside the circle contains strain energy, and in this simplified analysis it is assumed that the strain energy outside the circle is uniformly distributed at a level determined by the average stress, σ_0.

The analysis begins with the calculation of the energy released if each crack tip extends by ΔL, as follows. Eq. (4.4) shows that the stored energy per unit volume in a strained material can be calculated as:

$$\phi_{stored} = \frac{\sigma^2}{2E}.$$

$$(5.1)$$

If we multiply this energy per unit volume by the volume of the material inside the circle,

$$V_{circle} = \pi L^2 h,$$

$$(5.2)$$

where h is the thickness of the plate, which we assume is 1, we can calculate the amount of strain energy, W, missing from the unstressed circle, as:

$$W = \frac{\sigma^2}{2E} \pi L^2.$$

$$(5.3)$$

We can now differentiate this equation with respect to length, L, as:

$$dW = \frac{\sigma^2}{E} \pi L dL,$$

$$(5.4)$$

which indicates the incremental energy release if the crack length, L, grows by an increment dL.

Now we need to consider the energy required to break the bonds at the two tips of the crack as the crack length grows. This process is illustrated in fig. 5.1D, which shows one end of this double-ended crack. In growing this crack by length dL, we create two surfaces of area $dL \times h$. Again, with unit thickness, and with our double-ended crack, the increment of surface energy, dS, required to extend these two cracks (four surfaces) by a length increment, dL, is:

$$dS = 4\gamma_{surf} \, dL,$$

$$(5.5)$$

where γ_{surf} is the surface energy of the material that we used in our calculation of the theoretical strength of materials in eq. (4.6). Now, the solution to this analysis is the

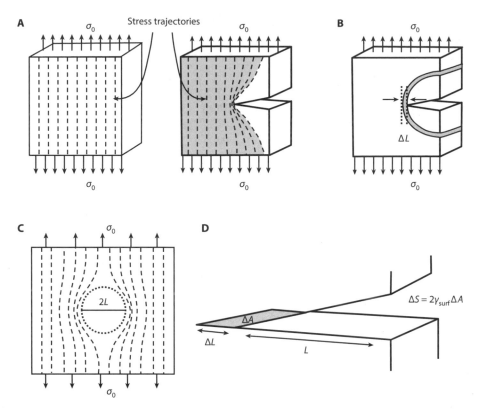

Figure 5.1. The energetics of crack propagation. **Panel A** illustrates the concept of stress trajectories in stressed objects. The *dashed lines* in the solid object to the left indicate the direction of stress transfer through this object from an external stress, σ_0, uniformly applied to the top and bottom surfaces. These stress trajectories are uniformly spaced to indicate that stress in this object is equal in all locations. The *arrows* at the top and bottom surfaces indicate force vectors. In the object to the right, which has a large notch cut into its right side, we see that stress trajectories away from the notch are equally spaced, but on the side with the notch the trajectories must divert around the notch and they bunch together tightly at the tip of the notch. Thus the notch, or flaw, acts as a stress concentrator that causes very large stresses to form in its vicinity. *Gray shading* indicates areas of the block that transfer stress through the material. **Panel B** illustrates the strain-energy release that will occur when the notch grows a small increment in length, ΔL. The *gray-shaded band* indicates the volume of strained material that will become unstrained when the notch increases in length by ΔL. **Panel C** presents the assumed distribution of stress trajectories (dashed lines) around an internal cut of length $2L$ in the center of a plate of a material test sample. This cut is indicated by the *solid dark line* inside the dotted circle at the center of the plate. The *dotted circle* indicates the part of the sample that contains no stress trajectories, and hence contains no stored elastic energy. If the external stress is sufficient to cause the crack to grow by a length dL, strain energy stored in a band of material surrounding the crack will be released, in accordance with eq. (5.4), and this energy will be available to "pay for" the creation of new crack surfaces in the processes of breaking the sample. **Panel D** illustrates the energy requirements for the elongation of a preexisting crack and the creation of new crack surfaces at each end of this flaw. The energy required at each end of this internal notch, ΔS, is equal to the surface energy of the material, γ_{surf}, times the surface area created, ΔA. Because two surfaces are created, ΔS at each end is $2\gamma_{surf}\Delta A$, or $4\gamma_{surf}\Delta A$ to extend the notch at both ends.

assumption that fracture will occur when the strain energy released with crack growth, eq. (5.3), is just equal to the strain energy required to extend the crack, eq. (5.4), as:

$$4\gamma dL = \frac{\sigma^2}{E}\pi L dL, \quad \text{or} \quad \sigma_{max} = \sqrt{\frac{4E\gamma_{surf}}{\pi L}}, \quad (5.6)$$

where σ_{max} is the failure stress for the material in this flawed plate. Griffiths used a mathematically exact analysis of the stress distribution around a sharp notch and came up with essentially the same result. The only difference is that the 4 inside the square-root term becomes a 2, and we will use this exact form,

$$\sigma_{max} = \sqrt{\frac{2E\gamma_{surf}}{\pi L}}, \quad (5.7)$$

from this point on.

3. THE REALIZED STRENGTH OF MATERIALS. This equation establishes the energy balance that is required to explain that the strain energy released with crack growth can pay for the creation of new fracture surfaces, and it also explains the catastrophic nature of fracture processes that can occur when this critical condition is reached. The fracture process is shown graphically in fig. 5.2, where we see at the left a notched block of material starting just as its first increment of crack growth releases the stored strain energy in the first increment of material volume indicated by the shaded band. This process starts because a rising applied σ_0 may break some bonds at the tip of the flaw, but if the stored strain energy released is not sufficient to break the next increment of bonds at the crack tip, then fracture will not occur, and σ_0 will continue to rise. The crack may then extend in a stable manner until the rising σ_0 reaches a level where the strain energy released is equal to or greater than the energy required to break bonds at the crack tip, and once this condition is reached the crack should grow rapidly because each additional increment of crack growth releases more and more strain energy to break bonds at the flaw tip. The center drawing of the notched block shows the test sample

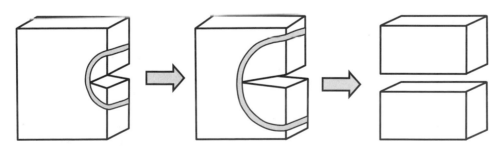

Figure 5.2. Progression of crack growth in the catastrophic fracture of a test sample. The drawing to the left indicates a small crack that has just reached the critical condition for crack growth, where stored energy released from strained portions of the test sample are just sufficient to "pay for" the creation of new crack surfaces through the extension of a preexisting crack. Once this first increment of crack growth has occurred, the next increment will release more energy, and hence the crack process will proceed catastrophically from the release of additional energy that is already stored in the strained test sample.

in the middle of this uncontrolled crack growth. Note that the longer crack length is releasing more energy, as the band of unloaded material is now much larger than in the initial state. Indeed, in brittle materials, which lack mechanisms to inhibit crack growth, fracture occurs catastrophically and may involve shattering of the test sample into small pieces. In this diagram the fracture is complete when the flaw has propagated fully across the sample, splitting it in two.

Griffiths tested his theory and analysis with experiments in which he determined the tensile strength of glass, a very brittle material. Based on his equation for the energetics of crack growth, he predicted that tensile strength would increase as the square root of one over crack length ($\sigma_{max} \propto L^{-0.5}$ in eq. [5.7]), and he tested this hypothesis by measuring the effect of glass-fiber diameter on its tensile strength. On the assumption that smaller fibers must have smaller flaws, he predicted that thinner fibers would be stronger, and fig. 5.3 shows that this is indeed the case. He saw a 50-fold increase in strength as fiber diameter decreased from $100\,\mu m$ down to about $3\,\mu m$. His tensile-strength data vary with fiber diameter, d, as $\sigma_{max} \propto d^{-0.7}$, which is close to the prediction, but it appears that the actual flaw size relative to fiber diameter changes somewhat in thinner fibers.

The energy-balance analysis above shows that strain energy released from distant sites in a test sample must be sufficient to pay for the creation of new surfaces, but it does not explain the mechanism by which this stored strain energy is transferred to the very local region at the tip of a flaw where bonds actually break. The mechanism for this transfer of energy occurs through the concentration of stress that occurs around the tips of cracks and other kinds of flaws because stress trajectories must divert around the flaw. Fig. 5.4 shows an elliptical "flaw" adjacent to a small portion of a crystal lattice of atoms that are being pulled apart by some distantly applied average stress, σ_0. For linearly elastic materials it is possible to calculate the local stress at the tip of a crack, σ_{crack}, which can be orders of magnitude greater than the average stress, as:

$$\frac{\sigma_{crack}}{\sigma_0} = 1 + 2\frac{a}{b}, \quad \text{or} \quad \frac{\sigma_{crack}}{\sigma_0} = 1 + 2\sqrt{\frac{a}{r_m}}, \qquad (5.8)$$

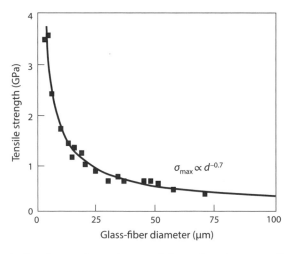

Figure 5.3. The slimmer the fiber, the stronger it is. This shows the results obtained by A. A. Griffiths in tests of the tensile strength of glass fibers as a function of fiber diameter. His model of fracture mechanics, embodied in eq. (5.7), predicts that the tensile strength of rigid materials, σ_{max}, should be proportional to crack length, $L^{-0.5}$. The test data he obtained clearly indicate that as glass-fiber diameter decreases the strength of the fibers increases, and, as indicated in the figure, the measured strength was proportional to diameter, $d^{-0.7}$ power, in close agreement with his model.

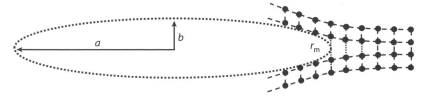

Figure 5.4. Stress concentration. Stress concentration explains how strain energy re-leased from distant locations in the strained test sample is able to concentrate on the small number of chemical bonds at the tip of a notch or flaw where crack growth oc-curs. The crucial mechanism is the process of stress concentration that occurs at sharp notches and edges in the surfaces of flaws. This diagram shows an elliptical flaw that exists inside a material sample, and at the right end of the ellipse there is a portion of the atomic lattice of the material. An elliptical flaw is used here because it is possible to calculate the stress concentration that exists at the tip of the ellipse, as indicated by eq. (5.8). For very elongated ellipses and for long, sharp flaws, the stress at the flat tip can be orders of magnitude greater than the average stress in the material at sites distance from the flaw. a, major axis of the ellipse; b, minor axis of the ellipse; r_m, radius of curvature at the end of the major axis.

where a and b are the major and minor axes of the ellipse, and in this equation r_m is the radius of curvature at the end of the major axis. Thus, for this ellipse, where $a/b = 5$, the stress concentration is 11. (Note, $r_m = b^2/a$.) Longer, sharp cracks will create higher stress concentrations, and the measured and theoretical tensile strengths listed in table 4.1 can give us some insight into the magnitude of stress concentrations in real materi-als. Since the theoretical strengths of materials can be hundreds of times greater than the measured strengths of macroscopic samples, we can infer that the stress concen-trations that exist in materials can be of the order of 100 or more. In summary, this analysis of fracture indicates that the magnitude of strain energy released from flaws is directly proportional to length of existing flaws, L_{init} (eq. [5.4]), but the shape of the flaws and hence the stress concentration achieved at the tip of the flaws will affect the outcome, with sharp flaws initiating catastrophic failures at lower stresses than rounded flaws. Now look at the drawing of the crystal lattice at the tip of this elliptical crack. The stress concentration in this illustration exists at atomic dimensions because the flaw itself is only a few nanometers long. In this case, we are seeing strains in the bonds between atoms, and a test at this level would likely produce strengths at or near the theoretical strength of the material. More typical flaws would exist at much larger size scales, and the critical flaw size is something that can be inferred from Griffiths's equation (eq. [5.7]).

Experiments like Griffiths's tests on glass fibers have been repeated for other brittle materials with similar results, and it is safe to assume that the fracture analysis is es-sentially correct, but there is one major addition that needs to be made. The Griffiths analysis of glass fibers assumes that the surface-energy term was the actual surface energy for the creation of an atomically flat surface, as indicated in fig. 5.1D, but in reality even brittle materials like glass and inorganic crystals will fracture with some surface roughness. Thus, the surface-energy term must be determined experimentally in fracture tests, and when this is done we find that the "surface energy" for brittle materials like glass is somewhat higher than the true surface tension. More importantly, the surface energy for tougher materials can be many orders of magnitude greater than the surface tension. For example, large objects made from steel, like cars, do not shatter when they crash; they crumple. Trees provide a similar example from nature; if you attempt to cut down a tree with an axe, it does not shatter like glass. The axe bites into the tree, creating an obvious crack-like flaw, but this crack does not propagate catastrophically across the tree, causing it to fall with a single blow. This strongly suggests that the surface-energy term must include energy costs for deforming and breaking bonds well beyond the plane of crack elongation in materials that strongly resist fracture. Indeed, this whole concept of fracture requires that we redefine our concept of toughness from the very simple definition given previously as the energy per volume to break, determined from the area under the stress-strain curve of a material tested to failure (eq. [2.5]). What we see now is that we must measure the energy required to extend flaws in a material sample to quantify its toughness, and hence make predictions about the conditions that will cause a material to break catastrophically.

4. FRACTURE TOUGHNESS. The field of fracture mechanics defines two additional properties, with units of $J \, m^{-2}$, but they are measured under specific test conditions that make it possible to quantify the total energy required to extend a preexisting crack. The theory and practice of fracture mechanics are very complex, and readers who wish to learn more should look to the literature for details (e.g., Anderson, 2005). At a very minimum, however, it is useful to at least recognize the names of the experimentally determined properties that are used to replace the surface-energy term, γ_{surf}, in the Griffiths equation. One property, called the strain-energy release rate, G, applies for materials with essentially linear elastic behavior to fracture. (Note that the symbol G, as used here, is not the shear modulus.) A second property, J, called the J-integral, is used to quantify the energy required to extend cracks in materials that undergo large-scale plastic deformation as cracks grow. Thus, the Griffiths fracture equation (eq. [5.7]), which includes $2\gamma_{surf}$ for the cost of crack elongation, becomes:

$$\sigma_{max} = \sqrt{\frac{EG}{\pi L}}, \text{ or } \sigma_{max} = \sqrt{\frac{EJ}{\pi L}}, \qquad (5.9)$$

where G and J are no longer restricted to quantifying the energy cost of creating an atomically flat surface. Rather, they quantify the energy required to extend cracks that form in real materials. Note that the 2 in the Griffiths equation disappears because G and J quantify the energy to extend the crack's length, which involves estimating the energy absorbed on both sides of the growing crack.

Table 5.1 provides some approximate numbers for fracture toughness for a variety of materials, ranging from very brittle glass to tough natural and engineering materials. Note that the table includes a column for L_{crit}, which is a calculated value for the critical flaw size in a material that accounts for the measured values of toughness (G or J), stiffness (E), and strength (σ_{max}). That is, eqs. (5.9) above are solved for L_{crit} with these inputs. This value predicts the approximate size of a flaw or defect that will create a catastrophic fracture event in samples of each material. It is perhaps useful to consider these predictions in light of what we intuitively know about the materials listed,

Table 5.1. Values for the Fracture Toughness of Structural Materials

Material	Toughness, G or J (kJ m⁻²)	Stiffness, E (GPa)	Strength, σ_{max} (GPa)	Critical Flaw Length, L_{crit} (mm)	Stress Intensity Factor, K (MPa m^½)
Glass, bulk	0.01	80	0.07	0.015	0.84
S-glass fiber	0.01	80	4.7	0.00004	0.84
Bone	1.5	20	0.16	0.5	5
High-tensile steel	10	210	1.5	0.6	45
Mild Steel	100–1000	210	0.4	80–800	140–450
Wood	10	10	0.1	12	12
Horse-hoof keratin	23	3	0.04	27	8

because these predictions should tell us useful things about the kinds of objects we find these various materials in.

Note first that bulk glass shows a critical flaw length of about 15 μm, which is a plausibly small defect that could account for the fact that both large and small glass objects shatter when they are dropped. L_{crit} for glass fibers made from a particular type of glass (S-glass) is 0.00004 mm, or 40 nm, well below the dimensions of this 10 μm diameter fiber. L_{crit} for bone is about 0.5 mm, which is small relative to the size of a whole bone, but comparable to the size of some bone substructures, as we will see later. Both high-tensile and mild steel are capable of significant plastic deformation before fracture, but mild steel can undergo extremely large-scale plastic deformation. As a consequence, L_{crit} for mild steel can be in the range of a meter. So cars, made from mild steel, don't fracture; they crumple. Fracture toughness of this magnitude is crucial, however, in the construction of very large-scale structures, such as bridges and ships, which may have stress-concentrating features on the scale of meters. Good design requires that these features not include sharp notches where stress concentrations can be high, or the large structures might fracture. Finally, wood and horse-hoof keratin are tough biological composites that have critical-flaw sizes on the scale of tens of millimeters. Cellular substructures in wood are well below this magnitude, but "defects" created by knots in lumber do indeed contribute to fracture in wood beams. It is worth noting that farriers routinely attach metal horseshoes to horses' hooves with nails that are pounded through the hoof edge, and the horses are able to run actively over rough ground without having these nail holes initiate failure of the hoof wall by fracture because the diameter of nail holes is about an order of magnitude smaller than the critical-flaw size.

Before we look at design strategies for creating strong and tough materials, we need to return to the Griffiths equation one more time to define an additional way to quantify the fracture resistance of materials, namely the stress intensity factor, K. To this point we have looked at the energy required to extend flaws, quantified as G and J, which are in the top term under the square root in eqs. (5.8). So increasing G or J will increase σ_{max}, and will increase L_{crit} if we rearrange eqs. (5.8) to solve for that parameter. Note also that E, the elastic modulus of the material, is also in the top term, and thus increasing both material properties specified in eqs. (5.9) will increase strength and toughness. It is likely very clear by now why increasing G and J increases toughness, because they determine how much energy is required to make flaws grow. The role of E in the fracture process is to determine the magnitude of the stored strain energy in a sample, and hence influence its rate of release in fracture. This may sound backward, but increasing E decreases the strain energy stored in a sample at any level of applied stress, σ_0, because increasing E decreases strain. That is, the area under a stress-strain curve falls as E increases, for any given level of stress. The stress intensity factor, K, is calculated as:

$$K = \sqrt{EG}, \text{ or } K = \sqrt{EJ}, \tag{5.10}$$

and will have the units of MPa m$^{1/2}$. For materials with equal toughness (as measured by G or J), the ability to resist fracture and the critical size of stress concentrators will increase as the modulus of the material increases.

The final column in table 5.1 gives estimates of the stress intensity factor, K, and we see some interesting differences between these materials. Both bulk glass and S-glass fibers have the same K value. They are the same "material"; they just have different flaw sizes. Both kinds of glass will fail with an abrupt, catastrophic drop in stress from a linear stress-strain line, just as seen for the left stress-strain plot in fig. 2.3C. The only difference between the bulk-glass and S-glass-fiber stress-strain behavior will be in the stress at which the failure occurs, which may differ by almost a factor of 100. All the other materials in table 5.1 have stress-strain curves that show features similar to those of the stress-strain curve to the right in fig. 2.3C. Here we see an initial linear rise in stress, followed by a yield, where the slope of the curve falls. However, they will show variations on this kind of yield behavior, where postyield stress may rise slowly, remain constant, or even fall before the fracture event that causes the sample to break in two. In some cases—such as bone, wood, and steel—the yield process arises from the onset of processes that ultimately lead to the onset of fracture, and in these cases the best interpretation of the yield process is that stiffness falls because of microdamage within the material that is associated with the elongation of preexisting flaws through stable crack growth. With continued strain, however, cracks elongate and may combine, presumably increasing stress concentrations until the conditions are met for the initiation of a catastrophic fracture event. In other cases, such as hoof keratin or hair as shown in fig. 2.2, the yield behavior is caused by the unfolding of helical protein fibers, and postyield strain hardening leads to failure at much higher stresses.

Chapter 6

The Molecular Origins of Soft Elasticity

When we think about the structural basis for soft materials it is important to realize that we cannot think in terms of the deformation of chemical bonds. Chemical bonds are exceptionally stiff and strong, and as we have seen in chapter 4 it is the strong bonding forces that create stiff and strong materials. It is interesting, however, that soft elastic materials require the presence of chemical structures that are held together by strong chemical bonds, but the primary bonding in soft solids forms linear chains of atoms, rather than having primary bonds organized in a three-dimensional lattice, as illustrated in fig. 4.2 for rigid materials That is, soft elastic materials are based on polymers, which are typically huge molecules made from linear arrays of repeating monomer units. Polymers can also be branched, but structural polymers are almost always linear polymers, as we will soon see. In addition, the primary bonds in polymers are typically covalent bonds, which means that the primary bonds in the polymer backbone are very strong and directional.

1. FLEXIBLE LINEAR POLYMERS. In biology, there are three main classes of polymers: nucleic acids, proteins, and polysaccharides. Nucleic acids (DNA and RNA) function in the storage and translation of genetic information. Proteins function in a wide variety of metabolic and regulatory mechanisms, but they are also present in a variety of structural materials. Polysaccharides are used as energy stores (e.g., starch) but they are also used extensively in structural materials. Indeed, cellulose, a linear polymer of the sugar glucose, is likely the most abundant polymer on earth, as cellulose is the major component of plant cell walls. Artificial structural polymers are created using a broad range of chemical monomers, but most are based on hydrocarbons of various types. In general, structural polymers are extremely long, containing hundreds or thousands of monomer units. Nucleic acids are amongst the longest of natural polymers, and the DNA molecules in eukaryotic chromosomes are likely the largest linear polymers known.

Linear polymers can exist in one of two basic forms. If the backbone of a structural polymer has a compact shape and exists in a highly regular repeating structure, the polymer has a high probability of exhibiting crystalline organization, in which extended or perhaps helical polymer conformations allow the polymers to pack tightly together, and these compact structures are stabilized by interchain secondary bonding forces. Fig. 6.1A shows the repeating two-carbon monomer unit of polyethylene, which is very compact and regular indeed because the only side groups on the carbon backbone are two hydrogens on each of the carbons. Note also that the carbon backbone is made entirely of carbon-carbon single bonds, which in principle should make this a very flexible polymer chain. However, polyethylene forms tight crystalline structures, where the backbone of the chain is extended and packed into parallel crystalline arrays that are held together by very weak bonding forces between the hydrogen atoms that extend out from the carbon backbone. That is, the lateral bonding forces that create the polyethylene crystal are the weakest of the bonding forces we recognize in organic molecules; namely, van der Waals induced-dipole bonds. It is really quite remarkable that this simple polymer can actually crystallize, but it does because the small size of the hydrogen side groups allows extremely close packing of extended polyethylene chains, bringing the hydrogens close enough that they form bonds with sufficient strength to stabilize the crystal at room temperature. Crystalline polymers are typically quite stiff and strong if they are loaded in a direction parallel to the covalent backbone of the polymers, and they exhibit bond-energy elasticity, as described in chapter 4. It is also quite remarkable that polyethylene can be made into a superfiber if the length of the polyethylene chain is sufficient and if the fiber is spun under conditions that fully extend the flexible chain parallel to the long axis of the spun fiber before it is allowed to crystallize. Spectra fiber, an ultra-high-molecular-weight polyethylene fiber, has a tensile strength of about 3 GPa. We will see a great deal more about crystalline polymeric materials in section III, on tensile materials.

If a structural polymer backbone has bulky side groups that prevent it from packing into compact crystalline arrays, the polymer will most likely take on a disordered and highly folded shape, and these polymers are often called amorphous polymers. Depending on the chemistry of the polymer backbone, the temperature, and the presence of swelling solvents, these folded polymers can exist as a rigid polymer glass, as a soft elastic solid, or as a viscoelastic fluid. Figs. 6.1B and 6.1C show monomer units for two synthetic polymers that adopt noncrystalline, amorphous chain conformations. Fig.6.1B shows the monomer for polymethyl methacrylate (PMMA), also known as Plexiglas or Lucite, which is clearly more bulky than the polyethylene monomer. Again, the monomer unit has only two carbons in the main chain, but one carbon has two side

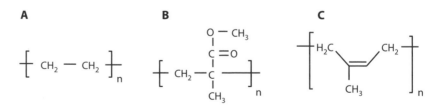

Figure 6.1. Repeating units of three common polymeric materials. **Panel A,** polyethylene; **panel B,** polymethyl methacrylate; **panel C,** *cis*-polyisoprene. These polymers provide useful examples to illustrate the basic principles of polymer mechanics required in our discussion of rubberlike elasticity.

groups, a methyl group on one side and a methacrylate group on the other. These side groups make close packing into crystalline arrays awkward, but the polar groups in the methacrylate side chain introduce the possibility of hydrogen bonding between adjacent side chains. The net result of this chemistry is that PMMA at room temperature is a polymer glass, and hence quite rigid, but its molecular structure is entirely random. However, the interchain hydrogen bonds are relatively weak, and as a result PMMA glass will melt at a glass-transition temperature, T_g, of about 105°C, turning it into a viscoelastic fluid. The key difference between the glass and the fluid state is that in the glass state the folded polymer molecules maintain a fixed shape and a fixed relationship with neighboring molecules, whereas above T_g the polymer chains can change their shape, under the driving force of thermal agitation, and with external loading the polymer chains can slide relative to their neighbors like particles in a fluid.

Fig. 6.1C shows the monomer structure of natural rubber, which is called *cis*-isoprene. That is, natural rubber, the polymer that we harvest commercially from the sap of the rubber tree, *Hevea brasiliensis*, is chemically identified as *cis*-polyisoprene. This polymer's chemical structure appears to be intermediate between polyethylene and PMMA. The four-carbon monomer contains a *cis* double bond in the chain backbone, and there is a single small methyl side group. This structure creates a polymer with a very flexible carbon backbone, but with two features that restrict crystal formation: the methyl side group and the *cis* double bond. Thus, at room temperature this polymer is both folded and highly mobile. In fact, it is so flexible that its glass-transition temperature, T_g, is −69°C. This molecule provides the starting point for our analysis of soft elasticity, often called rubberlike elasticity.

2. THE THERMODYNAMICS OF RANDOM-COILED MOLECULES. Fig. 6.2 shows a drawing of a single *cis* polyisoprene molecule, which we can imagine as being immersed in a suitable solvent. The arrowheads identify the ends of the molecule, and the drawing indicates the folded shape that results from the flexibility discussed above. What is not shown, however, is the fact that this drawing represents a snapshot in time of a dynamic structure that is changing its shape rapidly with time. The driving force for these shape changes comes from thermal agitation of all parts of the molecule and from the impact of rapidly diffusing solvent molecules surrounding the molecule and filling spaces within the molecule. We call this type of molecule a random coil, and if we imagine that we take hold of the two ends of this molecule and stretch it, as indicated in the drawing, the molecule will extend. Then, if we release the ends, the molecule will return to its initial state. If we were able, we could measure the increasing force that is required to stretch the molecule, and we could see this force decline as we allowed the molecule to recoil back to its starting state. Indeed, it is now possible, using atomic-force microscopy, to carry out this thought experiment on real molecules. The question is, what is the origin of the force generated when we stretch this molecule, and what makes it recoil elastically? To address this question, we need to turn briefly to thermodynamics to understand the energetic origins of elastic behaviors. The Helmholtz free energy, $\mathcal{F}$, is most useful to consider elastic processes, which most often occur at constant volume, V, and pressure, P. Thus,

$$\mathcal{F} = U - TS, \tag{6.1}$$

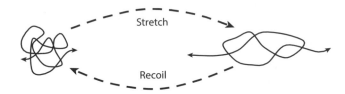

Figure 6.2. The tendency toward disorder drives elastic recoil. A drawing of a single molecule of *cis*-polyisoprene, natural rubber, in its resting and stretched states. This drawing represents a snapshot in time of a molecule that is very flexible and is writhing in space under the influence of thermal agitation. We call this type of molecule a random coil. If we could grab the ends of this molecule and stretch them apart we would see that on average the molecule would take on a more elongated set of shapes, and if we then let go of the ends, the molecule would recoil back to its more compact starting shapes.

where U is the internal energy, S is entropy, and T is absolute temperature (measured on the Kelvin scale). For a process taking place at constant temperature, the change in the Helmholtz free energy becomes:

$$d\mathcal{F} = dU - TdS. \tag{6.2}$$

So this relationship should apply to the extension of our random-coil molecule if it is extended with no change in volume and at constant temperature and pressure. The first law of thermodynamics states that the change in internal energy, dU, is:

$$dU = dQ + dW, \tag{6.3}$$

where Q is the heat absorbed by a system and W is the work done on the system by some external process. The second law of thermodynamics states that the change in entropy in a reversible process is:

$$TdS = dQ, \tag{6.4}$$

and combining these equations gives:

$$dU = TdS + dW. \tag{6.5}$$

Now, combining this relationship with eq. (6.2), we get:

$$d\mathcal{F} = dW, \tag{6.6}$$

for a reversible process that occurs at constant temperature.

3. ENTROPY ELASTICITY. Next, for an elastic system as shown in fig. 6.2, the work done by the application of an external force, F, extending our elastic object by a length, L, is:

$$dW = FdL - PdV, \tag{6.7}$$

where the PdV term accounts for any work associated with changes in pressure-volume work. Since, for a rubber, the volume changes with extension are quite small, we can usually ignore the pressure-volume term, so that:

$$dF = dU - TdS = FdL. \tag{6.8}$$

Finally, for a change in length at constant temperature and constant volume:

$$F = \left[\frac{\partial F}{\partial L}\right]_{T,V} = \left[\frac{\partial U}{\partial L}\right]_{T,V} - T\left[\frac{\partial S}{\partial L}\right]_{T,V}. \tag{6.9}$$

This relationship shows that the force required to create a reversible deformation of an elastic object can arise from two kinds of processes, changes in internal energy and changes in entropy. We have already considered the first possibility, that elasticity can arise from changes in internal energy, in our discussion of the bond-energy elasticity of rigid materials in chapter 4. Indeed, the changes in bond energy in eq. (4.1) are the same as the changes in internal energy in eq. (6.9). The important conclusion for our current discussion is that elasticity can also arise from changes in entropy, and it is this possibility that explains the mechanical behavior of materials that are made from random-coil molecules. We will model the behavior of random coils with random-walk statistics. Fig. 6.3 shows a three-dimensional random walk of n steps, with each step of equal length, ℓ. Note that the distance between the start of the walk, at the origin of the graph, and the end—the distance indicated by the dotted arrow labeled r—is much shorter than the extended length of the walk, $n\ell$. Gaussian statistics can be used to describe the dimensions of this random walk, as follows. The probability density, $p(x,y,z)$, of finding the chain end in a specific small volume element, dx,dy,dz, at a location (x,y,z) in three-dimensional space, which is indicated by the small box in fig. 6.3, can be calculated as:

$$p(x,y,z) = \frac{b^3}{\pi^{3/2}}\exp(-b^2r^2), \tag{6.10}$$

where:

$$b = \left[\frac{3}{2n\ell^2}\right]^{\frac{1}{2}} \quad \text{and} \quad x^2 + y^2 + z^2 = r^2. \tag{6.11}$$

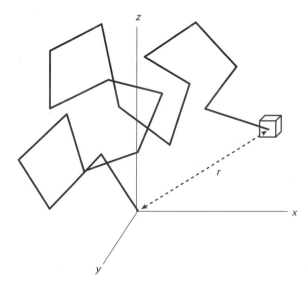

Figure 6.3. The random configuration of a polymer molecule. A diagram of a random walk of a number of steps of equal length, which provides the theoretical basis for our understanding of rubber elasticity. Statistical models are used to calculate the probability density of finding a chain in a small volume of space, eq. (6.10), or the most probable chain-end separation distance, r, eq. (6.11).

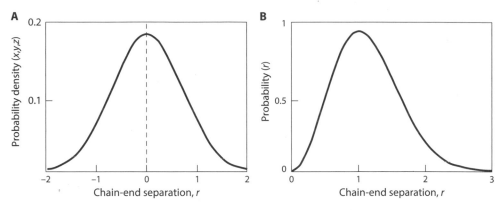

Figure 6.4. Probability versus probability density. The probability density forms a "bell" curve centered at zero separation between the chain ends (**panel A**). However, when probability density (essentially, energy per volume) is multiplied by the volume available at a given end-to-end separation, the probability of the molecule having a given separation is as shown in **panel B**. This probability has a peak at a distance from the chain origin. These functions allow us to describe the size, shape, and elastic properties of kinetically free random-coil polymers that exhibit rubberlike elasticity.

This probability-density function, $p(x,y,z)$, describes the well-known "bell" curve that is used to describe what are known as normal distributions of data. Fig. 6.4A shows the $p(x,y,z)$ function for a three-dimensional random walk, which indicates that the most probable location per unit volume for a random walk is at a chain-end separation of zero. The result is similar in two dimensions, and we can therefore think of an actual walk by a person on a flat surface, where each step proceeds in this random manner. The result of this walk will be the same; the highest probability density, $p(x,y,z)$, will occur at the origin of the walk.

However, the probability, $P(r)$ of observing the walker or a chain end being a distance r away from the origin in any direction, as indicated by the arrow in fig. 6.3, will not be determined by the probability density alone, but is calculated by eq. (6.10) multiplied by the volume of a spherical shell, V_r, of thickness dr, located at distance r away from the starting point of the random walk, as:

$$V_r = 4\pi r^2 dr. \tag{6.12}$$

The result is a probability function that gives us a more appropriate indication of the actual dimensions of a random-coil molecule with a shape that is equivalent to a random walk, as:

$$P(r)\,dr = \left(\frac{4b^3}{\pi^{1/2}}\right)r^2 \exp\left(-b^2 r^2\right) dr. \tag{6.13}$$

Eq. (6.13) is plotted in fig. 6.4B, and we see that the probability of a chain-end separation $r = 0$, is zero, not because the probability density is zero, but because the volume at the origin goes to zero. The probability $P(r)$ rises to a maximum at about $r = 1$, and we can also use eq. (6.13) to calculate the most probable chain-end separation, r_{mp}, as:

$$r_{mp} = \left(\frac{2n\ell^2}{3}\right)^{\frac{1}{2}},$$ (6.14)

and the mean-square end-to-end separation can be calculated, as:

$$\overline{r^2} = n\ell^2.$$ (6.15)

It turns out that the mean-square end-to-end separation is the most useful parameter describing the dimensions of random-coil molecules.

This random-walk model for the behavior of a random-coil molecule provides us with the basis for understanding the origins of the elastic behavior illustrated in fig. 6.2. The link to elastic behavior and the probability function for the shape of random-coil molecules comes from Boltzmann's definition of entropy, S, as:

$$S = k \ln N,$$ (6.16)

where k is the Boltzmann constant, 1.38×10^{-23} J K^{-1}, and N is the number of micro-states in a system. In our case N will be the number of different shapes that the random coil is capable of adopting, and we will use the probability-density function, $p(x,y,z)$ (eq. [6.10]) to determine the value of N as a function of chain-end separation. Substitution of eq. (6.10) into eq. (6.16) yields:

$$S_c = C - \left(kb^2 r^2\right),$$ (6.17)

where S_c is the entropy of the random chain and C is a constant. Differentiating with respect to r gives us the change in entropy with extension, r, as:

$$\frac{dS_c}{dr} = -2kb^2 r.$$ (6.18)

Substituting this equation into eq. (6.9), assuming that chain-end separation, r, is equivalent to sample length, L, in eq. (6.7), and that temperature and internal energy are constant, yields:

$$F = -T\left(\frac{dS}{dr}\right)_{T,V} = 2kTb^2 r.$$ (6.19)

This quite amazing result means that a single random-coil molecule will exhibit linear force-extension behavior, and the force required to stretch it is determined by kT, which is effectively the intensity of thermal agitation at the molecular scale. This also means that the random coil will become stiffer as it is heated, a rather unusual property for a material. Keep this in mind, because the effect of temperature on the mechanics of random-coil-based materials is one of the diagnostic features of this kind of elasticity.

Looking at fig. 6.4A we see that at an average chain-end separation $r = 0$, there is a very high probability density, which means that there are many shapes available to a random coil at zero extension, and this means that entropy is high and the molecule is in a highly disordered state. As r increases, the probability density falls, and this indicates that pulling on the ends to elongate the random chain reduces its entropy, in effect imposing order on the random chain. This fall in entropy provides the mechanism for the rise of force with the extension of a random coil, and for rubberlike materials in general. Elastic recoil occurs because ordered states tend to move spontaneously to less ordered states, and it is the thermal agitation of the random coil itself and the particles that surround and collide with it that drive it back to its initial dimensions and high entropy. Random coils are entropic springs!

We are almost there; we now understand rubber elasticity at the level of ideal random-coil molecules. We now need to translate this molecular behavior into real polymer molecules, and then consider how polymer molecules can be constructed into macroscopic samples that are part of the skeletal system in a whole organism. The random-walk model assumes that the junction point at the end of each "step" will allow complete rotational freedom, so that there is equal probability of the step being in any direction. This degree of rotational freedom is simply not allowed by the chemical bonds present in real polymer molecules. Carbon-carbon single bonds, for example, allow considerable rotational freedom, but the tetrahedral bonding of carbon single bonds restricts rotational freedom to a fixed arc that provides only a small part of the rotational freedom needed for behavior that is equivalent to the assumed random step. However, we can consider several single bonds in series to provide sufficient rotational freedom to comprise an "equivalent random link." The condition is that the mean-square end-to-end distance for the real molecule must equal the mean-square end-to-end distance for the theoretical random walk, $\overline{r^2} = \overline{r_e^2}$, and the extended length of the real and the equivalent molecule must also be equal, $n\ell = n_e\ell_e$. Calculations based on the known structure of polyethylene chains in a noncrystalline random state indicate that the length of the equivalent random link is 2.4 times greater than the real carbon-carbon bond length in polyethylene, and that the number of random links in the equivalent random chain is one-third of the number of carbon single bonds in polyethylene's backbone chain. For more natural polymers, like those found in rubberlike proteins, the structural complexity makes calculation of the dimensions of equivalent random links very difficult, and experimental values for chain properties are required, as will be discussed below. However, at this stage it is safe to say that the structure of the protein backbone is sufficiently different, and the size of side groups sufficiently larger than those in polyethylene, that equivalent random lengths in proteins are many times larger than those in polyethylene.

4. THE EFFECTS OF CROSS-LINKS. To produce a macroscopic rubber sample, we require many random-coil molecules, each of which exhibits the elastic behavior expressed in eq. (6.19), but they must be linked together so that when we apply a force to the molecules, all will experience equivalent deformations and hence develop equivalent forces. This is achieved by the simple process of introducing fixed linkages between the individual polymer molecules. Fig. 6.5A illustrates this concept with a cube of rubber, where the lines represent random-coil polymer molecules, and the dots indicate the location of chemical cross-links that tie the molecules together into a continuous network. In this drawing, and in most real rubber networks, the original polymer chains are much longer than the contour distance along the chains between cross-links, although polymer chemists also make rubber networks that are end linked. Fig. 6.5B shows the cube after it has been stretched uniaxially by 100%; note that there is a large decrease in the lateral dimensions because the sample deforms at constant volume. The extension ratios that describe this deformation are indicated beside the stretched sample. λ_1 is the longitudinal extension, and at 100% extension $\lambda_1 = 2$. The lateral contraction will be equal in both the y and z directions,

A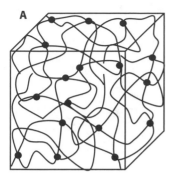

B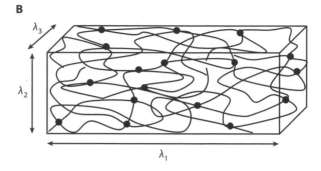

Figure 6.5. The rubber network. The organization of a rubber network made from ki-netically free polymer chains that are chemically cross-linked into a three-dimensional solid, where the cross-links are indicated by the *black circles*. **Panel A** represents the un-stressed resting state, and **panel B** represents a stretched state. Each polymer segment between cross-links behaves as a single random-coil chain, and exhibits rubberlike elasticity. Calculation of the entropy changes takes account of the three-dimensional shape changes that occur when this solid block is stretched to extension ratios λ for the three dimensions of the sample, as explained in the text.

and they will fall to a value that maintains constant volume for the sample, which will be $\lambda_2 = \lambda_3 = 0.707$. Now we have a continuous material, and deformation at the sample surface will be transferred to all chains within. Thus, we have a macroscopic network of molecular springs, and we can calculate the mechanical properties of the rubber networks.

The analysis of network elasticity is based on the following assumptions:

1. The network chains are defined as the length of polymer between cross-links, and we can therefore define the number of network chains per unit volume, N.
2. The mean-square end-to-end distance of the network chains is the same as for the free polymer.
3. The network deforms at constant volume, and the network chains in the interior of the network deform in proportion to the changes in the external shape of the rubber sample (this is called the affine-network assumption).
4. Finally, and most important, the entropy of individual chains in the network is the same as the entropy of an isolated chain under the same conditions (eq. [6.17]), making the total entropy of the network equal to the entropy per chain times the number of chains, N.

Now we need to apply these assumptions to the calculation of the mechanical proper-ties of rubber networks.

Starting with eq. (6.17), we can express the entropy of an unstrained chain, S_0, as:

$$S_0 = C - kb^2 r^2 = C - kb^2 (x_0^2 + y_0^2 + z_0^2), \tag{6.20}$$

in a form that allows us to specify deformation in all three dimensions (x_0, y_0, z_0). We can also express the entropy of chains that have been strained to new dimensions (x, y, z), where $x = \lambda_1 x_0$, and similarly for the y and z directions, as:

$$S = C - kb^2 r^2 = C - kb^2 (\lambda_1^2 x_0^2 + \lambda_2^2 y_0^2 + \lambda_3^2 z_0^2). \tag{6.21}$$

In this equation the λ terms indicate the extension ratio of individual random chains, but with the affine-network assumption they also describe the macroscopic deformation of the cross-linked network. The change in entropy with deformation, $\Delta S = S - S_0$, is obtained by subtracting eq. (6.20) from eq. (6.21), and in this process all the constant terms cancel. Then the result can be summed over all N chains in a unit volume to give us the total change in entropy of a macroscopic sample of these cross-linked random chains (Treloar, 1975) as:

$$\Delta S = -\frac{1}{2} Nk(\lambda_1^2 + \lambda_2^2 + \lambda_3^2 - 3). \tag{6.22}$$

This is as far as we will go with the detailed derivation of rubberlike-elasticity theory, because with this equation it is possible to translate macroscopic deformations as indicated by easily measurable extension ratios of a cross-linked network into changes in the conformational entropy of random chains at the molecular scale. With this information it is possible to calculate a number of macroscopic material properties, based entirely on the Gaussian statistical model for random-coil molecules, as follows. Starting with the assumption that the change in internal energy is zero, $\Delta U = 0$, the work of extension, which is equal to the change in free energy, is $\Delta W = \Delta \mathcal{F} = -T\Delta S$, making the work of extension:

$$W = \frac{1}{2} NkT(\lambda_1^2 + \lambda_2^2 + \lambda_3^2 - 3) = \frac{1}{2} G(\lambda_1^2 + \lambda_2^2 + \lambda_3^2 - 3), \tag{6.23}$$

where k is the Boltzmann constant and T is temperature. Further, we can define an elastic modulus for a sample of rubber, $G = NkT$, and interestingly this term, G, turns out to be the shear modulus of elasticity of the rubber network. So we can now start to quantify the stiffness of this block of rubbery material. In addition, it is possible to calculate the average molecular weight of the network chains between cross-links, M_c, as:

$$G = NkT = \frac{\rho RT}{M_c}, \tag{6.24}$$

where ρ is the density of the polymer (kg m^{-3}) and R is the molar gas constant. That is, if we measure the shear modulus we can predict the size of the polymer chains between cross-links in the rubber network.

Finally, with minor adjustments to these equations, we can determine the tensile and compressive properties of a rubber, as follows. However, instead of calculating the work of deformation, W, we will calculate the stress. For properties expressed in engineering stress, force per unit initial cross-section area:

$$\sigma = G\left(\lambda - \frac{1}{\lambda^2}\right) \frac{\rho RT}{M_c} \left(\lambda - \frac{1}{\lambda^2}\right). \tag{6.25}$$

Fig. 6.6 shows eq. (6.25) plotted for a rubber with a shear modulus $G = 0.33$ MPa. It is important to note that the zero-stress line is in the middle of the graph because this plot shows data for both tension and compression, and that zero stress occurs at extension ratio $\lambda = 1.0$. That is, positive stress is for tension and negative stress is for

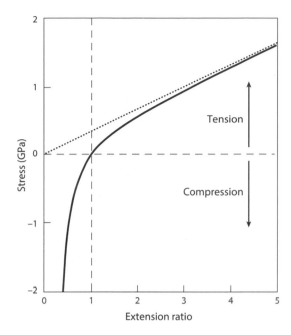

Figure 6.6. Tension and compression properties of a cross-linked rubber network. The form of the tension and compression properties of a cross-linked rubber network according to eq. (6.25). The shear modulus, G, for this sample is 0.33 MPa, and the initial tensile and compressive stiffness at extension ratio = 1.0 will be 1 MPa. The dotted line starting at the origin and running tangent to the plotted curve has a slope of 0.33 MPa, so at extension ratios beyond about 3 the material's stiffness is equal to the shear modulus.

compression. Note that the stress-strain curve for rubber is very nonlinear, and this is primarily because of the large changes in cross-sectional area that occur at the very large strains imposed in this experiment. The dotted line starting at the origin of the graph, which merges with the curve at high extension, has a slope equal to G, the shear modulus. The slope of the stress-strain curve at extension ratio = 1.0 is equal to the initial tensile modulus, $E = 3G$, as well as the initial compression modulus. Thus, in this case $E_{init} = 1.0$ GPa.

For true stress, the relationship that emerges from the random-coil model is:

$$\sigma_{true} = G\left(\lambda^2 - \frac{1}{\lambda}\right) = \frac{\rho RT}{M_c}\left(\lambda^2 - \frac{1}{\lambda}\right), \tag{6.26}$$

and this calculation largely eliminates the strong curvature seen in fig. 6.6. It might be a useful project to plot this equation to see the predicted true-stress behavior of rubber.

One final aspect of rubber-network properties that remains is what happens when a random-coil molecule or cross-linked network is extended to the point where the molecular chains are approaching full extension. Eq. (6.25) above, based on Gaussian statistics, predicts that the stress-extension curve in tension will continue with a constant slope indefinitely, but this is clearly not possible. At some point the random chains will be fully straightened, and the covalently bonded polymer backbone will begin to resist further extension very strongly. In fact, well before the chain is fully extended, stiffness will start to rise because chain entropies will rise more rapidly than predicted by the Gaussian analysis described above. Non-Gaussian statistical analysis of chain entropy allows us to quantify this increase in stiffness. An important feature of non-Gaussian behavior is that the extension at which the rise in stiffness starts is determined by the number of "equivalent random links" that exist in the polymer chains, and this effect provides us with a method to estimate the size of the equivalent random links in a rubberlike polymer.

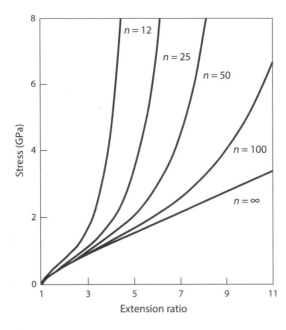

Figure 6.7. Effect of chain length on the shape of stress-strain curves for rubber networks. Each line is drawn for networks formed from random chains made from different numbers, n, of equivalent random links. The line labeled $n = \infty$ is for very long chains, which will have properties as described by eq. (6.25). The other lines show the properties of short-chain networks, with 100, 50, 25, and 12 random links in the chains between cross-links.

Fig. 6.7 shows non-Gaussian predictions for the tensile behavior of random-chain networks. The data are plotted as engineering stress versus extension ratio curves that are based on an analysis by Treloar (1975). The curve at the bottom, labeled $n = \infty$, is the Gaussian analysis, eq. (6.25), which effectively assumes infinitely long chains. The curves above it are based on non-Gaussian models for chains with different numbers of equivalent random links in the network chains. A network with 100 random links tracks the Gaussian prediction to about $\lambda = 4$, and the deviation from the Gaussian prediction occurs at smaller extension as the number of random links drops. At 12 random links per chain the non-Gaussian prediction departs from the Gaussian analysis before the material has been extended by 100%. We will see when we discuss the mechanics of rubberlike proteins from animals that this kind of analysis will allow us to estimate the inherent flexibility of protein polymers by comparing their extension curves with the models plotted in this figure.

5. EXPERIMENTAL MEASUREMENTS. The final issue that we will consider in this analysis of rubber elasticity is the assumption that the elastic mechanism is due entirely to the change in entropy of kinetically free random chains. Our assumption has been that there are no changes in internal energy with extension, $\partial U / \partial L = 0$, and we can test this assumption by investigating the effect of temperature on the elastic behavior of rubbers. To do this we need to revisit briefly some of the thermodynamic relations we used previously. Starting with eq. (6.9), which is repeated below as eq. (6.27):

$$F = \left[\frac{\partial \mathcal{F}}{\partial L}\right]_{T,V} = \left[\frac{\partial U}{\partial L}\right]_{T,V} - T\left[\frac{\partial S}{\partial L}\right]_{T,V}. \tag{6.27}$$

Maxwell's relations allow us to conclude that:

$$\left(\frac{\partial F}{\partial T}\right)_{L,V} = -\left(\frac{\partial S}{\partial L}\right)_{T,V}, \tag{6.28}$$

and substituting eq. (6.28) into eq. (6.9) (or [6.27]) gives:

$$F = \left(\frac{\partial U}{\partial L}\right)_{T,V} + T\left(\frac{\partial F}{\partial T}\right)_{L,V}. \tag{6.29}$$

Note that this equation now contains notation that specifies that the partial differentials are at constant volume, because one of our assumptions for eq. (6.2) was that extension of the random coils should occur with no change in volume. Therefore, eq. (6.29) allows us to calculate the force arising from changes in internal energy, F_U, and the force arising from changes in entropy, F_S, as follows:

$$F_U = \left(\frac{\partial U}{\partial L}\right)_{T,V} \quad \text{and} \quad F_S = T\left(\frac{\partial F}{\partial T}\right)_{L,V}. \tag{6.30}$$

This result forms the basis for a simple experimental protocol that can be used to measure the relative contribution of changes in internal energy (bond-energy elasticity) and entropy (rubber elasticity) to the elastic force. Eq. (6.29) is a linear equation, where the first term on the right, F_U, is a constant, and the second term on the right, F_S, is a variable in T, where the variable is the change in elastic force, F, with change in temperature, at constant volume and length.

The experiment involves stretching an elastic object to a fixed length and then measuring the effect of changing temperature on the magnitude of the elastic force. This is called a thermoelastic experiment. Fig. 6.8A shows a graph that indicates the results of this type of experiment. The graph plots three values for the elastic force at several temperatures, indicated by the crosses at the top right. The slope of these data is the value in the brackets of the second term on the right of eq. (6.29), and this slope is extrapolated to its intercept at 0° on the temperature axis, because when $T = 0$ the second term is zero, and the force equals the internal-energy component alone. That is, the intercept in eq. (6.29) is equal to the fraction of the force due to changes in internal energy, F_U, and $1 - F_U$ is the fraction of the force due to changes in entropy. This experiment can be repeated at a range of extensions, and the results will look like those in fig. 6.8B. This diagram shows the kind of results that would be obtained for natural rubber tested at constant length, and note that it is assumed that the volume of the sample remains constant as temperature is changed. It can be seen that the data labeled F_S track the measured force closely, but appear to be shifted significantly below the force curve. In addition, the data labeled F_U are well above zero. Together, these results suggest that there is a significant contribution from changes in internal energy, which suggests that bond-energy elasticity makes a fairly significant contribution to the energetics of natural rubber.

However, the experiment described above assumes that the rubber sample remains at constant volume as its temperature is changed, and it is observed that rubber exhibits quite a large coefficient of thermal expansion, about 10^{-4} per °C. Thus, the experiment as described does not maintain constant volume, and eq. (6.29) does not really apply. It can be very difficult to maintain constant volume in this type of experiment, but it

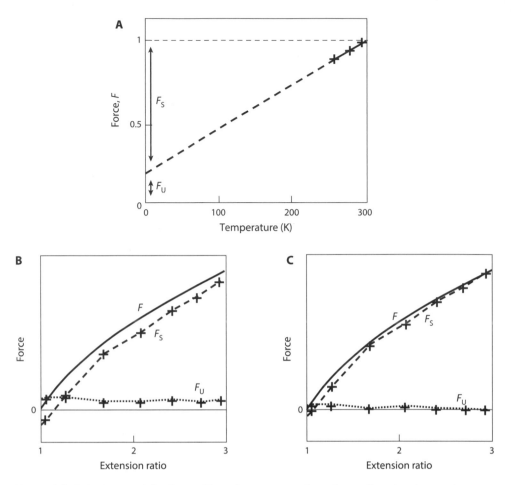

Figure 6.8. Relative contributions of bond-energy and conformational entropy changes to elasticity of rubberlike materials. The results of thermoelastic experiments to determine the relative contributions of bond-energy and conformational-entropy changes to the elastic mechanism of rubberlike materials. **Panel A** shows how a thermoelastic experiment works. A rubber sample is stretched to a constant length at high temperature, and the elastic force is measured as the sample is cooled. The data are represented by the three crosses in the upper-right corner of the graph, and the force axis is scaled to the force value at the highest temperature. The trend of these data is extrapolated to 0 Kelvin, and the intercept on the force axis indicates the fraction of the elastic force that arises from changes in internal energy, F_U. The remainder of the scaled force gives the fraction of the force that arises from changes in the conformational entropy of the random chains in the rubber network, F_S. **Panel B.** Typical data obtained for rubber networks formed from natural rubber, *cis*-polyisoprene. The graph plots the force-extension curve (F) as well as plots of F_U and F_S at a number of different extensions. The expected trend is seen, as the majority of the elastic force appears to arise from changes in entropy, but, interestingly, at low extensions the entropy component is negative and the internal-energy component dominates. However, as indicated by eqs. (6.29) and (6.30), this experiment requires that sample volume remain constant as temperature changes. **Panel C.** If the force-temperature data are corrected for thermal expansion, giving the force at constant extension ratio, then the calculated F_U and F_S values are essentially in agreement with the theory of rubber elasticity. That is, the elastic force arises primarily from changes in the conformational entropy of random-coil polymer chains.

is possible make an approximate correction for volume changes by measuring F_S at constant extension ratio, as:

$$F_S \approx T \left(\frac{\partial F}{\partial T} \right)_{P,\lambda},$$

(6.31)

which means that we can carry out the experiment at constant pressure, P, as long as we maintain constant extension ratio, λ, of the sample in the experiment. This is actually relatively easy to do, and it provides a reasonable correction for temperature-dependent volume changes, as illustrated in fig. 6.8C. This plot shows that F_S now essentially accounts for all of the elastic force, and F_U is essentially zero, indicating that changes in the conformational entropy of random-network chains are indeed responsible for rubberlike elasticity.

Chapter 7

The Molecular Origins of Viscoelasticity

In the last chapter, we described experimental approaches to studying the mechanical properties of polymeric materials. Now, with an understanding of the behavior of kinetically free random-coil molecules, we have a basis for analyzing the molecular origins of polymer viscoelasticity.

1. DIFFUSION AND ENTANGLEMENT. Starting first with simple condensed fluids composed of atoms or small molecules, we know that the particles that make up a fluid are constantly moving owing to thermal agitation in the process of diffusion. Einstein's theory of diffusion defines the diffusion coefficient, D, as:

$$D = \frac{kT}{f_0},\tag{7.1}$$

where kT (k is Boltzmann's constant, 1.38×10^{-23} m^2 kg s^{-2} K^{-1}; T is temperature in K) is the intensity of thermal energy, and f_0 is the molecular friction coefficient (kg s^{-1}). Molecular friction in a fluid arises from the weak secondary bonding forces that act between fluid particles and are essential for the maintenance of the fluid state. These weak bonds include hydrogen bonds, dipole-dipole bonds, and induced-dipole bonds, depending on the chemistry of the particles involved, and it is the strength and number of the bonds present in any particular fluid particle that determine the rate of diffusion in the fluid as well as the temperature at which the fluid becomes a gas. If thermal energy (kT) is high, individual particles will fly apart and become a gas because the attractive forces between particles are not sufficiently strong to maintain the fluid state. At lower temperature, however, bonding forces are sufficient to maintain a condensed liquid state, but thermal energy still plays a key role in driving the diffusion of fluid particles.

What is required for an individual fluid particle to move relative to its neighbors? Because fluid particles are relatively tightly packed together, the movement of any individual particle will be determined by the presence of a "hole" of sufficient size for that particle to move into. Because holes are created by the breaking of the weak bonds between neighboring particles, there is an activation energy (ΔH_{act}) requirement for the formation of a hole, and we can describe the effect of temperature on the diffusion constant as:

$$D(T) = D_0 \exp\left(-\frac{\Delta H_{act}}{kT}\right), \tag{7.2}$$

where D_0 is a reference diffusion coefficient. This relationship provides a mechanistic explanation for the molecular friction coefficient that may be helpful in thinking about the molecular origins of viscoelasticity.

With viscous flow in fluids, bulk fluid motion is created by the action of an external force, F, which arises from a shear stress applied to a volume of liquid, as we saw in fig. 2.5C. The viscous flow of fluid particles relative to one another requires an input of external work to create this movement. However, a relationship similar to eq. (7.2) holds because the forced movement of any fluid particle relative to its neighbors will also require the creation of a hole that it can move into. Thus, fluid viscosity, η, will also be related to an activation energy for the creation of that hole, as:

$$\eta(T) = A \exp\frac{\Delta H_{act}}{kT}, \tag{7.3}$$

where A is a constant. So the viscous processes that occur in the deformation of viscoelastic materials are resisted by molecular friction processes similar to those that determine the diffusion coefficient.

When we translate these ideas to amorphous polymers, we can see similar phenomena at two length scales. We will do this by analyzing the behavior of a synthetic polymer, polymethyl methacrylate (PMMA; Plexiglas), because of the extensive data available for this common industrial material. In later sections we will see how the behaviors described here for PMMA relate to a variety of natural materials. Consider a single molecule of PMMA, a polymer that is a rigid glass at room temperature. However, we will start with this molecule dissolved in a solvent so that at room temperature the molecule becomes kinetically free and behaves like a molecular random walk. We will describe this polymer molecule as being constructed from a series of polymer segments, where the term "segment" is a more convenient name for "equivalent random link" that we used in chapter 6 in our discussion of rubber elasticity. We can now consider the diffusion and the forced movement of polymers and polymer segments through their solvent and amongst their neighboring polymers and polymer segments. The force, F, required to move a single particle at unit velocity, u, through its environment is:

$$F = uf_0, \tag{7.4}$$

where f_0 is the molecular friction coefficient defined above, and if we have a polymer molecule that contains n polymer segments, then f_0 for the polymer will be:

$$f_0 = nf_s, \tag{7.5}$$

where f_s is the segmental friction coefficient, and the force, F_p, to move the polymer at velocity u will be:

$$F_p = unf_s. \tag{7.6}$$

2. VISCOSITY AND CHAIN LENGTH. Polymer molecules in dilute solution behave independently, and they exert a relatively small effect on solution viscosity. However, as their concentration or their size is increased, they begin to interact, and their impact on solution viscosity increases dramatically. Experiments to measure the effect of polymer molecular weight on the viscosity of concentrated polymer solutions demonstrate this phenomenon clearly. In the experiment described below, the weight concentration of the polymer is held constant, but the molecular weight of the polymer (i.e., chain length, or segment number) was varied, and solution viscosity was measured. The form of the results, which is quite general for a wide variety of amorphous polymers, is shown in fig. 7.1A and is based on data from Bueche et al. (1963) for PMMA dissolved in a compatible solvent at a weight concentration 0.25 kg polymer per liter. The graph plots log viscosity, $\log \eta$, against the log of polymer molecular weight, $\log M$, of different batches of polymer with molecular weights that span from 10 to about 300 kD. What we see is that at low molecular weight (small segment number, n) the viscosity of the solution rises as molecular weight to the first power, $\eta \, \mu \, M^1$. However, as molecular weight increases beyond a threshold at about 60 kD, viscosity increases as molecular weight to the 3.5 power, $\eta \, \mu \, M^{3.5}$. The breakpoint at the intersection of the two lines occurs at a molecular weight, M_b, where entanglement between two polymer chains begins to occur, and at this point on average each polymer molecule is entangled with one other molecule, effectively doubling the size of the molecule. But what do we mean by entangled?

Fig. 7.1B provides a drawing of two random-coil polymer molecules, one drawn with a solid line and the other with a dotted line, that are entangled, and what we see is that the two polymer chains overlap in space and interact with each other very closely. The individual chains interpenetrate one another completely because a segment of one molecule is chemically indistinguishable from one in the other molecule. Recall that these molecules are in solution, so the empty spaces are also filled with solvent, which keeps the polymer chains from bonding together. Thus, both chains are kinetically free and are changing their shapes owing to the thermal agitation of the solvent molecules and their component polymer segments. Individual molecules will exhibit rubberlike elasticity, as we have described above, and in the polymer solution at the entanglement breakpoint, M_b, the chains will be entwined together sufficiently that if a force is applied to a chain end on one molecule it will be transmitted through the entangled chains to the other molecule. Fig. 7.1C shows the two molecules after the application of opposing forces at one end of each chain (arrows), and the chains have moved apart, revealing the single entanglement that holds the chains together. This is likely a reasonable description of average molecular interactions at M_b, and as molecular weight becomes larger more molecules will interact, forming larger aggregates with multiple entanglements. This process is what causes polymer viscosity to rise rapidly in the high-molecular-weight range. There are, however, no permanent chemical cross-links between the chains, and a static force on the molecules, as indicated by the arrows in fig. 7.1C, would decline over time to zero as segmental motion, driven by thermal agitation, allows the two molecules to relax back to their unentangled molecular shapes.

Fig. 7.1B could also represent two entangled chains in a solid sample—that is, one with no swelling solvent—and we could consider their behavior as a function of temperature. The situation described above, where the polymer chains are kinetically free

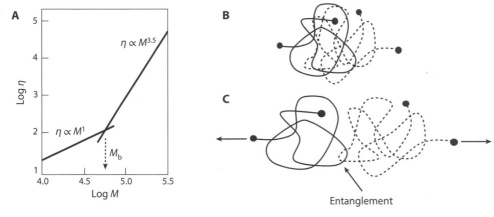

Figure 7.1. Effects of molecular entanglement. **Panel A** shows the results of experiments to investigate the effect of polymer molecular weight (M) on the viscosity (η) of concentrated polymer solutions. The breakpoint in this plot (M_b) indicates where high-molecular-weight polymers begin to interact to form entanglements, which form temporary "cross-links" between the molecules and allow the polymers to have a much larger effect on the viscosity of the solutions. Entanglements between polymer molecules can create time-dependent rubber elasticity in un-cross-linked polymer systems. **Panel B** shows two long-chain polymers that are entangled with one another. If we could grab hold of one end of each molecule and pull them apart at the arrows **(panel C)**, we would experience a rubberlike elastic force between the two molecules that would arise from changes in their conformational entropy. However, the force on the molecules would decline over time to zero as the two molecules to relax back to their unentangled molecular shapes.

and able to move and shed entanglements, requires that a solid sample made from PMMA would have to be well above its glass transition temperature, T_g; for PMMA, $T_g \cong 105°C$. However, we could design an experiment where we dynamically loaded the test sample at a fixed frequency, around 1 Hz, and then measure the dynamic stiffness or storage modulus, E', as a function of temperature. Typical results for this test are illustrated in fig. 7.2A. Starting from the left, we see the sample tested at about 10° above its T_g, and at this point the sample's stiffness is somewhat below 0.1 GPa. In this temperature range the dynamic stiffness is falling rapidly because of rapid changes in the mobility of segments in the coiled polymer molecules, which we will discuss next. For the moment, note that at about 130°C and above the modulus levels off in what is labeled the rubber plateau. This is the temperature range where the diffusion coefficient for the polymer segments is high and rising, according to eq. (7.1). Thus, the polymer chains are kinetically free, and they are becoming more mobile as temperature rises. In this temperature range the polymer should exhibit rubber elasticity, and we know the signature of rubberlike behavior; namely, that stiffness rises with increased temperature. This is exactly what we see for the dotted line, which represents the behavior of an amorphous polymer that is chemically cross-linked to form a permanent rubber network.

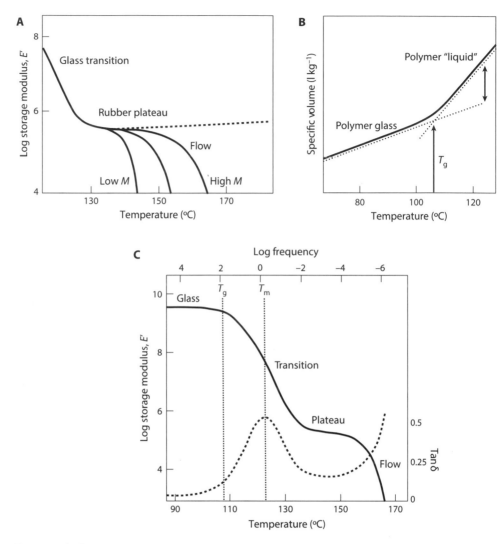

Figure 7.2. Polymer viscoelasticity—overview. **Panel A** shows the mechanical properties of polymethyl methacrylate (PMMA) in the solid state, without solvent. The property measured is the storage modulus, E', determined with dynamic mechanical testing at a frequency of about 1 Hz, over a range of temperatures somewhat above the glass-transition temperature of this polymer. The data are plotted as log E', because the range of stiffness is very large. M, molecular weight. **Panel B** illustrates the specific-volume method for measuring the glass-transition temperature (T_g) for an amorphous polymer. The double-headed arrow highlights the different slopes of the temperature response between the glass and liquid states, a difference that causes the two lines to intersect at T_g. **Panel C** shows the full range of viscoelastic behavior expected for un-cross-linked amorphous polymers, again plotted as the log of storage modulus, E' (*solid line*), and the damping term, tanδ (*dotted line*), against temperature. The properties shown are for PMMA, but they are characteristic of other amorphous polymers. The details of these behaviors are explained in the text. T_m, midpoint of the glass transition.

Now look at the three solid lines that continue at temperatures above 130°C. They illustrate the behavior of un-cross-linked PMMA with different molecular weights. In each case the stiffness levels off with increasing temperature, indicating that molecular entanglements are forming temporary "cross-links," but because there are no permanent cross-links these polymers will always show a drop in modulus that occurs because entangled polymer molecules slide apart as the material starts to flow like a fluid. As illustrated in the viscosity experiment shown in fig. 7.1A, the ability of polymer molecules to slide apart is strongly dependent on polymer molecular weight. So the sample that created the curve with the broadest plateau zone must have contained the highest-molecular-weight polymers, which is indeed the case.

3. THE GLASS TRANSITION. Now we can consider the glass transition itself, which is the process by which heating a rigid polymer glass causes segmental diffusion processes to increase rapidly. This transition occurs because the secondary bonding forces that hold the randomly coiled polymer molecules into fixed conformations are broken by the increasing intensity of thermal agitation (kT). For PMMA this process starts at temperatures below the glass-transition temperature, T_g, and it is useful to understand how T_g is determined and what it tells us about the molecular processes involved in the glass transition. T_g can be measured in several ways, but most commonly it is measured with either thermomechanical measurements to track the change in specific volume of the polymer (liters per kilogram of polymer), or with differential scanning calorimetry to measure the change in the heat capacity of the polymer across the transition. It will be useful for us to consider the specific-volume method, as it provides important information about the change in free volume in the polymer, which is directly related to the change in material properties seen in this transition.

Fig. 7.2B shows a typical plot of polymer specific volume against temperature for an amorphous polymer at its glass transition. The temperature scale has been adjusted so that this graph roughly describes the behavior of PMMA. The experiment involves heating the polymer sample to well above the glass point in a fixed-volume chamber. The sample is then cooled at a controlled rate, typically a few degrees per minute, and the specific volume is plotted against temperature. As the temperature falls in the "liquid" state, which would correspond to the heat-softened polymer melt, there is a rapid decline in specific volume. As the sample approaches T_g, there is a discontinuity, and over a narrow temperature range the slope of the fall in specific volume is reduced to about one-third of the value above the transition. The intersection of the two linear extrapolations provides a discrete value for T_g, in spite of the fact that the process of transforming a rigid polymer glass into a fully mobile sample of kinetically free polymer chains can span more than 50°C. In addition, the exact temperature of the intersection that defines T_g varies with the rate of cooling, with faster cooling rates giving higher values for T_g. So the glass-transition temperature of a polymer is not nearly as well defined as melting point of a crystalline material, where the melting occurs over a very narrow range of temperature.

There is additional information available in fig. 7.2B from the magnitude of the changes in specific volume that occur at the glass transition, specifically the threefold increase in thermal expansion seen above T_g. The increased thermal expansion produces a dramatic change in the rate at which free volume is created in the polymer melt. The volume coefficient of thermal expansion for PMMA below the glass transition is approximately 10^{-4} per °C, and above T_g the expansion coefficient will be of the order of 3×10^{-4} per °C. This means that over the roughly 50°C span of the glass transition there will be an additional 2% increase in volume over the expansion that would have occurred in the rigid glassy state of this polymer, as indicated by the double-ended arrow in fig. 7.2B. Much of this additional volume contributes to the "free volume" that is required to create the holes into which polymer segments move when a kinetically free random-coil molecule diffuses into new shapes. Referring back to eq. (7.2) for the diffusion coefficient of particles, or in this case of polymer segments, there are two factors that increase the diffusion coefficient. First, rising temperature, expressed in the kT term, reflects the rise in thermal agitation that drives segmental motion. Second, the magnitude of the activation energy, ΔH_{act}, falls as temperature rises because of the increase in free volume created by the rapid thermal expansion of the material above its T_g. Together, these factors make it more likely for a "hole" of sufficient size for segmental motion to occur as temperature rises above T_g. Amorphous-polymer systems like PMMA or the protein rubber elastin, described briefly in fig. 3.10, follow an empirical relationship, called the WLF equation (William, Landel, and Ferry equation) that provides a quantitative relationship for the change in ΔH_{act} over the glass transition of amorphous polymers, and those wanting to understand this process in greater detail should consult texts on polymer viscoelasticity, such as Shaw and McKnight (2005).

With this information in hand, we can now relate the discrete value of the measured T_g to the mechanical behavior of an amorphous polymeric material over the full process of the glass-to-melt transition. This is shown in fig. 7.2C, which shows a viscoelastic response curve for an un-cross-linked amorphous polymer such as PMMA loaded dynamically. Again, the data are plotted as log storage modulus, E', at a single frequency around 1 Hz, versus temperature, and in this plot an additional dotted line is included for the damping term, $\tan \delta$. Starting at the low-temperature end of the graph, we see the behavior of PMMA as a rigid glass, where at 90°C and below stiffness is constant and high, in the range of 3 GPa. In addition, in this glass state the damping, $\tan \delta$, is low, reflecting the fact that the randomly coiled polymer chains are not changing their shape appreciably because the secondary bonding forces are holding the coiled chains into fixed conformations. Thus, high stiffness arises from the deformation of the secondary bonding forces that stabilize the coiled molecule, along with a contribution from the rigid covalent bonds in the polymer backbone. It is worth noting that the actual magnitude of the stiffness in this polymer glass is quite a bit lower than that seen for other rigid materials that are stabilized by stronger and stiffer bonding forces. For example, inorganic glass, ceramics, and metals have stiffness values that range from 70 GPa for glass, to 200 GPa for steel, to 1220 GPa for diamond. The relatively low stiffness for amorphous polymers in their glassy state is due to the low stiffness of the secondary bonding forces that stabilize the randomly folded polymer chains.

The lack of damping in the glassy state reflects the virtual lack of segmental motion, which prevents energy loss through frictional processes. Somewhere near 100°C, however, the modulus begins to fall, and at about 105°C we reach T_g, as measured in the specific-volume experiment described above. At this point the stiffness has not fallen much, and thus the value of T_g measured in this way falls at the low-temperature end of the actual transition. Typically, the full transition involves a roughly 1000-fold drop in stiffness, and thus the measured T_g value is typically 10–20°C lower than the midpoint of the transition, T_m, as indicated by the peak of the damping factor, tanδ, which occurs very near to the midpoint of the fall in stiffness. At about 135°C the sample enters the rubbery plateau zone described above, where the fall in stiffness is reduced, and the damping falls but does not reach the low levels seen in the glass at low temperatures. This is because PMMA is not cross-linked into a permanent network like that found in rubbery materials; thus, the plateau arises from a temporary network created by entanglements between the long, intertwined polymer chains, which will slide apart as temperature rises because diffusion allows increased flow of entire molecules past one another. Thus, at the highest temperatures the material becomes a viscous liquid, where stiffness falls precipitously and damping rises again to very high levels. However, if the PMMA were cross-linked into a polymer network, the rubber plateau would extend to high temperatures with a rising slope, as indicated by the dotted line in fig. 7.2A.

It is important to note again that the properties illustrated in this diagram are based on cyclic loading at 1 Hz, and you should remember from the discussion of the time-temperature superposition principle in chapter 3 that the position of these curves on the temperature scale would also depend on the frequency of loading used to document the changes in mechanical properties. Thus, mechanical testing at high frequency (e.g., 10^4 Hz) will shift the curve to the right (higher test frequencies would see greater stiffness at any particular temperature), and lower test frequencies would shift the curves to the left. One could even do this entire test at a single temperature and document the dynamic mechanical properties by varying only the frequency. To make this relationship clearer, the horizontal axis at the top of fig. 7.2C shows the frequency scale that would be used if this dynamic mechanical test were conducted at a constant temperature of 105°C, and the dynamic test frequency were varied. The frequency axis is on a logarithmic scale because the range of viscoelastic behavior shown here spans about 12 logarithmic decades of frequency. Now, zero on the log frequency scale (1 Hz) is located at T_m, the peak of the damping curve and the midpoint of the glass transition. Note also that high frequency is to the left on this scale, because high-frequency deformation will require high forces to overcome the viscous forces within the material, and at low frequency the un-cross-linked chains will be able to flow apart.

4. AN EXAMPLE: ELASTIN. Before we conclude this chapter it will be useful to relate this analysis to the kinds of protein and polysaccharide polymers that exist in animals. Figs. 3.8 and 3.10 show dynamic test data for the rubberlike protein elastin, both in small-amplitude dynamic tests (fig. 3.8) and in constant-strain-rate tensile tests to failure at strains as high as 2 (fig. 3.10). Elastin is one of very few biomaterials where we have a reasonably complete understanding of its viscoelastic properties, and it is appropriate for you to look at these figures again because they illustrate one particularly important feature of natural materials; namely, that in their normal function in live organisms they typically are in water-rich environments. Thus, solvent effects must be considered in the analysis of viscoelastic behaviors, in addition to the effects of

temperature and strain rate. Fig. 3.8 shows results from dynamic tests on fully hydrated elastin (solid lines) and on elastin with hydration reduced by about 10% (dashed lines), plotted as a function of dynamic test frequency. Elastin functions in many tissues in vertebrate animals, but one of its major roles is in the elastic arteries of the central circulation, where the contractions of the heart produce pulsatile flow in the arteries. During systole, when blood is forced into the arterial system, elastic arteries expand, and the efficient elastic recoil of these arteries provides energy to maintain blood flow to the tissues during the diastolic period when the heart refills. Cardiac frequencies are of the order of 1–3 Hz (at about zero on the log frequency scale), so fully hydrated elastin meets the requirements of circulatory dynamics because over this frequency range the storage modulus, E', is low, at about 1 MPa, and damping, $\tan\delta$, is also low, indicating high resilience, or efficient elastic recoil. Loss of a small amount of loosely bound water shifts the curves to the left, causing elastin's stiffness and energy loss to increase, and it is possible that shifts of this nature occur in diseased arteries where lipid accumulation and calcification of elastic tissues are present.

Elastin is also subject to static loading over very long time scales in arteries, owing to the constant diastolic pressure between systolic pulses, and in tissues such as skin that remain under constant tension over long periods. This aspect of elastin's function would be far to the left of the graph as shown, at very low frequencies. Because elastin is a cross-linked network, the values of E' and E'' established at the left of this graph should continue at the same level to very long times, barring any fatigue processes leading to eventual failure. On the other hand, the high-frequency behavior, documented up to frequencies of about 10 MHz in fig. 3.8, really has no biological significance, other than to indicate that elastin is a typical amorphous polymer, like PMMA, and it will go through a glass transition under appropriate conditions of time, temperature, and hydration. However, other polymeric biomaterials, such as spider dragline silk, function under conditions that truly span the full range of properties illustrated in fig. 7.2C. As we will see in later sections, spider dragline silk is formed from a hybrid protein polymer that contains elastin-like amino-acid sequence blocks that alternate with sequence blocks that crystallize into rigid, fibrous nanostructures. In the process of silk formation within the spinneret the silk proteins remain hydrated, and as the crystalline regions begin to develop, the hydrated, amorphous chain segments deform to allow the crystal domains to assemble and to become oriented along the silk fiber's axis. As the silk emerges from the spinneret it dries, and the amorphous regions become rigid and lock the crystalline domains into the ordered composite structure we know as spider dragline silk.

Interestingly, the stress-strain curves in fig. 3.10 give us some insights into the origins of spider dragline silk's remarkable mechanical properties. This figure shows typical stress-strain curves for elastin samples tested in constant-strain-rate tensile tests to failure under a wide variety of testing conditions. The curve labeled "rubber" shows the behavior of fully hydrated elastin tested at low strain rates, and we see behavior typical of the elastin in arteries. The stiffness is low, at about 1 MPa; tensile strength is low, at about 1.3 MPa; and extensibility is modest, at about $\varepsilon_{max} = 1.2$. In the region labeled "glass transition" we see a progression of properties where stiffness, strength, and extensibility all rise. These curves were obtained by increasing strain rate and decreasing temperature and hydration. The exact conditions are not important, but shifting elastin into its glass transition, where energy dissipation is high, provides viscoelastic mechanisms that increase the energy to break elastin, as calculated from the area under the stress-strain curve (J m^{-3}), by roughly 30-fold. Similar changes will occur in the amorphous segments of spider silk when the silk dries as it exits the spinneret. The combination of a rigid matrix with fibrous substructures in this composite-like material produces silk fibers with truly exceptional strength and toughness, as we will see in section III, on the mechanical design of tensile materials.

Chapter 8

The Design of Composite Materials

The details of failure mechanisms of biomaterials will be discussed in greater detail in sections III, IV, and V, devoted to the three classes of biomaterials, but it will be useful to consider some general features of design in composite materials, as essentially all structural biomaterials show some level of composite design. Fiber-reinforced-composite design in engineering materials takes advantage of the inherent strength of thin fibers, as we have seen for the S-glass fibers that are commonly used in fiberglass composites. But thin fibers are extremely flexible, and they are not able to provide the structural rigidity in compression, in bending, or in torsion needed in objects such as tennis rackets or limb bones. However, if glass fibers are bonded together with polymer resins to form a rigid composite structure, the many thin individual fibers deform as a single, much larger, unit that behaves very much like an object made from a single piece of solid glass. The main difference between a fiberglass tennis racket and a solid glass tennis racket is that the fiberglass racket is much stronger and will not shatter if it is dropped. There are a number of mechanisms that are responsible for the toughening of the glass in a fiberglass composite. First, as mentioned above, the use of thin glass fibers means that stress-concentrating flaws in the glass fibers are necessarily very small, so failure of individual fibers will happen at very high stress, perhaps even two orders of magnitude higher than the failure stress of bulk glass. In addition, interactions between the glass fibers and the matrix of organic plastics that bonds them together into a rigid solid provide mechanisms for controlling and limiting the development of cracks capable of initiating catastrophic fracture events. So we should expect that toughness will increase with fiber-reinforced composites.

1. FIBER AND MATRIX. However, there are compromises in properties that arise with composite designs, and let's start by looking at the stiffness of composites relative to the stiffness of the material that forms the reinforcing fibers (e.g., the glass in fiberglass). Fig. 8.1A shows a diagram of a fiber-reinforced composite, where the gray stripes represent reinforcing fibers, and the clear stripes represent the matrix. This diagram indicates a composite with a fiber volume fraction $V_f = 0.5$, as the matrix layers are the same width as the fiber layers in this diagram. Bulk glass and glass fibers have essentially identical stiffness, so the stiffness of a composite made with glass fibers will

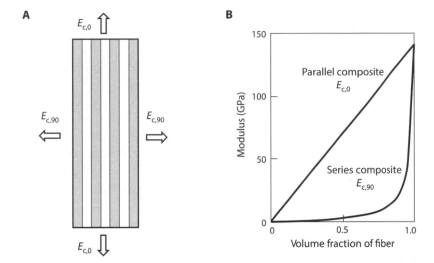

Figure 8.1. Design of composite materials I. **Panel A** is a diagram of a fiber-reinforced composite, where the gray stripes represent a reinforcing fiber, and the clear stripes represent the matrix. The diagram shows a composite with a fiber volume fraction $V_f = 0.5$. The *arrows* surrounding this diagram indicate the directions of loading if this composite were stressed parallel or perpendicularly to the fiber axis, allowing us to determine the magnitudes of the modulus of the composite under parallel and perpendicular stress, $E_{c,0}$ and $E_{c,90}$. **Panel B** shows the results of calculations for a composite of Kevlar fibers and an epoxy matrix at different fiber volume fractions, based on eqs. (8.1) and (8.2). The modulus of the parallel composite rises linearly from the matrix stiffness at $V_f = 0$ to a maximum value equal to the fiber stiffness at $V_f = 1$. The modulus of the series composite rises asymptotically between the same extreme levels, but over the midrange of volume fractions the series composite will be much less stiff than the parallel composite.

depend on the volume fraction of reinforcing fibers in the composite as well as the orientation of the fibers relative to the applied load. The arrows surrounding this diagram indicate the directions of loading if this composite were stressed parallel or perpendicularly to the fiber axis, allowing us to determine the magnitudes of $E_{c,0}$ (modulus of the composite under stress parallel to the fiber axis) and $E_{c,90}$ (modulus under stress perpendicular to the fiber axis). For stress applied parallel to the long axis of reinforcing fibers (fiber angle = 0°), the rule of mixtures applies, as shown in eq. (8.1), where E_f and E_m are the modulus of the fiber and matrix respectively, and V_f and V_m are their volume fractions. This equation is based on the assumption that fibers and matrix are strained equally, and because E_f is typically much greater than E_m, composite stiffness increases linearly with the fiber volume fraction, as:

$$E_{c,0} = E_f V_f + E_m V_m, \qquad E_{c,0} = E_f V_f + E_m (1 - V_f). \qquad (8.1)$$

However, if the applied stress is oriented perpendicular to the fiber axis (fiber angle = 90°), the situation is very different because the fiber and the matrix are arranged in series and experience equal stress:

$$\frac{1}{E_{c,90}} = \frac{V_f}{E_f} + \frac{V_m}{E_m}. \tag{8.2}$$

In this case, the matrix will dominate because of its lower stiffness, which means that matrix strain will be greater than fiber strain at most volume fractions. For example, if $E_f = 100$, $E_m = 1$, and $V_f = V_m$, then the parallel composite will have a stiffness $E_{c,0} = 50.5$, and the perpendicular composite will have a stiffness $E_{c,90} = 2$. Fig. 8.1B shows the results of calculations for a composite of Kevlar fibers and an epoxy matrix, based on eqs. (8.1) and (8.2). The modulus of Kevlar is 134 GPa, and the modulus of epoxy resins are of the order of 3 GPa, so in this case the reinforcing fiber is about 45 times stiffer than the matrix. The modulus of the parallel composite rises linearly from the matrix stiffness at $V_f = 0$ to a maximum value equal to the fiber stiffness at $V_f = 1$. The modulus of the series composite rises asymptotically between the same extreme levels, but over the midrange of volume fractions the series composite will be much less stiff than the parallel composite. Clearly there are considerable benefits of making V_f of composite materials as high as possible, and also of having loads applied parallel to the direction of the reinforcing fibers, but it is very difficult to assemble composites at very high V_f, and it is also difficult to organize the loading of composite structures, be they artificial or part of a living organism. In artificial composites, it is rarely possible to achieve V_f values greater than about 0.7, but for parallel composites this provides a material with 70% of the stiffness of the reinforcing fiber. At comparable V_f values the series composite is much less stiff, and this aspect of composite mechanics is a very serious concern in structural design.

2. THE EFFECTS OF FIBER ANGLE. We can see from fig. 8.1B that the properties of our uniaxial composite will vary dramatically with the orientation of an applied load, but it is not clear from this graph just how anisotropic a unidirectional composite will be. Fortunately, engineers have developed mathematical approaches to analyze composite properties, and fig. 8.2 shows the results of some of these calculations.

The dashed line in fig. 8.2A, labeled "single lamina," shows the stiffness of a single layer of a Kevlar-epoxy composite with a volume fraction of 0.65 that is oriented at varying angles relative to the test axis. Fig. 8.2B shows the arrangement of a single-lamina test sample in which the fibers are oriented at 15° relative to the test axis. The graph in fig. 8.2A shows the behavior of a full range of single-lamina fiber orientations, and it is clear that fiber reinforcement drops very quickly as the fibers are rotated away from the test axis. At 0° the applied stress is carried primarily by tensile stress in the fibers, with only a small fraction carried by tension in the matrix. However, with rotation composite stiffness falls rapidly, and at 15° the stiffness of the composite has fallen by 70% from the parallel composite stiffness.

Composite design in artificial structures typically involves the buildup of multiple layers (laminae) of a parallel-fiber composite, with the fiber orientation changing from layer to layer, creating a laminated composite structure. The benefit of this kind of composite construction, which is found frequently in both natural and artificial composites, is that building with parallel-fiber laminae makes it possible to produce complex laminates that have high volume fractions of reinforcing fibers. Then it is possible to organize these high-stiffness (and high-strength) laminae to create laminate structures

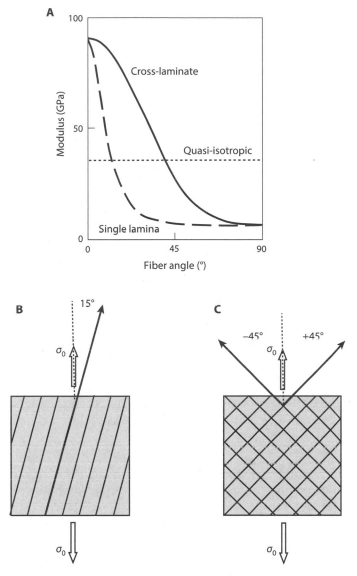

Figure 8.2. Design of composite materials II. **Panel A** shows the results of calculations of the effect of fiber orientation on composite laminate properties. The *broken line* shows the behavior of a single lamina, over a full range of fiber orientation. Note that the stiffness drops off at very small angles relative to the test axis, with stiffness falling to about 30% of the parallel composite at a fiber angle of 15°. The *solid line* shows the effect of fiber orientation on cross-laminate properties. In this case the stiffness drops off more gradually, and at 45° the stiffness is about 40% of the parallel composite. The *dotted horizontal line* shows the behavior of quasi-isotropic laminates, which can be created in laminate composites with multiple fiber orientations, as described in the text. In this case the laminate stiffness is equal in all directions at a level that is about 40% of the parallel fiber composite. **Panel B** shows the arrangement for a single-lamina test sample in which the fibers are oriented at 15° relative to the test axis; σ_0, applied stress. **Panel C** shows a cross-laminate arrangement in which the fibers in successive layers are oriented at +45° and −45° to the test axis.

with a wide range of properties. A common arrangement is to create balanced cross-laminates, where successive laminae of equal thickness are oriented at the same angles to the right and to the left of the test or anticipated-load axis. Fig. 8.2C shows a diagram of a cross-laminate arrangement in which the fibers in successive layers are oriented at +45° and −45° to the test axis, and the solid line labeled "cross-laminate" in fig. 8.2A shows composite stiffness for a Kevlar-epoxy composite ($V_f = 0.65$) assembled at different fiber angles, ranging from 0° (a parallel-fiber composite) to 90° (a series composite). It is clear that such cross-laminates improve the situation considerably over the single lamina because the stiffness falls much more slowly as the fiber angle increases. In this sample calculation the stiffness of the 45° cross-laminate is virtually identical to that for the 15° single-lamina sample for the same material. However, simple cross-laminate composites will fall to the series composite behavior as the fiber angles approach 90°.

The cross-laminate curve in fig. 8.2A was based on a calculation of fiber-reinforcement efficiency at angle θ, $\eta_{\text{ref},\theta}$, in composites (Bunsell and Renard, 2005), as this provides a useful approximation of fiber reinforcement in laminate structures. Readers who wish to understand the full mathematical analysis of laminate mechanics should consult the many texts (such as the one cited here) and technical monographs on fiber-reinforced-composite mechanics. The analysis presented here is based on the observation that fiber-reinforcement efficiency, $\eta_{\text{ref},\theta}$ for any one layer of reinforcing fibers in a balanced cross-laminate composite will contribute to the overall reinforcement as:

$$\eta_{\text{ref},\theta} \mu \cos^4 \theta, \tag{8.3}$$

where θ is the angle between the fibers in the lamina and the loading axis. By summing over all layers in the cross-laminate, one can calculate a reinforcement efficiency of the laminate structure, as:

$$\eta_{\text{ref},\theta} = \sum_i a_i \cos^4 \theta_i, \tag{8.4}$$

where a_i is the fraction of the laminate thickness occupied by any particular lamina, and θ_i is the orientation angle of the fibers in that lamina relative to the loading axis of the sample. The cross-laminate curve in fig. 8.2A was created by using eq. (8.4) to calculate the reinforcement efficiency for a range of cross-laminate fiber orientations ranging from 0° to ±90°, and the composite stiffness was calculated as:

$$E_c = \eta_{\text{ref},\theta}(E_{c,0} - E_{c,90}) + E_{c,90}. \tag{8.5}$$

Eq. (8.4) can also be used to demonstrate the possibility of making laminate composites with equal stiffness in all directions in the plane of the laminae, which are referred to as quasi-isotropic laminates. The simplest arrangement involves three lamina orientations that are 60° apart. That is, if one lamina is parallel to the applied load, the other two lamina orientations will be at +60° and −60°. If these angles are applied to eq. (8.4), with equal fractional thicknesses of the laminae, a_i, the value for $\eta_{\text{ref},\theta} = 0.375$ or 3/8, and the same result will be obtained for any combination of fiber angles as long as they are always 60° apart. Similar results will be obtained for a four-layer design with fiber angles of 0°, 90°, and ±45°. Note, however, that the benefit of equal stiffness in all directions comes at a cost; namely, that the magnitude of the stiffness is at about 40% of the parallel-composite stiffness. And, finally, it is important to note that in all these laminate designs, the stiffness through the thickness of the composite will always be equal to the series-composite stiffness (eq. [8.2]), regardless of the fiber angles in the laminae. That is, the stiffness through the laminate will always be dominated by the properties of the matrix.

A consequence of these properties of laminates is that fiber-reinforced laminates are most commonly used in tubular designs and shell-like structures in which the important mechanical loads run in the plane of the structure, rather than through the tube's or shell's thickness. Examples of engineering composite structures constructed in this way include kayak hulls, tennis rackets, and airplane fuselages/wings. Rigid composite structures in animals constructed in this manner include bones, feathers, mollusk shells, and insect cuticles, and similar designs are seen for soft composites such as arteries, skin, and the body walls of a huge variety of invertebrate animals.

It is also possible to create composites with fiber orientations in all three dimensions, although it is generally not possible to achieve high stiffness and strength in 3-D composites. One way to create a composite with equal fiber orientation in three dimensions is to disperse short fiber segments randomly in the matrix. This process, however, has two significant disadvantages. First, in a random 3-D composite it is very difficult to achieve a high fiber volume fraction, V_f, simply because random packing leaves large spaces that will be occupied by the softer and weaker matrix. Hence, stiffness and strength will necessarily be relatively low. That said, random 3-D composites will have equal stiffness in all directions, and at a given V_f the composite stiffness should be about one-fifth of the stiffness of a parallel composite made with the same reinforcing fibers.

The second disadvantage of using short fibers is that to this point we have assumed that our composites are constructed from reinforcing fibers that span the full length of the objects that are tested. That is, for a sample of a parallel-fiber composite, the fibers run from end to end, and there is no loss of stiffness or strength through the transfer of force from fiber to matrix and then to another fiber. Of course, as we apply stress in directions that differ from the fiber direction, stress transfer between fiber and matrix becomes increasingly important. We have seen this effect in fig. 8.1B and eqs. (8.1)–(8.4) for continuous-fiber composites. If we want to apply these relationships for laminates with discontinuous fibers, we must add another correction term to account for the loss of stiffness and strength in the transfer of stress from fiber to fiber through the matrix. The analysis of this process is quite complex, and readers should consult a composite-materials text for details (e.g., Daniel and Ishai, 2006; Bunsell and Renard, 2005), but it will be useful for us to develop a simplified analysis of discontinuous-fiber composites because these are seen commonly in biological systems.

Fig. 8.3A shows a diagram of a single short fiber surrounded by a large block of matrix. The matrix is loaded in tension at each end by an external stress, σ_0, and the dotted lines that run across the matrix block indicate the distribution of strains in the matrix created by the applied stress. At each end of the block the dotted lines are widely spaced, indicating large longitudinal strains in the low-stiffness matrix material because there is no reinforcement from the fiber. Moving inward toward the end of the reinforcing fiber, the dotted matrix strain lines become bowed because the stress in the matrix is being transferred to the fiber, and in this process the matrix becomes sheared, thus producing the curved lines. Starting where the matrix strain lines start to cross the fiber there is a gradual reduction in both the curvature and the spacing of the dotted matrix strain lines. In the center of the fiber the strain lines are straight and closely

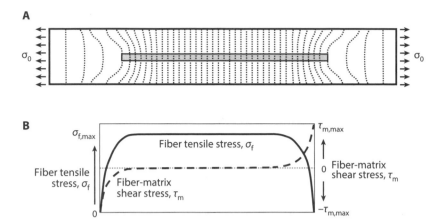

Figure 8.3. Design of composite materials III. This illustrates the mechanics of discontinu-ous-fiber composites. **Panel A** shows a single short fiber surrounded in a block of matrix material, where stress is applied at the ends of the matrix block. In this simple system, tensile stress at the ends of the matrix block is transferred as shear stress in the matrix that surrounds the fiber into tensile stress in the fiber. **Panel B** illustrates the pattern of this stress transfer between matrix and fiber. The plot shows how fiber tensile stress, σ_f, and fiber-matrix shear stress, τ_m, vary along the length of the fiber. At the fiber ends the fiber-matrix shear stress is high, whereas the fiber tensile stress is maximal through the central portion of the fiber.

spaced, indicating that essentially all of the stress applied to the matrix at the end is now transferred into a tensile stress in the fiber. This visual indication of the transfer of matrix shear stress into a tensile stress within the fiber is indicated quantitatively in the graph in fig. 8.3B, which shows what is happening in the composite at the correspond-ing position in the diagram directly above. This graph presents two of the outputs of the short-fiber-composite model that will be described below, the fiber-matrix shear stress, τ_m, and the fiber tensile stress, σ_f. The graph shows clearly that near the fiber ends there is a dramatic fall in fiber-matrix shear stress, τ_m. and a dramatic rise in the fiber tensile stress, σ_f. In the central region of the fiber, however, σ_f reaches a constant high value, and shear stress, τ_m, at the fiber interface is zero, indicating that all of the applied external stress has been transferred through the matrix into the fiber.

Fig. 8.3 shows output for a Kevlar-epoxy composite with $V_f = 0.75$, with a fiber as-pect ratio (total length, L_{tot}, divided by fiber diameter, d) of about 30. For composite systems with similar fiber aspect ratios but different fiber/matrix compositions, the stress transfer from matrix to fiber would not be as complete as that indicated in fig. 8.3B. Thus, calculations based on this theory can give useful estimates of the reinforc-ing process in short-fiber composites, such as those found in natural materials. The short-fiber-composite model requires four basic inputs: fiber aspect ratio, s; fiber vol-ume fraction, V_f; fiber tensile modulus, E_f; and matrix shear modulus, G_m. Fiber aspect ratio, s, is defined as:

$$s = \frac{L_{0.5}}{r} = \frac{L_{tot}}{d},\tag{8.6}$$

where d is the fiber diameter, $L_{0.5}$ is the average half-length of the fiber, L_{tot} is the average full length of the fiber, and r is fiber radius. Fiber volume fraction, V_f, will determine the ratio of the radius of the fiber, r, and the thickness of the matrix layer that surrounds the fiber, r_0. This term is dependent on the packing geometry of the fibers in the matrix, but for a generic solution we will assume that the ratio, r_0/r can be calculated as:

$$\frac{r_0}{r} = \frac{1}{\sqrt{V_f}}. \tag{8.7}$$

3. REINFORCEMENT EFFICIENCY. The theory can be used to calculate a reinforcement-efficiency term, $\eta_{ref,0}$, that can be used to calculate the stiffness of parallel composites made with discontinuous fibers. In this analysis we define a scaling-factor term, n, which provides input for the volume fraction of the composite, the shear modulus of the matrix, and the tensile modulus of the fibers, as follows:

$$n = \sqrt{\frac{2G_m}{E_f \ln \frac{r_0}{r}}}. \tag{8.8}$$

With this value as input for the model, we can calculate the magnitude of the fiber-matrix shear stress, τ_m, along the length of the fiber, as well as the tensile stress, σ_f, that develops along the length of the fiber, as follows:

$$\tau_m = \frac{1}{2}nE_f\varepsilon_c\left[\frac{\sinh(nx/r)}{\cosh(ns)}\right], \tag{8.9}$$

$$\sigma_f = E_f\varepsilon_c\left[\frac{\cosh(nx/r)}{\cosh(ns)}\right]. \tag{8.10}$$

In these equations the term E_f is the fiber modulus; ε_c is the strain of the composite; r is the fiber radius; x is the distance along the length of the fiber, starting from the midpoint and proceeding to the "+" or "−" end of the fiber; and s is the fiber aspect ratio. These are the equations that are used to plot the fiber-matrix shear stress and fiber tensile stress in fig. 8.3B.

Based on these equations, it is possible to calculate the reinforcement efficiency, $\eta_{ref,0}$, that a particular fiber-matrix system will impart to a composite, keeping in mind that the value of n from eq. (8.8) contains information about the fiber and matrix modulus as well as the volume fraction of fiber and matrix in the composite. The value of $\eta_{ref,0}$ is calculated as:

$$\eta_{ref,0} = 1 - \frac{\tanh(ns)}{ns}. \tag{8.11}$$

The sinh, cosh, and tanh terms in these equations are the hyperbolic sine, cosine, and tangent. Those not familiar with these terms will be pleased to know that many spreadsheet programs will calculate them for you. The reinforcement term provided by eq. (8.11) can be put into the formula for the modulus of a parallel composite (eq. [8.1]), giving:

$$E_{c,0} = \eta_{ref,0}E_fV_f + E_m(1 - V_f). \tag{8.12}$$

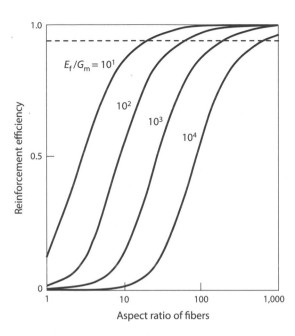

Figure 8.4. Design of composite materials IV. This shows the results of discontinuous-fiber-composite calculations, based on eq. (8.11). The graph shows four plots of the reinforcement efficiency, $\eta_{ref,0}$, against the fiber aspect ratio, where each plot is for a different magnitude of the modulus ratio, as indicated. The dashed line at the top of the plot indicates a reinforcement efficiency of 95%, which is generally taken as the level at which a discontinuous-fiber composite is functionally equivalent to a continuous-fiber composite of the same composition. E_f, fiber tensile modulus; G_m, matrix shear modulus.

This equation indicates that $\eta_{ref,0}$ acts to scale the fiber contribution to the composite, and it is important to note that the two main determinants of the reinforcing behavior of discontinuous fibers are the fiber aspect ratio, s, and the ratio of the matrix shear modulus, G_m, and the fiber tensile modulus E_f. Fig. 8.4 shows four plots of the reinforcement efficiency, $\eta_{ref,0}$, against the fiber aspect ratio, where each plot is for a different magnitude of the modulus ratio. The dashed line at the top of the plot indicates a reinforcement efficiency of 95%, which is generally taken as the level at which a discontinuous-fiber composite is functionally equivalent to a continuous-fiber composite of the same composition.

The curve on the far left has a modulus ratio, E_f/G_m, of 10, and each successive curve to the right increases the modulus ratio by a factor of 10. In all four cases E_f is held constant at 100 GPa, and G_m is adjusted to alter the modulus ratio. The curve for modulus ratio = 100 gives a reasonable approximation of the behavior of the Kevlar-epoxy composite described in fig. 8.1 because the E_f for Kevlar is indeed about 100 times G_m for the epoxy matrix. Note that this curve crosses the 95% line at an aspect ratio somewhat above 30, which means that a discontinuous-fiber composite for Kevlar-epoxy will require fibers with aspect ratios of 30 or greater to achieve a stiffness that approaches that of the continuous-fiber composite. Increasing the modulus ratio to 10^3 and 10^4, by reducing matrix stiffness by a further 10- and 100-fold, shifts the curves to the right, indicating that much higher aspect ratios are required to achieve stiffness levels that approach those of a continuous-fiber composite.

We see in this example the key function of the matrix in stress transfer between discontinuous fibers in a composite. Low matrix shear stiffness will increase the composite strain by allowing the fibers to slide relative to one another, and this can only be prevented either by having much longer fibers or by increasing the stiffness of the matrix.

Decreasing the modulus ratio to 10 increases G_m to one-tenth of E_f, and this shifts the curve to the left of the Kevlar-epoxy composite. Now we see that the 95% stiffness level is achieved at fiber aspect ratios of the order of 10. Volume fraction, V_f, also plays a role in determining the properties of discontinuous-fiber composites, with higher volume fraction increasing the reinforcing efficiency, although there is generally less scope for adjusting properties with volume fraction.

Engineers usually use fabrics woven from continuous fibers to make composite laminates in crucial structural components such as those found in airplanes. However, in less crucial structures, such as boat hulls, it is not uncommon to find fiberglass composites made from relatively short discontinuous fibers. The situation in biology is rather different because in most instances the synthesis of reinforcing fibers occurs at the cellular level, by the self-assembly of nanoscale subunits, and it is likely difficult for living organisms to produce continuous reinforcing fibers that run the full scale of an animal or a plant. As a consequence, it is very likely that the theory described here will play an important role in the design of many composite biomaterials.

4. THE STRENGTH OF COMPOSITE MATERIALS. To this point we have concentrated on the stiffness of composite materials, to develop an understanding of how the properties and volume fraction of stiff reinforcing fibers can be used to create rigid structural materials that can be matched to the requirements of the structural systems in living organisms. However, we started thinking about composite materials in the context of creating strong composites from relatively brittle materials such as glass. Recall that brittle glass becomes extremely strong when it is formed into very thin fibers, and surrounding parallel glass fibers with a rigid matrix to form a composite has the potential to increase the strength of the composite to magnitudes that approach the strength of the thin fibers. The fracture of composites is very complex, and there are many modes of failure that ultimately contribute to the failure of a composite. However, there are some relatively straightforward principles that can give us insights into the design of biological materials.

For parallel-fiber composites we can use mixture equations, like eq. (8.1) for the stiffness of these composites, to predict the strength of composite materials. But there are some additional issues to deal with. We will consider the situation for composites in which the breaking strain of the fiber is less than the breaking strain of the matrix, as this is likely the most common situation in natural materials. As a result, composite failure should be initiated by failure of the fibers. The logic here is that in a parallel composite the matrix and fiber are at equal strain under load, and therefore fibers should reach their breakpoint before the matrix, as indicated in fig. 8.5A. Note that the dashed vertical line indicates the stress in the matrix that occurs when the fibers break, σ_m^*, and with values for the failure stress of the fibers, $\sigma_{max,f}$, we can calculate the breaking stress of the composite as:

$$\sigma_{max,c} = \sigma_{max,f} V_f + \sigma_m^* (1 - V_f). \tag{8.13}$$

This equation is plotted as the line labeled "eq. (8.13)" in fig. 8.5B, and this relationship works at high fiber volume fractions, V_f. At low V_f, however, the situation is quite different because when there are very few fibers present, the individual fibers can all

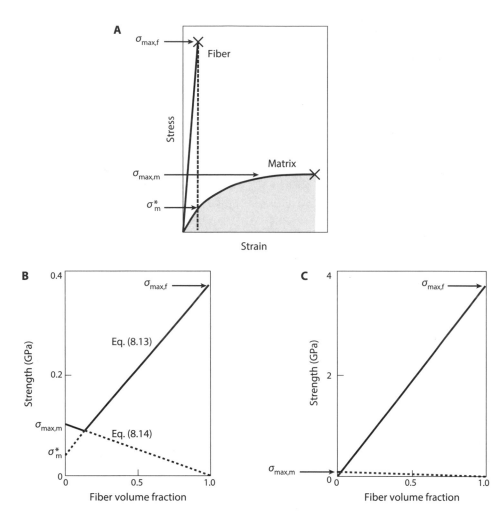

Figure 8.5. Design of composite materials V. **Panels A, B, and C** show how composite strength varies with fiber volume fraction. The systems presented here are assumed to have reinforcing fibers that fail at smaller strains than the matrix material, as illustrated in the stress-strain diagram in **panel A**. In this diagram, a key parameter is the stress in the matrix that occurs when the fibers break, σ_m^*. At low V_f, the strength of the composite is determined by the properties of the matrix. Over a small range of volume fractions, increasing the magnitude of V_f will decrease the strength of the composite, as shown in **panel B**, which has been calculated for a composite in which the matrix strength is one-third that of the fiber. However, in composites where the fiber strength is much larger than the matrix strength, as is the case in **panel C** for Kevlar-epoxy composites, this weakening effect is essentially negligible. See the text for the details of this calculation. $\sigma_{max,f}$, failure stress of the fibers; $\sigma_{max,m}$, failure stress of the matrix.

break and leave the matrix intact. Thus, at low values of V_f the composite strength will be determined by the matrix breaking stress, $\sigma_{max,m}$, and the volume fraction of the matrix, V_m. Since the broken fibers still occupy part of the volume, V_m will be less than 1, and failure strength will be less than the failure strength of the matrix alone. So as V_f increases from 0, the strength of the composite will actually fall because more fiber means less matrix to support the applied load. Thus, at low V_f composite strength should follow the relationship:

$$\sigma_{max,c} = \sigma_{max,m}(1 - V_f), \tag{8.14}$$

and this relationship is plotted in fig. 8.5B labeled "eq. (8.14)." Where these two lines intersect, the composite will exhibit its minimum strength, which, as shown, can be considerably less than the strength of the matrix alone. Note that parts of these two lines are solid and parts are dotted. The solid lines indicate the relationships that actually apply to the strength of the composite. Thus, at low V_f eq. (8.14) applies, and at high V_f eq. (8.13) applies.

The graph in fig. 8.5B has been drawn in a manner that illustrates this relationship clearly by using strength values of $\sigma_{max,f} = 0.4$ GPa, and $\sigma_{max,m} = 0.1$ GPa. That is, the fiber is only four times stronger than the matrix. The matrix strength is reasonable for an epoxy, but 0.4 GPa is very small for a reinforcing fiber, so this relationship has been redrawn in fig. 8.5C with $\sigma_{max,f} = 4.0$ GPa, a value close to the strength of S-glass or Kevlar. Now it is clear that when the fiber strength is much larger than the matrix strength, the dip in strength at low V_f is actually quite minimal, and eq. (8.13) is appropriate at essentially all useful volume fractions.

The analysis above is interesting, but it does not tell us anything about the toughness of fiber-reinforced composites. We have assumed that all fibers fail at the same stress, which is the stated value for $\sigma_{max,f}$, but our discussion above of the fracture process and the role of flaws in initiating catastrophic fracture of materials tells us that some fibers in a composite will have larger flaws and will fail at lower stresses than others. In addition, the analysis implies that once a fiber breaks in a high-V_f composite, the flaw will propagate across the whole sample, causing catastrophic failure of the sample. This need not be the case because the rather complex structure created in composites provides mechanisms that can stop cracks from proceeding into adjacent components in the composite. Consider the drawing in fig. 8.6A, which shows a small area of a fiber-reinforced composite. Note the circle highlighting a large notch or crack in one of the fibers. This notch will create a stress concentration at its tip that will likely initiate the growth of the flaw. If this were a flaw in a macroscopic sample of a brittle material that is loaded in tension, it would initiate catastrophic crack growth and cause the object to break in two, or perhaps even to shatter into many pieces. However, in this composite, crack growth through the glass fiber is limited by the thickness of the fiber, which in the case of S-glass fibers would only be 10–20 μm. A crucial feature of fiber-reinforced-composite design is that the process of separating the high-stiffness and high-strength components (i.e., thin fibers) with matrix materials that have very different properties can dramatically alter the growth of cracks.

The remaining diagrams in fig. 8.6 illustrate several of the "mechanisms" that can occur in composite materials to stop crack growth and hence increase the toughness of these materials. In the discussion of fracture mechanics we considered the role of flaws in generating stress concentrations at the tip of a crack, and we used eqs. (5.8) to calculate the magnitude of the stress level that would act to pull apart the atoms at the tip of an elliptical flaw. Fig. 8.6B illustrates an interesting feature of the stress concentration existing at the tip of an elliptical crack that was pointed out by Cook and Gordon (1964). This diagram shows the tip of an elliptical flaw, and it also shows a series of stress trajectories, as dotted lines, to represent the stress concentration at the tip of the ellipse that is created when a uniform external stress, σ_0, is applied at the top and bottom of this sample. The vertically oriented arrow that is placed at the tip of the flaw indicates the peak value of σ_y, the local vertical stress that acts to break vertically oriented bonds that are right at the tip of the flaw. It is this stress that will cause the crack to grow to the right in this diagram.

Note also that there is a shorter, horizontally oriented, arrow placed a small distance to the right of the ellipse, which is labeled σ_x. This arrow represents a horizontally oriented stress in the material that is maximal at a distance in front of the crack tip of about r_m, which is the radius of curvature of the ellipse at the tip of the major axis. This smaller stress will load bonds that are a small distance in advance of the growing crack and that are oriented parallel to the direction of crack growth. In linearly elastic materials, σ_x is about one-fifth of σ_y, and the lengths of the two arrows have been adjusted to reflect this ratio. Cook and Gordon suggested that if the interface between the fiber and

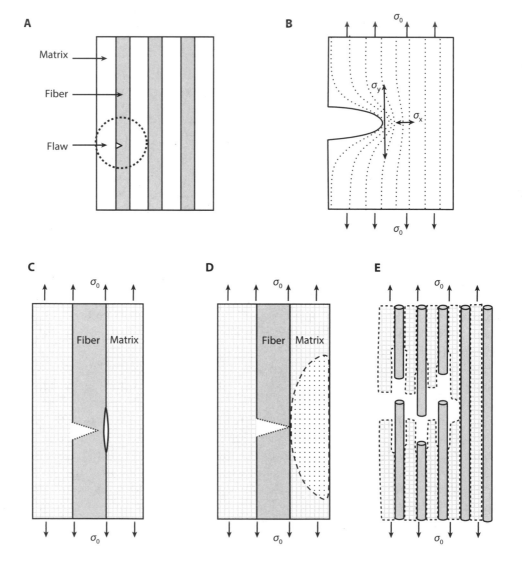

the matrix is relatively weak, say less than one-fifth of the strength of the fiber, then σ_x may be sufficiently large to cause the fiber and matrix to separate, as illustrated in fig. 8.6C. In this case the interface is weak enough that σ_x has separated the interface, leaving a narrow gap. Note that the surface of this gap is almost flat, and this means that the local radius of curvature that exists when the growing flaw merges with the gap will be very large, and the local stress concentration will fall dramatically, as expressed in eqs. (5.8). Also, because the local stress is more broadly distributed in the material around the flaw, the strain energy released by any additional crack growth will necessarily be distributed over a much larger volume of material. In this case, the growing flaw will most likely be stopped, and this is one of a number of crack-stopping mechanisms that make fiber-reinforced composites and other tough materials difficult to break.

Fig. 8.6D shows a similar situation, but in this case the fiber-matrix interface is sufficiently strong that the interface has not opened, and the local stress concentration and the strain energy released at the crack tip will be transferred into the matrix as the crack tip reaches the fiber-matrix interface. If the matrix has properties similar to those of the fiber, then it is possible that the growing flaw will release sufficient energy to continue growth of the flaw into the matrix phase and beyond. In this case, it is likely that catastrophic failure of the material sample will occur. However, if the matrix properties are sufficiently different from those of the fiber, it is possible that the conditions for crack propagation are not met, and crack growth will stop. Suppose that the fiber and matrix properties are similar to those illustrated in fig. 8.5A. The fiber is very stiff

Figure 8.6. (*left*) Composite fracture. **Panel A** introduces the concepts of fracture toughness in composite materials, as it illustrates a composite structure with a significant flaw in a fiber, which could form the origin of a fracture process. The key question for this system is whether a crack propagating from this flaw will proceed through the fiber-matrix boundary to initiate the catastrophic failure of the entire composite sample. An important feature of composite mechanics is that composites provide mechanisms that can stop crack growth, and panels B–E illustrate some potential crack-stopping mechanisms. **Panel B** shows a plot of stress concentrations around the tip of an elliptical flaw, as calculated by Cook and Gordon (1964). The vertically oriented *double arrow* at the tip of the flaw, labeled σ_y, indicates the stress that will act to extend the crack in the reinforcing fiber. A small distance in advance of the crack there is a horizontally oriented double arrow that is labeled σ_x, and this stress is about one-fifth of the magnitude of σ_y. **Panel C.** If the interfacial strength of the fiber-matrix boundary is relatively low, as shown here, the horizontally oriented stress may cause the fiber-matrix interface to open, leaving a much blunted crack tip that will lower the stress concentration and may stop crack growth. **Panel D** indicates how matrix extensibility, as illustrated in panel B, may allow the matrix to stop crack growth by simply straining and absorbing the energy released by the failure of a fiber. **Panel E** shows that when strong fibers break and stress is transferred to adjacent matrix, energy may be required to disrupt the fiber-matrix interface near the break and to pull the fiber out of the matrix hole that encased it in the intact composite, absorbing additional energy that may stop crack growth.

and quite strong, but it fails at small strain, which results in a relatively small energy to break (i.e., the area under the stress-strain curve). The matrix is less stiff and less strong, but it fails at large strains. In this case the energy to break is higher for the matrix than for the fiber, as indicated by the difference in the areas under their respective stress-strain curves. So in spite of the fact that a sharp crack arrives at the matrix layer with the potential to create a high stress concentration that will break bonds in the matrix, the energy released to this crack tip may no longer be sufficient to break the bonds in the matrix material. First, because the matrix material is less stiff and more extensible than the fiber material, the shape of the notch tip will change its geometry, attaining a larger radius of curvature at the crack tip and lowering the stress concentration on the bonds that must fail for the crack to proceed. Another way of saying this is that high matrix strain will redistribute the incoming strain energy over a larger volume of matrix material, as indicated by the dark, cross-hatched area in the matrix adjacent to the crack tip, and thus the crack may not propagate through the matrix layer. In addition, matrix materials are often polymeric in nature, both in artificial and in natural composites, and polymeric materials can be viscoelastic. That is, they can dissipate mechanical energy as heat in load-unload cycles, and thus reduce the energy available for crack growth. Thus, strain energy released in distant sites in the test sample may in part be consumed in frictional processes in this transfer. More importantly, the high matrix strains that will occur in the immediate vicinity of the flaw will likely dissipate large amounts of strain energy in viscoelastic processes. Energy dispersed or dissipated in these processes represents energy not concentrated on the key bonds at the crack tip where crack growth must proceed.

Fig. 8.6E gives a different view of a fiber-reinforced-composite fracture zone. Assume that this is the crack tip of a flaw that extends to the left by some unspecified distance. We can ask a number of questions about the structure of this defect and about its origins. First, the crack morphology suggests some additional mechanisms for increasing the energy required to extend a crack. Clearly, energy is required to break the fibers, but many reinforcing fibers used in industry are made from inherently brittle materials, like glass, and the energy required to break them is relatively small because the fracture surfaces are relatively smooth. However, when a strong fiber breaks and stress is transferred to adjacent matrix layers, energy may be required to disrupt the fiber-matrix interface near the break and to pull the fiber out of the matrix hole that encased it in the intact composite. These energies will contribute to the fracture toughness of the composite as a whole.

Next consider this crack tip in a test sample that has been loaded to some intermediate stress level that was not sufficient to initiate a catastrophic fracture, but was sufficient to break a few of the fibers at this crack tip. That is, the crack-tip morphology was created by stable crack growth, likely because the overall crack length is small. Now, if the test sample is loaded and unloaded repeatedly to this subcritical stress level, components at the crack tip will fatigue, and over many load-unload cycles this flaw will gradually elongate. This process will occur at other locations in the test sample, and over time neighboring flaws will become sufficiently close that they amalgamate to form larger defects. At some point in time a flaw will initiate a catastrophic failure process that causes the fracture of the test sample. This process is known as fatigue, and it is another crucial material property that plays as significant a role in biology as it does in engineering. Fatigue testing, like other material-testing procedures, is quite complex, but in principle it involves tracking the loading history of numerous test samples that are loaded either cyclically or statically at specified stress or strain levels until the sample fractures. Unfortunately, there is not a very extensive literature on the fatigue behavior of biomaterials, but we will see some information of the fatigue lifetime of tendon and bone in the following chapters.

Finally, it is important to realize that the defects that initiate fracture and fatigue processes in composites need not be at the microlevel illustrated in the diagrams in fig. 8.6. In composite laminates the interface between laminae with differing fiber orientations will provide a matrix-dominated layer that may be a weak link in the composite structure. These interfaces may initiate the failure process, particularly in loading regimes that involve compression, bending, or shearing, and hence create additional modes of composite failure. We will see these kinds of effects in the hierarchical structures seen in complex biomaterials such as bone.

SECTION III

THE MECHANICAL DESIGN OF TENSILE MATERIALS

Chapter 9

The Structural Design of Collagen

Tendons and Ligaments

1. CRYSTALLINE POLYMERS AND TENSILE FIBERS. The design of tensile materials in animal structures is based on the properties of polymer molecules that spontaneously form crystalline arrays, as opposed to the amorphous polymers described in the previous section. A brief look at some synthetic polymer fibers will provide insights into this class of polymeric materials. A number of these polymers have a sufficiently regular chemical structure that they spontaneously form ordered crystalline arrays, and these polymers are used commercially in the production of a broad range of rigid fibers that can be used in the construction of tensile materials and structures. Figs. 9.1A and 9.1B show the repeating monomer units for nylon and polyethylene (PE), well-known examples of crystalline polymers that can be formed into fibers that are stiff and strong in tension, and can therefore be used to form threads and ropes. However, the production of high-stiffness fibers from these polymers requires extensive processing to ensure that the polymer crystals are rich in extended chains, and that the crystals have a high degree of preferred orientation parallel to the long axis of the fibers.

Polyethylene provides an excellent example of this. It has been known for many years that polyethylene in dilute solution will crystallize, under appropriate conditions, to form thin, sheetlike crystals, which have been observed under the electron microscope. However, the polyethylene molecules in these crystals take on a folded-chain organization, as shown in fig. 9.1C. Here we see a sheetlike crystal that is only 10–20 nm thick, and electron diffraction has established that the polymer backbone runs perpendicularly to the plane of the sheet, and it folds back and forth on itself through the crystal's thickness. Because PE is made entirely from aliphatic carbon atoms, these crystals are stabilized exclusively by very weak induced-dipole bonds. A variety of methods can be used to unfold these crystals and impose a degree of orientation parallel to the axis of fibers that are drawn from the bulk polymers.

Nylon is a more complex polymer, and the structure of nylon 6–6 in fig. 9.1A reveals a polymer that shares important features with PE and also with proteins. Note that the monomer repeat of nylon contains two amide groups that are separated by four and six aliphatic carbons. The amide groups are identical to the peptide groups in proteins, where they are separated by a single alpha carbon that carries the amino acid's side chains. The aliphatic carbon chains in nylon are identical to the main chain of PE. So, nylon is a very flexible molecule, but it has polar groups in the polymer backbone that can form stronger hydrogen bonds with neighboring chains to form more robust crystals. Figs. 9.1D and 9.1E show diagrams that illustrate the kind of transformations that can occur in fiber processing with nylon. Fig. 9.1D shows the kind of lamellar organization that exists in bulk nylon, and fig. 9.1E shows how this structure can be changed in the extrusion and draw processing stages that are used to produce high-tensile-strength

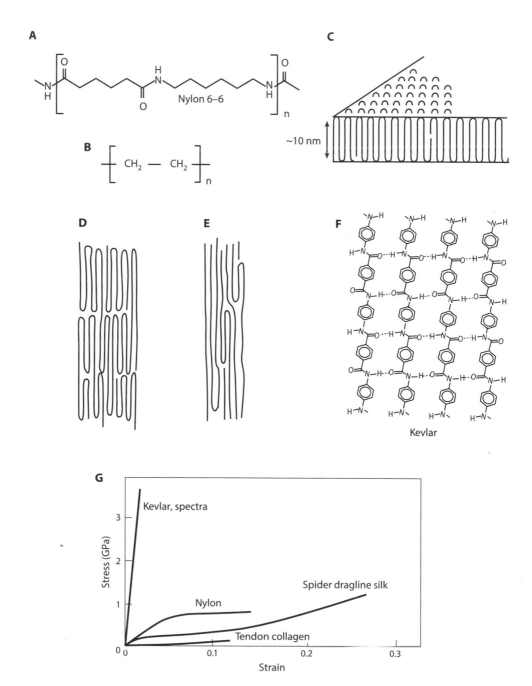

nylon. Typically, bulk nylon is heated and extruded through a tapered spinneret, and the extruded filament is then pulled around a series of rotating drums, each successive drum rotating more rapidly than the previous drum, to gradually extend the fiber. This process is carried out in hot air to facilitate the reorganization of the polymer molecules, and after multiple stages the drawn fiber is wound onto a spool, where the fibers cool and solidify. In this process, the lamellar organization is pulled apart, and extended-chain conformations are enhanced, but the process does not produce fully extended-chain crystalline organization, which is ultimately the desired end product.

There are, however, synthetic fibers that achieve exceptional mechanical properties through gel-spinning processes, in which high-molecular-weight polymer molecules are dissolved at high concentrations in an appropriate solvent. Extensive draw processing and solvent removal can dramatically increase polymer-crystal volume

Figure 9.1. (*left*) Tensile materials. **Panel A.** The polymer-repeat structure of nylon 6–6. Note the presence of multiple amide groups that allow the formation of this linear polymer. The amide linkages, which are chemically identical to the peptide bonds in proteins, provide the potential for extensive networks of hydrogen bonds between adjacent polymer chains that run parallel or antiparallel to each other. **Panel B.** The polymer repeat for polyethylene. This polymer is very flexible because its covalent backbone is made entirely from carbon-carbon single bonds that allow rotational freedom, and because the hydrogen side groups on the backbone are very small. In spite of this flexibility, polyethylene is able to form stable polymer crystals at room temperature. **Panel C.** When polyethylene is allowed to crystallize from dilute solutions, it typically forms thin, sheetlike crystals in which the polymer backbone folds back upon itself into thin, regular crystals. However, if high-molecular-weight polyethylene is swollen in good solvents and subjected to draw processing, the folded chains can be extended, and uniaxial arrays of polymer chains will create materials that are exceptionally stiff and strong in the direction of the fiber. **Panel D** shows the kind of lamellar organization that is found in bulk nylon, and **panel E** indicates the changes in molecular organization that can occur in the draw processing of nylon, or of other synthetic polymers such as polyethylene. **Panel F** shows the basic structure of the crystalline units in Kevlar. The repeating polymer unit is based on a benzene ring, and these planar rings are interconnected through amide linkages to form the polymer. Kevlar forms planar sheets in which fully extended polymer molecules are interconnected with hydrogen bonds between the amide groups, similar to those seen in nylon and in proteins. **Panel G** shows typical stress-strain curves for a number of tensile fibers formed from synthetic and natural polymeric materials. The line labeled "Kevlar, Spectra" provides a clear indication of the high stiffness and strength that is achieved in fibers with highly oriented crystals of fully extended polymer molecules. Drawn nylon shows considerably less stiffness and strength than Kevlar and Spectra. Spider dragline silk is nylon-like in its structural chemistry, but it is actually a hybrid molecule that combines crystal-forming sequence blocks with more extensible sequence motifs that provide extensibility. Tendon collagen appears inferior to the other polymer fibers shown in this graph, but its utility in animal skeletal materials cannot be underestimated.

fraction and alignment in fibers. One such material, Spectra, is made from ultrahigh-molecular-weight polyethylene, with molecular weights for the PE of the order of 5 million Da. These huge molecules entangle extensively in the solvent-gel, and they allow the swollen material to be draw processed to achieve a very high degree of extension and crystallization. Another class of material is based on aromatic-polyamide polymers, such as Kevlar, fig. 9.1F, which are much less flexible than nylon or PE because the planar benzene ring that connects the amide groups makes the polymer backbone quite rigid. Kevlar chains can interact with neighboring chains both through an extensive hydrogen-bond network between the numerous amide groups, and through the stacking of the benzene rings, a process called pi-stacking. These interactions make the polymer crystals in Kevlar exceptionally stable. Kevlar is spun from a gel formed in concentrated sulfuric acid, where it forms a liquid-crystalline structure. Extrusion and drawing, with removal of the acid, completes the process, which produces a fiber with a very high crystal volume fraction and high alignment in the fiber.

Fig. 9.1G shows typical stress-strain plots for the synthetic polymers discussed above, along with curves for two biological materials. The line labeled "Kevlar, Spectra" shows a typical stress-strain curve for these materials, which have quite similar properties. The curve for nylon is for a highly drawn, high-tensile-strength nylon fiber. The comparison between the commercial fibers and the natural fibers, spider dragline silk and mammalian tendon, provide some interesting insights for our analysis of tensile biomaterials. Spider dragline represents the strongest of the natural fibers, with tensile strength generally greater than 1 GPa. This means that its strength is about one-third that of Kevlar and Spectra, but is somewhat greater than that of high-strength nylons. It is also considerably more extensible than these synthetic fibers, suggesting that it is likely tougher than all of them. Tendon collagen, on the other hand, does not look particularly impressive, as it is about an order of magnitude less stiff and strong than spider silk. This is misleading, however, because, as you will soon see, collagen is a truly remarkable biomaterial.

This is an interesting comparison, but it misses an important difference between artificial fibers and fibrous proteins made inside living organisms. Industrial processes occur on a huge scale, with polymer synthesis occurring in reaction vessels that are of the order of 10^4 liters in volume, and fiber spinning produces continuous fibers that may be multiple kilometers in length. In living systems, polymer synthesis occurs on the scale of individual cells, with sizes measured in tens of micrometers, and volumes of the order of 10^{-14} liters. In spite of the small scale of cellular synthesis systems, animals can produce robust tensile structures such as tendons that can be as long as a meter or more. Thus, it is truly remarkable that animals can produce structural materials of the quality of spider silk and tendon, and they do this by employing molecular mechanisms of self-assembly, whereby the detailed sequence design actually preprograms the assembly process that will occur spontaneously under appropriate physical-chemical conditions. In addition, these processes occur at different scales to produce hierarchical structures, with nanoscale molecular structures combining into microscale aggregates, and then the microscale structures assembling into macroscale objects. The collagen-based tensile structures called tendons provide one of the best understood examples of this hierarchical organization in natural materials. In this chapter and the next, we will consider in detail the molecular and structural designs of collagen and spider silk,

including the processes involved in the synthesis and assembly of the constituent polymers, as well as their mechanical properties and functional attributes. That is, these
materials will serve as case studies that illustrate the principles of tensile-materials design in animals.

2. THE EVOLUTION OF COLLAGEN. Collagen is the most widely used tensile
fiber in the animal kingdom (the Metazoa). It is found in tensile structures, such as
tendons and ligaments, but it is also a major component of rigid materials such as bone,
and it is a dominant component in pliant tissues, such as skin, cartilage, artery wall, and
the body-wall connective tissues of many invertebrate animals. The evolutionary origin
of collagen is now known to reside in the sister group of the metazoans, the choanoflagellates. As the name suggests, these single-celled organisms have an actin-based collar
that surrounds a single flagellum, which they use for aquatic propulsion. Choanoflagellates are morphologically very similar to the choanocyte cells found in sponges, which
are regarded as amongst the most primitive of the metazoans. Sponges are multicellular
organisms in which the various cell types are held together by an extracellular matrix
(i.e., skeletal structure) that contains a very primitive kind of collagen, called spongin,
and recent genome-sequence analyses of choanoflagellates and primitive metazoans by
Richard Hynes (2012) indicate that the collagen structure in metazoans was derived
from proteins present in choanoflagellates. That is, collagen-like repeat-sequence elements are present, as are other sequence elements found in the collagen genes of modern
metazoans. However, the choanoflagellates do not actually make collagens. True fibrillar
collagen does not appear until the eumetazoans, in the cnidarians (corals, jellyfish, sea
anemones), and fibrillar collagens exist in all of the metazoan phyla that exist today.

The structural organization and material properties of collagen and collagen-
containing structural tissues have been extensively studied and documented over many
years, largely because of the importance of these materials in medical science. There
have been numerous research articles written about collagen, too many to list compre-
hensively here. However, two books on the biology and biomechanics of collagen provide access to more detailed coverage of this topic (Brinkmann et al., 2005; Fratzl, 2008).

In vertebrates, we recognize a collagen gene superfamily of at least 28 collagen types,
but in this discussion we will deal primarily with the fibrillar collagens, which are collagen types I, II, III, V, and XI. However, most of our treatment of collagen structure
will focus on type I collagen, which is the most abundant and best documented. We
will start with a brief introduction to the organization and processing of the collagen
gene product, and the secretion of the protein into the extracellular space where collagen molecules assemble into the filamentous substructures that ultimately become
tendons and ligaments. The type I collagen is encoded by two genes, but the sequence
structures of these two gene products are very similar. The genes encode the amino-
acid sequence of alpha chains (α chains), which are designated $\alpha 1$(I) and $\alpha 2$(I). Type II
and III collagen are made from a single gene, and type V and XI collagen are encoded
in multiple genes.

3. TROPOCOLLAGEN, THE COLLAGEN MOLECULE. It has been known for
many years that it is possible to isolate small amounts of individual collagen molecules
in a fully functional state from young or rapidly growing tissues by simple extraction

with dilute saline solutions. These molecules have a very interesting structure that gives us insights into the functional design of this molecular system. Solution studies and electron microscopy of individual collagen molecules in their native structure reveal that this structure is dramatically different from the structures of the vast array of proteins found in living systems, which function as catalysts (enzymes), hormones, receptors, ion channels, and other related functions. These kinds of proteins typically take on a "globular" conformation, where the polypeptide chain folds back and forth to form a compact, globular structure with a diameter that is much smaller than the extended length of the protein chain. Structural proteins, like collagen, have very different structures that relate directly to their mechanical function. Similar to the crystalline polymer structures discussed above for Kevlar and Spectra, high stiffness and strength are achieved with crystalline structures that hold the polymer chains in extended conformations so that the inherent stiffness and strength of the covalent backbone of the polymer is manifest in the stiffness and strength of the structural materials formed. As a result of this arrangement, the collagen molecule has a structure that is essentially that of a molecular filament. Fig. 9.2A shows a diagram of a single collagen molecule, called tropocollagen, which has the form of a long, thin, and reasonably rigid rod, with a length of 300 nm and a diameter of about 1.5 nm. The shape of the line in this figure has a length/diameter ratio that just about matches these dimensions, so it is a reasonable representation of the shape of a native tropocollagen molecule. Interestingly, if one heats a solution of native tropocollagen molecules, they will denature and come apart into individual disordered protein chains, as indicated in the figure. These are the three α chains that combine to make tropocollagen, and in the case of type I collagen there are two $\alpha 1(I)$ chains and one $\alpha 2(I)$ chain in tropocollagen.

The amino-acid sequence for the full collagen protein encodes the rodlike molecule described above, along with large extensions at the amino- (N-) and carboxyl- (C-) terminal ends that play crucial roles in the transport of the molecule through the cells, called fibroblasts, where collagen is synthesized, as well as enabling the assembly of the three polypeptide chains into the rodlike structure. The full polypeptide is called preprocollagen, and it contains 1487 amino acids (this number is based on bovine collagen, and it is representative of other mammalian collagens, although some minor variations

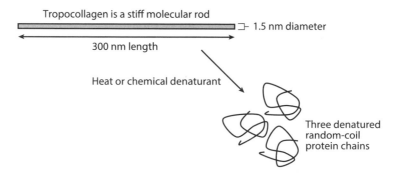

Figure 9.2. Structure of tropocollagen. Tropocollagen is a molecular rod, with a length of 300 nm and a diameter of 1.5 nm. When this molecular rod is denatured, it collapses into three separate randomly coiled polypeptide chains.

exist). At the N-terminal end of this molecule there is a signal sequence that directs the molecule to be synthesized by endoplasmic-reticulum (ER)-bound ribosomes and hence targeted for packaging and secretion from the cell into the extracellular space. The signal sequence is cleaved from the protein as it is transported into the ER (which eventually buds off to form the Golgi apparatus), and this produces the procollagen molecule, which contains 1437 amino acids. The portion of this molecule that will become collagen is in the center of the procollagen peptide, and is 1056 amino acids long. It is flanked by the N-terminal (pN) and the C-terminal (pC) propeptide domains that function in collagen-fiber assembly. Several processes occur as the procollagen peptide is formed, the most important being the chemical modification of a large fraction of the prolines and some of the lysines. Many of these amino acids in collagen are hydroxylated by specific enzymes present in the ER. As we will soon see, hydroxyproline is crucial for the stabilization of the final rodlike collagen structure, and hydroxylysine plays a key role in the chemical cross-linking of collagen molecules together in fibrous aggregates called collagen fibrils. In addition to these modifications of proline and lysine, the peptide chain becomes glycosylated by the addition of short sugar polymers at key locations along collagen chain. The pN and pC propeptide domains are globular, and they assist in the assembly and translocation of the collagen molecule into the extracellular space. The pC domains from three procollagen chains join together, and this initiates the assembly of the collagen rod, which involves the winding of the three chains around each other to form a triple-helical structure, which will be described in detail below. When the helical rod is formed, it remains capped by the globular pN and pC terminal domains. This creates a dumbbell-shaped molecule that prevents individual collagen-rod domains from associating with their neighbors and assembling into fibrous aggregates inside the cell. The fully processed and assembled procollagen molecule then proceeds from the Golgi apparatus into secretory vesicles and is released into the extracellular space, where enzymes cleave off the pN and pC domains. This releases the rodlike tropocollagen molecule that is ready to self-assemble into the filamentous structures that ultimately become tendons and ligaments.

The self-assembly of the tropocollagen rod from three polypeptide chains and the assembly of these rods into collagen fibrils is largely programmed into the amino-acid sequence encoded in the collagen gene. So it is now appropriate to consider the details of collagen's amino-acid sequence. The three chains formed with the removal of the propeptide terminal domains are 1056 amino acids long, and 1014 of these amino acids are arranged into 338 repeating tripeptide units (glycine–X–Y), with glycine in every third position for the entire 1014 amino acids. This central tripeptide sequence block will ultimately form the triple-helical-rod domain of tropocollagen, and this will be capped by short N-terminal (16 amino acids) and C-terminal (26 amino acids) sequence blocks called telopeptides.

Glycine is the smallest of all the amino acids in proteins because its α-carbon side group is a single hydrogen, and the presence of glycine in every third position of the tripeptide repeat is the key feature that directs the assembly of the rodlike tropocollagen molecule. The next most striking feature of the amino-acid sequence is that imino acids proline and hydroxyproline (OH-proline) are found in exceptionally high levels in collagen, where they may account for more than 20% of the total. In addition, proline and OH-proline show very strong preferences for their position in the repeating tripeptide unit, where proline is almost always in the X position and OH-proline almost always located in the Y-position. For example, in the bovine $\alpha 1$(I) collagen chain, proline and OH-proline account for about 24% of the amino-acid composition of the rod domain, with 97% of the proline in the X position and 99% of the OH-proline in the Y position. Thus, polypeptides with repeat sequences of glycine–proline–OH-proline, (gly-pro-OHpro)$_n$, spontaneously assemble into rodlike molecules with diameters of 1.5 nm and lengths determined by the number of repeating tripeptide units. However, in real collagen molecules other amino acids in the X and Y positions play important roles in the assembly of tropocollagen molecules into collagen fibrils and in the cross-linking of tropocollagens together in these fibrils. Before looking at these aspects, we need to consider the details of the 3-D structure of the three polypeptide chains in the tropocollagen molecule.

The structural organization of tropocollagen has been worked out in considerable detail, largely through X-ray diffraction studies and electron microscopy. Without going into detail, the three protein chains exist in the form of extended helices, and these three helices wind around each other to form a rope-like triple helix that holds the chains in an extended molecular unit that is crucial to the creation of structural fibers with great tensile stiffness and strength. The organization of this triple helix is shown in fig. 9.3. In panel A we see short segments of the three polypeptide chains (the three α chains) arranged in parallel, and it should be apparent that each α chain is organized into a helical conformation, which is best described as a left-handed helix. The "N" and "C" labels at the bottom and top of this diagram indicate the N-to-C-terminal direction of these polypeptide chains, and you can tell that this is a left-handed helix because the helix spirals in the direction of the fingers of your left hand when held with the thumb pointing upwards. These helices take on the form that is characteristic of a polyproline-II type of helix, in which all of the peptide bonds are in the *trans* conformation. The rise per amino acid in this helix is typically 0.286 nm and the rise per turn of the helix is about 0.9 nm, as indicated in the diagram. This makes it a highly extended helical structure, as opposed to the other common helical polypeptide structure found in proteins, the α helix, where the rise per turn is only 0.5 nm. The extended nature of the polyproline-II helix is a key feature that allows the collagen molecule to have high axial stiffness.

Figure 9.3A shows each polyproline-II helix wrapped around a vertical arrow that indicates the central axis and the N-to-C direction of this helix, and the arrows are dotted, dashed, and solid lines to distinguish the three individual helices. The positions of amino acids in this diagram are indicated by the solid or open circles, where the small solid circles indicate the α carbons of glycines, and the open circles indicate the α carbons of the amino acids that occupy the X and Y positions in the sequence. You should note that the glycine residues are not arranged directly across from each other in this diagram. Rather, each chain is offset vertically by one amino acid, so that there is a single glycine at every level in the collagen molecule. This is a key feature of collagen because it allows the three left-handed helices to wrap around each other to form a very tightly wound right-handed triple helix, as shown in fig. 9.3B. In this diagram the individual helices are indicated only by the dotted, dashed, and solid arrows that show the vertical axis of each helix. The pitch (distance for one turn) of the triple helix is about 9 nm, so there are 10 turns of the left-handed polyproline helices per turn of the right-handed triple helix.

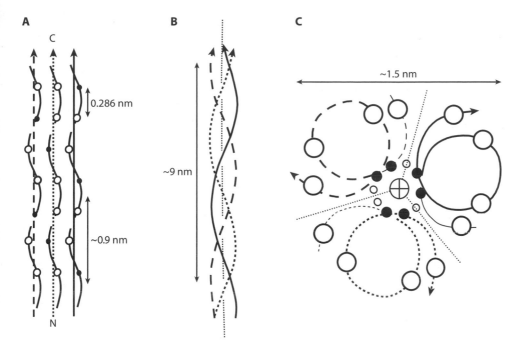

Figure 9.3 Hierarchical organization of collagen fibers. **Panel A** shows the lowest level of structural organization in tropocollagen. Each of the three polypeptide chains forms a left-handed helix with three amino acids per turn. The *circles* indicate the positions of the α carbons of individual amino acids in these polypeptides, with the smaller *black circles* indicating those of glycines, which occur at every third position. The *open circles* indicate the locations of the α carbons of amino acids in the X and Y positions of the (gly–X–Y) triplet repeat motif. **Panel B** shows the basic organization of the tropocollagen triple helix. In this diagram the individual peptide helices from panel A are represented by the *solid, dashed,* and *dotted arrows* that twist around a central axis to form a right-handed triple helix. **Panel C** shows the organization of these two helical systems in greater detail. In this diagram we are looking down the long axis of the tropocollagen triple helix, and the 1.5 nm arrow indicates the approximate diameter of tropocollagen. The drawing only shows structural detail for a very short length, so it is possible to see the organization of the individual peptide helices within the triple helix. The three *dotted lines* radiating out from the center of this diagram indicate the domains of the three peptide helices, which show about two turns of each. Note that there are six *dark circles* at the center of this structure. These indicate the position of the glycine residues at the center of the triple helix, and the adjacent small *open circles* indicate the position of an additional four glycines that would exist above and below this section of the triple helix and which would complete the central ring of 10 glycines per turn of the triple helix.

As mentioned above, the importance of glycine in every third position in the sequence is that this allows the three helices to twist together very closely. This arrangement is shown more clearly in fig. 9.3C, which shows a cross-sectional view of the triple helix. The central cross indicates the vertical axis of the triple helix as it emerges from the plane of the page toward you, and the three dotted lines radiating outward from the cross separate the domains of the three helical chains at this level, but remember if we plotted 10 turns of the α-chain polyproline helix it would wrap one full turn around the central axis of this diagram. In this view you see only about 1½ turns of each α-chain helix spiraling toward you in the N-to-C-terminal direction, so that it is possible to distinguish each of the three left-handed helices. The small black circles indicate the positions of glycine α carbons, and the larger open circles indicate the positions of the α carbons of the amino acids in the X and Y positions of the triplet repeat. It should be clear that the glycines are placed at the center of the triple helix, and that this placement allows the three coiled polypeptides to twist together very tightly. The glycines in each α-chain helix are offset by 36° around the central axis of the superhelix. This rotation is due to the fact that there are 10 turns of the α-chain helices for one turn of the superhelix. A drawing with 2.5 turns for each α-chain helix would have shown that there is a circle of 10 glycines at the center of the triple helix, which together form the central core of glycines that allows the very tight winding of the three polypeptide chains. However, this drawing would be more complex and less useful because the three helices would begin to overlap. In place of this, four small open circles are interspersed between the six solid black circles to indicate the positions of the four additional glycines, positioned above or below the plane of this diagram, which together would complete the central column of glycines at this level in the tropocollagen molecule. Finally, it is important to note that the circles in this diagram do not indicate the true size of the atoms or bond lengths in the real proteins. Using the 1.5 nm arrow as a scale, the black dot for the glycine is about 0.05 nm in diameter, which is less than the actual diameter of the single α-carbon atom (the van der Waals radius of carbon is 0.17 nm), and it is much smaller than the carbon with its attached hydrogen. Space-filling models indicate that the protein chains in tropocollagen are very tightly packed together, creating a very stiff molecular rod.

The stabilization of the triple helix involves several key features. First, the all-*trans* form of the peptide group is the minimum energy state for the polypeptide chain, and the triple-helical structure locks the three peptide chains into the all-*trans* form. Second, the position of these chains relative to their neighbors allows a single hydrogen bond per tripeptide unit to form between adjacent chains in the triple helix. In addition, the hydroxyl group on the OH-proline, when it is present in the Y position of the repeating tripeptide unit, provides additional stabilization, possibly through a hydrogen bond to a bound water molecule that bridges to a peptide group on a neighboring chain. Indeed, it has been known for many years that the tropocollagens of ectothermic animals (animals whose body temperature varies with environmental temperature) may denature at lower temperatures than mammalian collagens, because of differences in the OH-proline content of collagens from different animals. That is, there appears to be a strong correlation between the OH-proline content and the maximal body temperature experienced by animals in nature. Thus, ectothermic animals that live in cold environments have collagens with lower OH-proline content, and their tropocollagen molecules denature at lower temperatures than endothermic birds and mammals with body temperatures of the order of 35°C to 40°C.

4. THE ASSEMBLY OF COLLAGEN FIBRILS. With this understanding of the structural organization of tropocollagen we can begin to analyze the assembly of this molecule into crystalline filamentous structures that will ultimately lead to the formation

of macroscopic tensile structures such as tendons. We will now use a simple arrow, as shown in fig. 9.4A, to represent individual tropocollagen molecules, where the arrow points in the N-to-C-terminal direction of the polypeptides within the tropocollagen. We will also use this arrow as a scale bar, as it represents a length of 300 nm. The question is how this molecule is organized into crystalline fibrils. Electron microscopy provides a clear indication of the organization of collagen fibrils, as shown in panels B–D of fig. 9.4. Fig. 9.4B shows the appearance of a negatively stained collagen fibril, as viewed under the electron microscope. Note first that the width of this fibril is about one-third the length of the tropocollagen drawn next to it, so this makes the fibril about 100 nm in diameter, which is actually a very appropriate size for a typical collagen fibril. In general, however, collagen fibrils can range in diameter from 15 to 250 nm in different tissues and for different collagen types. The other important feature of this collagen fibril is that it is drawn with a prominent pattern of dark and light bands, which indicates the presence of a repeating structural motif, with a dimension that has been measured in hydrated collagen fibrils as 67 nm. This banding pattern has been called "D-spacing," and it arises from the pattern of overlapping tropocollagen molecules in the lattice structure of the fibril. Interestingly, X ray diffraction clearly identifies 67 nm longitudinal spacing in collagen fibrils, indicating that this dimension is a prominent characteristic of the fibril crystalline structure. Careful inspection of the amino-acid sequence of collagen α chains indicates that there is a characteristic repeating distribution of the non-proline/OH-proline amino-acid residues in the X and Y positions of the collagen tripeptide motif that is 67 nm in length, and the distribution of this motif is indicated by the crossbars drawn on the tropocollagen arrow in fig. 9.4A. Since 300 divided by 67 is somewhat less than 4.5, we can conclude that there are about 4.5 of these repeating sequence motifs along the length of tropocollagen, and it seems plausible that they direct the assembly of the tropocollagen molecules into crystalline fibrils.

The generally accepted pattern for the crystalline assembly of tropocollagen in collagen fibrils is illustrated in fig. 9.4C, where we see that tropocollagen molecules are aligned in parallel, but they overlap each adjacent molecule by one D-spacing of 67 nm. Because there is not an integer number of D-spacings in a tropocollagen molecule, logic implies that there must be a gap present between the C-terminus of one tropo collagen molecule (arrowhead) and the N-terminus of the next in line. This leaves a pattern of overlapping molecules with a gap between the ends of each tropocollagen. Note also that the gaps line up across this array at every fifth tropocollagen, leaving a transverse gap zone across the fibril. It is important to note that the diagram in fig. 9.4C does not show the full width of a 100 nm diameter collagen fibril, which would contain about three times as many tropocollagens at its widest point.

The banding pattern shown in fig. 9.4B arises from negative staining. The mechanism of negative staining is that the pH of an electron-dense metallic stain (e.g., phosphotungstate) is adjusted to the point where it has low net charge (at a pH of about 7 for the phosphotungstate ion), so the stain will not bind to the charged amino-acid side chains on the collagen molecules in the fibril. This is the condition for negative staining, and the staining reaction for the collagen fibril under these conditions comes from the accumulation of the stain in the open space in the crystal lattice provided by the gap zone. If you look at the arrangement of gap zones in fig. 9.4C, you will see that they line up with the dark-staining bands on the negatively stained collagen fibril shown in fig. 9.4B.

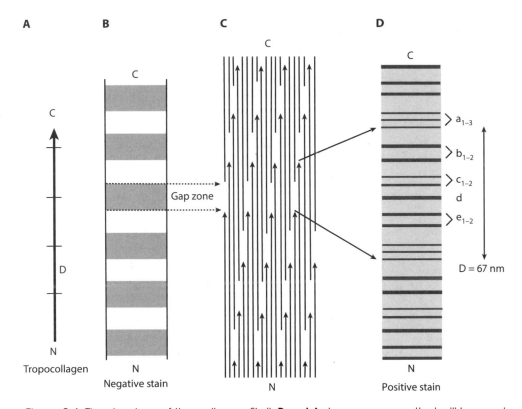

Figure 9.4. The structure of the collagen fibril. **Panel A** shows an arrow that will be used to represent a single tropocollagen molecule. The arrow is divided into five sections to reflect the pattern of overlap that forms when tropocollagen molecules assemble into a fibril. The distance, D, for one of these sections is 67 nm, and it represents the overlap length of adjacent tropocollagens in the crystalline lattice of the collagen fibril. **Panel B** shows the appearance of a negatively stained collagen fibril. **Panel C** shows the pattern of overlapping tropocollagen molecules in a fibril that accounts for the appearance of the negatively stained fibril. Because the 67 nm overlap length is not an integer fraction of the tropocollagen length, tropocollagen molecules do not assemble end to end; rather there is a gap of about 37 nm between the aligned tropocollagens. **Panel D** shows details of the positive-staining pattern observed for collagen fibrils. These bands indicate the various zones along the tropocollagen where hydrogen bonds, ionic bonds, and hydrophobic bonds between adjacent tropocollagens stabilize the crystalline structure of collagen fibrils. The labels a, b, c, d, and e are the standard designations for the particular bands revealed by positive staining.

However, at a pH of about 4 the phosphotungstate ion carries a net negative charge, and it will bind to positively charged lysine and arginine side chains, in a process called positive staining. This reveals a characteristic distribution of clusters of positive charges along the length of the overlapping tropocollagen molecules in the form of a set of fine, dark lines that repeat every 67 nm. In combination with another heavy-metal stain, uranyl acetate, which carries a positive charge at this pH and will stain the negatively charged side chains of glutamic acid and aspartic acid, it is possible to reveal the distribution of all charged amino acids within the D-period repeat of the collagen fibril. This pattern is illustrated in fig. 9.4D, where we see identified the 10 most prominent bands produced by positive staining of type I collagen. Note that the size scale in this panel has been expanded to reveal the banding pattern in greater detail, as indicated by the arrows connecting one repeat unit in fig. 9.4C to one positive-stain banding pattern repeat in fig. 9.4D. This pattern is repeated regularly every 67 nm along a collagen fibril over distances of many thousands of micrometers, because the overlapping pattern of tropocollagens is repeated over the full length of these crystalline fibrils. Close inspection of the amino-acid sequence of the tropocollagen molecule indicates that sequence blocks of 234 amino acids have charge distributions that very closely match the banding pattern observed in positive-stained fibrils (fig. 9.4D). So the D-spacing repeat in the collagen fibril actually corresponds to 4.45 repeats of this 234-amino-acid sequence block along the full length of the tropocollagen molecule.

There is one final design feature that we need to consider before we leave the structural level of individual collagen fibrils; this is the formation of covalent cross-links within and between the collagen fibrils described above. To this point we know that the tropocollagen molecule is a triple-helical coiled-coil molecule in which the covalent backbones of the three α chains are nearly fully extended, making tropocollagen a rigid tensile structure, and we know that this structure is stabilized by secondary bonding forces such as hydrogen bonds. Similarly, in the collagen fibril, where tropocollagens are organized into crystalline arrays of overlapping tropocollagens, the structure is stabilized by an array of secondary bonding forces (ionic bonds, hydrogen bonds, hydrophobic bonds) all along the length of adjacent tropocollagens. What is not clear from the information in this figure is whether these secondary bonds are sufficient to create the stiffness and strength in collagen fibrils that are required to create functional macroscopic tensile materials. The short answer, however, is that additional covalent bonds between the α chains, both within tropocollagen molecules and more importantly between α chains in adjacent tropocollagen molecules in fibrils, are required for collagen fibers and fiber bundles, as found in tendons, ligaments, and other collagenous tissues, to achieve functional levels of stiffness and strength.

Collagen cross-links are derived from lysine side chains, and in most cases from lysines that have been hydroxylated at the C-3 position on the side chain. The variety of chemical structures found in collagen cross-links is quite large, and details are available in Brinkmann et al. (2005) and Fratzl (2008). The overview presented here will summarize the general types of cross-links that are found in mammalian systems. Collagen α chains contain a large number of lysines, but only four are involved in the formation of primary cross-links. During the processing of procollagen in the cell secretion pathway these four lysines are hydroxylated to varying degrees into OH-lysine. Two are in the triple-helical portion of the α chains: one near the N-terminus, at position 89, and one near the C-terminus at position 930. In addition, there are two lysines in the telopeptides that participate in cross-linking, one each in the N-terminal and C-terminal telopeptides, and these are also hydroxylated.

The key reaction that initiates cross-link formation is the oxidative deamination of the lysine or OH-lysine side chains to form a reactive free aldehyde, called allysine or hydroxyallysine, depending on whether it is formed from lysine or from OH-lysine.

This reaction occurs in the extracellular space and only with tropocollagens that are incorporated into fibrillar structures. That is, the enzyme that catalyzes this process, lysyl oxidase, binds only to appropriate sites on fibrillar structures, not on individual tropocollagens in solution. As a result, only the lysines present on the telopeptides are accessible to this enzyme, and only these are converted into reactive aldehydes. Once these reactive aldehyde groups are formed, they are available to spontaneously form divalent cross-links with neighboring lysines or OH-lysines. The lysine side chain is quite long, with four methyl groups plus a terminal amino group, so the reacting lysines can project a reasonable distance out from the triple-helix, and the allysines present on the nonhelical telopeptides likely can project even further. It should be clear, however, that reactions will only occur between tropocollagens if the reactive groups are bought together by precise alignment of tropocollagens in the fibrillar structure. This is indeed the case. The tropocollagen overlap structure, described in the D-spacing pattern in fig. 9.4, precisely aligns the helical lysines and OH-lysines with the telopeptide allysines and OH-allysines, allowing the spontaneous formation of the immature covalent cross-links, as shown in figs. 9.5A and 9.5B. These are divalent cross-links because they form from the reaction of two amino-acid side chains, and they link together two tropocollagen molecules. Fig. 9.5A shows the reaction of a telopeptide allysine with a helix OH-lysine to form a Schiff-base compound called dehydro-hydroxylysino-norleucine (ΔHLNL). An alternate pathway is indicated in fig. 9.5B, where a telopeptide OH-lysine reacts with a helix OH-lysine to form a slightly different divalent cross-link called hydroxylysino-ketonorleucine (HLKNL). These are two of several intermolecular divalent cross-link structures that can form between two adjacent tropocollagen molecules. In addition, intramolecular cross-links can form by aldol condensation of adjacent allysines or OH-allysines in the telopeptides of adjacent α chains in an individual tropocollagen molecule. Fig. 9.5C shows how the four lysine residues identified above could be involved in a pair of divalent cross-links at the overlap between two adjacent tropocollagens in the fibril overlap structure.

The divalent cross-link structures in figs. 9.5A and 9.5B differ in one significant way; namely, that the Schiff-base cross-link, ΔHLNL, in fig. 9.5A is chemically quite unstable. It will degrade with modest heating or in acid or alkaline conditions, but it can be converted into a stable compound by reduction with sodium borohydride. HLKNL (fig. 9.5B), on the other hand, is a very stable compound. The unstable cross-link provides the basis for a simple but elegant experiment, conducted by Allen Bailey (1968) that clearly documents the importance of intermolecular covalent cross-linking in the mechanical properties of collagen fibers. He followed the maturation of collagen cross-linking in the development of rat tail tendon collagen, where the immature cross-links are predominantly ΔHLNL. He used changes in pH to destroy this chemically unstable cross-link, which is the dominant cross-link in young animals. The experiment involved measuring the tensile strength of tendons from animals of increasing age that had been exposed to a range of pH levels. The mechanical tests, however, were all carried out at pH 7. Tendons from young (three-week-old) animals lost virtually all their tensile strength when treated at pH 4 and below, as well as at pH 10 and above, but they had full strength at physiological pH (pH 7–8). As animals increased in age, the pH effect declined, suggesting that initial ΔHLNL cross-links may become more stable over time. In addition, tendons from young animals that had been treated with

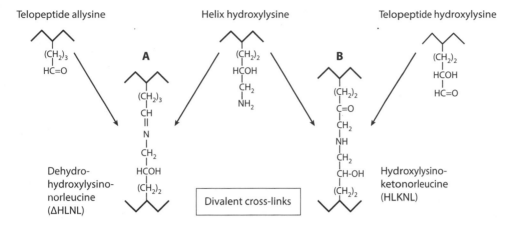

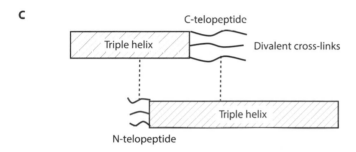

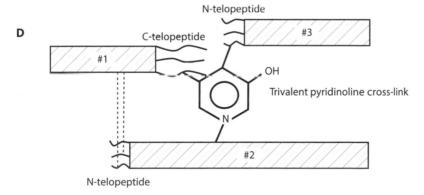

Figure 9.5. The cross-linking chemistry of collagen fibrils. **Panels A and B** show the structure of divalent cross-links formed from the reactions of telopeptide allysine and hydroxyallysine with a helix lysine. **Panel C** shows the typical organization of the divalent cross-links at the overlap of the C-terminal and N-terminal ends of two adjacent tropocollagens. **Panel D** shows a possible structure for one of the mature cross-links that form from the reaction of pairs of divalent cross-links. This cross-link is drawn to illustrate the possibility that the mature cross-links have the potential to connect three tropocollagens.

sodium borohydride to stabilize the ΔHLNL cross-links did not show any observable decline in tendon strength over the full range of pH. The result is clear: without covalent intermolecular cross-links in place, collagen fibers in tendons lose essentially all of their mechanical integrity. It is also clear that over quite long time periods there are changes in collagen cross-link chemistry that increase the mechanical integrity of collagen fibers in tendons and similar collagenous structures. This leads us to the mature cross-links that arise from further reactions involving the divalent cross-link structures described above.

The mature cross-links in collagens also form spontaneously, and many of them form from reactions between pairs of neighboring divalent cross-links as illustrated previously. Fig. 9.5D shows the basic form of a trivalent cross-link that forms from the reaction of a pair of HLKNL divalent cross-links that are in close proximity within a collagen fibril. The ring compound shown here is called a pyridinoline. This figure was drawn to illustrate two possibilities for the placement of this type of cross-link. The first is illustrated by pair of HLKNL cross-links, indicated as dotted lines, running vertically between the helix of the tropocollagen #1 downward to the N-terminus of tropocollagen #2. These two cross-links will be sufficiently close that they could easily react to form a pyridinoline cross-link that joins these two tropocollagen molecules together. To the right of the diagram we see the actual chemical structure of the pyridinoline cross-link, and in this case it is located in a position that would allow it to connect three tropocollagen molecules together. This would require an HLKNL cross-link running from the C-terminus of collagen #1 to the helix of collagen #2, as well as a second HLKNL cross-link running from the N-terminal telopeptide of collagen #3 to the helix of collagen #2. Tropocollagens #1 and #2 are drawn as they would exist in the plane of the overlap pattern shown in fig. 9.5C. Tropocollagen #3 is drawn slightly above and to the side of #1 to indicate that it is likely located either above or below the plane of the overlapping molecules, where it presumably had established a divalent cross-link from its C-telopeptide to the helix of the tropocollagen #2. It is not clear what proportion of the trivalent cross-links interconnect three tropocollagens. Regardless, these mature cross-links create a strong network of interconnections that stabilize the structure of collagen fibrils.

5. THE STRUCTURAL ORGANIZATION OF COLLAGEN FIBERS IN TENDONS AND LIGAMENTS. With this information about the structural organization of collagen fibrils, we can begin to think about their organization into macroscale tensile structures, such as tendons and ligaments. The next level of the hierarchical structure of a tendon is the collagen fiber, and we will start with the organization of collagen fibrils within a fiber. Electron microscopy easily documents the fact that collagen fibers are made from parallel arrays of collagen fibrils, as indicated in fig. 9.6. This drawing shows the basic organization observed when collagen fibers are sectioned in a plane perpendicular to the fiber's long axis and observed under the electron microscope. Here we see what looks like a parallel-fiber-reinforced composite material with a fiber volume fraction of the order of 0.7. The collagen fibrils look quite circular in cross-section, and typically they are observed to have a range of diameters, usually spanning from 50 to 250 nm, although in certain collagen types the diameters can be smaller and/or larger. This drawing does not provide any information about the nature of the material, if any,

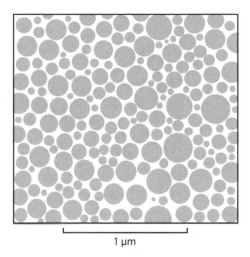

1 µm

Figure 9.6. The microstructure of a collagen fiber. A collagen fiber contains many collagen fibrils arranged in parallel, which are surrounded by and embedded in a matrix made of a proteoglycan gel. In this drawing we are looking at a cross-section of the fiber, where the fibrils project out of the page. The matrix fills the white spaces between the fibrils.

that fills the spaces between the fibrils, or about the length of the fibrils themselves. However, it is clear that there must be some connections in this system at some level. Tendons in large mammals, such as horses, can be of the order of one meter in length, and although it is theoretically possible that individual collagen fibrils span this full length, it is very unlikely. However, if there is a matrix material that fills the spaces between fibrils, then it is possible that collagen fibers function as a discontinuous fiber composite material, as described in chapter 8. In order to understand the mechanical design of tendon fibers, we need to understand the components of this composite and to establish their structural and mechanical properties.

Collagen-fibril length in tendon collagen has been evaluated in two studies, and the results suggest that collagen fibrils can reach lengths of the order of 10 mm. Since collagen-fibril diameters are of the order of 100 nm, the aspect ratios of collagen fibrils are of the order of 10^4 to 10^5. In one study, by Craig and colleagues (1989), the collagen-fibril length in rat tail tendon was estimated from observation of the distribution of fibroblasts, the cells that synthesize collagen, during the development of the tendon. The fibroblasts were found in the tendon in columns of varying length that ran parallel to the long axis of collagen fibers, and electron microscopy revealed that these cellular columns were the sites of collagen synthesis. That is, cell-membrane invaginations were observed to form extracellular compartments that provided a controlled environment for the assembly of growing collagen fibrils. Thus, the authors reasoned that the two ends of a growing fibril could be identified by the longitudinal distance along collagen fibrils between two cell columns. However, it is not possible to identify unequivocally the opposite ends of a single fibril from the large number of cell columns observed at any stage in development. They measured the changes in cell-column length and volume fraction in the tendon with age from 5 days to 6 months, and then used this information in a statistical model that allowed them to estimate the length of collagen fibrils at 6–12 mm in the oldest tendon. Mean fibril diameters were estimated for each age in the study, and, as shown in the far right column of table 9.1, fibril aspect ratios were found to be of the order of 10^5 in tendons from older animals. In addition, John Trotter (Trotter and Wofsy, 1989) analyzed serial sections of rat extensor digitorum

Table 9.1. Structural Organization of Tendon Collagen Fibrils as a Function of Age

Age	Crimp Angle (θ)	Crimp Length, l_0 (μm)	Strain to Extend Crimps, Measured (Calculated as $1 - \sec\theta$)	Estimated Fibril Length (mm) (Age)	Estimated Fibril Aspect Ratio (Age)
Rat Tail Tendon					
Term fetus	40°	6	(0.305)	0.025–0.045 (5 days)	410–750 (5 days)
2 weeks	20°	21	0.14 (0.060)		
3 weeks	19.4°	46	0.08 (0.058)	0.31–0.62 (22 days)	3100–6200 (22 days)
2 months	18.6°	64	0.07 (0.055)		
5 months	17.5°	86	0.05 (0.055)	3.1–7.6 (3.5 months)	1.9×10^4–3.8×10^4 (3.5 months)
13 months	15.4°	92	0.035 (0.037)	6.4–12.7 (6 months)	3.5×10^4–7×10^4 (6 months)
29 months	12.5°	110	0.024 (0.024)		
Horse Leg Tendon					
Fetal (ca. 210 days)	27°	30	(0.122)		
Young (2–4.5 years)	14.6°	20	(0.033)		
Mid-age (5–7 years)	13°	19.5	(0.026)		
Old (10–15 years)	10.5°	17.5	(0.017)		

The data for rat tail tendon presented in this table have been taken from three sources. The data for rat crimp angle, crimp length, and strain to extend crimps are taken from Diamant et al. (1972). The horse data for these parameters come from Patterson-Kane et al. (1997), and the values for estimated collagen-fibril length and aspect ratio come from Craig et al. (1989).

longus tendon using electron microscopy, and in this process traced 5639 fibrils for lengths about 750 nm, or for a total length of 4.2 mm. Four free ends were observed, and statistical analysis indicated the 95% confidence limits for fibril length spanned about 1–6 mm, which with a measured average fibril diameter of 85 nm gave fiber aspect ratios of between 10^4 and 7×10^4. These quite similar estimates of aspect ratio give some confidence that at least the order of magnitude of the aspect ratio, 10^4–10^5, is a reasonable estimate for the tendons of mature animals.

Clearly, collagen fibrils are a dominant component of the fiber-composite structure that makes up a collagen fiber, but what is the matrix? It turns out that the matrix is provided by a class of molecules called proteoglycans, which, as the name suggests, are polymeric complexes made from proteins and sugar-based polymers. Proteoglycans are a class of macromolecules that is found in a wide range of collagen-containing connective tissues, such as cartilage, skin, arteries, and other tissues. Proteoglycans are characterized by a very large molecular size because they function as highly hydrated, space-filling macromolecules that provide a viscoelastic matrix for the widely dispersed collagen and elastin fibers that dominate the mechanical properties of these tissues. The spaces between collagen fibrils in tendon collagen fibers are quite narrow, of the order of 50 nm, and the proteoglycans in tendons are accordingly smaller. In tendons the dominant proteoglycan is called decorin, but it is accompanied by smaller amounts of biglycan and other small proteoglycans. Before we look at the structural role of these molecules we need to know more about their chemical structure.

The protein components of proteoglycans are not significantly different from many other "globular" proteins, having a fairly compact structure, as opposed to the extended structure of tropocollagen. Decorin contains a core protein made from a single poly-peptide chain that contains 359 amino acids in the full human protein. However, it exists in a number of splicing variants. The full-length protein contains a series of 12 leucine-rich sequence motifs that contain parallel-beta structures interconnected by helical units. Together, these motifs form a relatively flat shape that takes on an arch-like form. It appears that this arch provides a surface through which the decorin core protein can bind to single tropocollagen molecules, and hence to the surface of collagen fibrils. Indeed, it has been established by electron microscopy that decorin binds specifically to the d-band of the positive-staining pattern illustrated in fig. 9.4D. Decorin bound to tropocollagen is thought to regulate the assembly of collagen fibrils, and its binding to collagen fibrils likely makes it the major component of the matrix that holds collagen fibrils together in a collagen fiber.

In addition, the protein core contains a single polysaccharide chain that contains on average about 50 disaccharide units, which would be about 125 nm long when fully extended. This sugar polymer is very different from the polyglucose polymers found in cellulose and in starch. The accepted name for the sugar polymers in proteoglycans is glycosaminoglycan, and the abbreviated name that we will use is GAG. GAGs are linear polymers, and their chemistry is notable mainly because they contain negatively charged groups on their sugar rings. These groups transform them from uncharged polymers into highly charged polyelectrolytes, which at physiological pH carry a large number of negative charges. Their polymer backbone is based on repeating disaccharide units, and the structures of several GAG repeat units are shown in fig. 9.7. The decorin GAG chain is known to contain both dermatan sulfate and chondroitin sulfate components, in amounts that vary with the particular type of collagenous tissue where it is found. Note that each repeating disaccharide unit contains a six-member ring that looks like a glucose ring. However, in sugars the ring carbons all have attached hydroxyl groups, and there are important differences in the side groups of these GAG polymers. In dermatan sulfate and the two forms of chondroitin sulfate, the ring on the left contains a carboxyl group at carbon 6, making this a uronic acid rather than a sugar. In chondroitin sulfate this ring is glucuronic acid, with the carboxyl group above the ring,

A   Dermatan sulfate

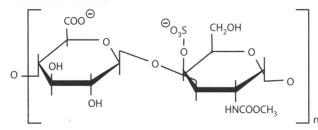

B   Chondroitin-4-sulfate

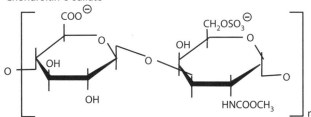

C   Chondroitin-6-sulfate

Figure 9.7. Repeating chemical units of several of the main glycosaminoglycan polymers. These glycosaminoglycan (GAG) polymers are key components of the proteoglycans that make up the matrix of tendons and other collagen-based structural materials. The key feature of these GAGs is that they have multiple charged side groups attached to their sugar rings, making linear polymers of these repeating structures function as highly expanded polyelectrolytes.

and in dermatan sulfate it is iduronic acid, with the carboxyl group below the ring. The second sugar ring remains a sugar, but it has an N-acetyl group on carbon 2, making it N-acetyl-D-galactosamine. In addition, there is a sulfate group attached to carbon 4 of dermatan sulfate and of chondroitin-4-sulfate, and, as you might expect, the sulfate group of chondroitin-6-sulfate is on carbon 6. These carboxyl and sulfate groups carry the negative charges that establish the characteristic behavior of these polysaccharides.

The presence of negative charges on every sugar unit in these GAG polymers creates a very high negative-charge density, and the repulsive forces between the negative charges makes these chains take on highly extended conformations. Thus, we expect that decorin molecules bound to the surface of a collagen fibril will have their GAG chains pointing outward toward neighboring fibrils, where they may interact with other decorin GAG chains emerging from the neighbor, or they may bind directly to the surface of the neighboring fibril. Fig. 9.8 shows a possible arrangement for the decorin

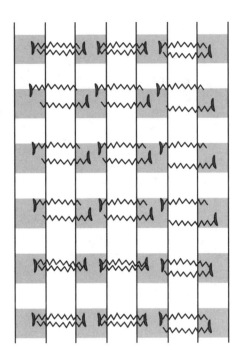

Figure 9.8. The basic fiber-reinforced-composite structure of the collagen fiber. The banded rods represent the crystalline collagen fibrils, now seen from the side. The proteoglycan molecules that bind to the fibril surfaces and interact with each other in the space between the fibrils create a soft matrix that binds the fibrils together as a fiber-reinforced composite.

matrix, in which GAG chains from opposite collagen fibrils overlap, and hence interact. Please note that the space between fibrils is drawn much wider than actually exists in tendons, and it is likely that the long GAG chains, which should be quite flexible, could coil around each other and interact through bridging cations that bind the GAG chains together. In this way, decorin could provide a matrix that holds the fibrils together in a fiber to form a fiber-reinforced composite system.

Experiments to test this hypothesis have had varying results. An electron microscopic study by Liao and Vesely (2007) of GAG-chain morphology in mitral valve chordae tendineae under load revealed interesting changes in the organization of the GAG chains. The study used a cationic stain, cupromeric blue, that is specific for GAGs under appropriate conditions, to stain the decorin GAG chains, and they observed dramatic changes in the organization of the decorin network between collagen fibrils in this tendon. Specifically, in longitudinal sections of the tendon at zero load the GAG chains were seen to run essentially perpendicularly to the collagen fibrils, but as load was increased to levels that approached the failure load, the GAG chains rotated to angles of about 45°, indicating that the proteoglycan matrix between the fibrils achieved quite high shear strains. Interestingly, cross-sections of the tendons showed that the spacing between collagen fibrils in the tendon was also dramatically decreased. These results provide strong evidence for decorin's role in force transfer between collagen fibrils within fibers. Experiments to remove decorin from tendons provide some support for this hypothesis.

Danielson and colleagues (1997) used a decorin-knockout mouse to investigate the role of decorin on collagen-fiber formation and on the material properties of skin. They saw extensive disruption of fiber morphology, and they also observed a fourfold reduction in skin strength, confirming decorin's role in fiber assembly and material properties. Interestingly, Robinson and his colleagues (2005) measured the mechanical properties of tendons from decorin-knockout mice and from biglycan-knockout mice, and they saw no change in mechanical properties for tail tendon from either of the knockouts, a small increase in modulus for patellar tendon in the decorin knockout, and a modest decrease in strength and stiffness of the flexor digitorum longus tendon in the biglycan knockout. More recently, Lujan et al. (2009) used chondroitinase-B, an enzyme that selectively cleaves dermatan sulfate GAGs, on human medial collateral ligament, and they were not able to see any changes in ligament properties, in spite of the fact that 90% or more of the dermatan sulfate GAGs were removed. Similarly, LeAnn Dourte et al. (2012) used decorin-knockout homozygote and heterozygote mouse strains to test the role of decorin in patellar tendons, and although they did detect small changes in some properties, most properties were not changed. We will reconsider this issue when we discuss the mechanical design of the collagen-fiber composite structure later in this section, but it appears that these knockout studies do not provide very strong support for the role of proteoglycans as matrix molecules for the collagen fiber's composite organization.

Collagen fibers are typically found with diameters of the order of a few micrometers, and these fibers are collected into bundles, called fascicles, with diameters of the order of 1 mm or less. Interestingly, the fibers in fascicles are surrounded by a fibrous sheath, made largely from collagenous tissue, and mechanical tests of force transmission between adjacent fascicles suggest that they essentially act as independent tensile units. So, even though a tendon is typically made up of a number of fascicles, presumably with distinct attachment sites on muscle and/or bone, the fiber bundles within a fascicle have generally been taken as the highest level of the tendon structural hierarchy that we need to consider in our analysis of tendon/ligament design.

Polarized-light microscopic observation of collagen-fiber bundles reveals an interesting morphological organization; namely, that the collagen fibrils within a fiber are not straight—rather they run in a zigzag pattern usually referred to as the collagen crimp. This pattern is defined by a crimp length, l_0, of the order of $20\,\mu\mathrm{m}$ or longer, and by crimp angles, θ, that range from about 10° to 25°, as shown in fig. 9.9. The crimp pattern varies with development; typically with higher angles in younger tendons. Indeed, this is the case for both rat tail tendon and for the superficial digital flexor tendon in wild, untrained horses as shown in table 9.1. Crimp length also changes with development, but the patterns are different for the rat tail tendon, which shows crimp length increasing almost 20-fold with age, and the horse tendon, which shows crimp length decreasing by about a third with age. The result of these patterns, particularly those in crimp angle, is readily apparent in the mechanical behavior of tendons and ligaments, as we will soon see. Briefly, initial loads straighten the crimps, giving reduced stiffness in the initial stages of extension. However, when the crimps have been pulled straight, tendons enter into a linear stress-strain region where stiffness is maximal. It is likely that the length of the low-stiffness region of stress-strain curves will increase with crimp angle.

The origin of this crimp pattern is not well understood, although crimping at the level of individual collagen fibrils has been observed by electron microscopy for many years. It is clear that there are abrupt deviations in the direction of adjacent collagen fibrils within individual collagen fibers, and these local crimps in the fibrils are lined

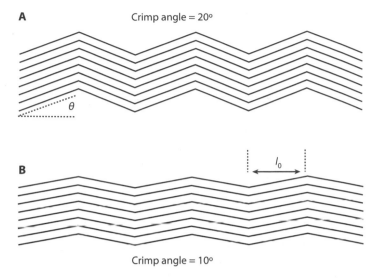

Fig. 9.9. The typical crimped structure of collagen fibrils in collagen fibers in a tendon. Crimp length, l_0, is usually of the order of 20 μm or longer, and crimp angles, θ, range from around 10° to 25°. The pattern varies with development.

up across the full width of collagen fibers. Initially, it was thought that these crimps were entirely planar structures, but careful observation of collagen-fiber bundles using polarized-light microscopy indicates that in some collagenous tissues the fiber "crimp" actually follows a helical path. However, Marco Franchi (2010) and his colleagues have shown that even in collagen fibers that appear to have a planar crimp structure, there is a left-handed helical twist to the fibrils within the fiber. In the case of fibers that show the planar zigzag appearance, the helical rotation is compressed to a half turn of the helix located just at the crimp point of the fibril. The remainder of the fibril is straight.

6. MECHANICAL PROPERTIES: STIFFNESS, STRENGTH, RESILIENCE, AND TOUGHNESS. Mechanical testing of tendons and ligaments reveals a characteristic set of mechanical properties that arise directly from the hierarchical structures of collagen fibers described above. Fig. 9.10 presents a series of stress-strain curves that document the progression of mechanical properties during the development and growth of rat tail tendons with age. These stress-strain curves have been constructed from the data obtained by John Kastelic and Eric Baer (1980) with adjustments made for the strain required to straighten crimps from the study by Diamant and colleagues (1972), as given in table 9.1. Note that the most mature tendons, at ages of 1–3 years, show a characteristic, "J-shaped" stress-strain curve that starts with a low-stiffness region where modulus rises gradually at strains of 0.02–0.03. This is called the "toe region," where the crimps are straightened and before the load is applied directly to taut collagen fibers. This is followed by a "linear region" that continues up to a stress of about 20 MPa, over which the modulus is about 1.5 GPa. This is a region of fully reversible elasticity. With extension beyond this point, however, the slope of the stress-strain curve drops slightly as the sample enters what the authors called "yield region 1." This region extends for

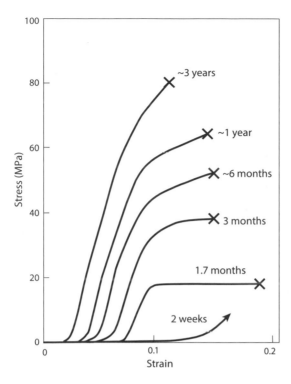

Figure 9.10. Changes in the mechanical properties of rat-tail tendon through development. The properties shown for the tail tendon of a 3-year-old rat are reasonably similar to those of mature tendons from other animals. However, in young animals the properties differ dramatically. In early development the tendons extend easily by more than 10% before they begin to stiffen, and they fail at quite modest stress. Over the life-span of a rat, the low-stiffness "toe region" of the stress-strain shortens and the tensile strength increases.

about another 0.02 strain or so, and strain in this range causes irreversible changes in material properties. That is, load cycles in this range will cause gradual drops in stress and stiffness. "Yield region 2" follows where the slope of the stress-strain curve declines rapidly, leading to the rupture of the sample. The oldest tendon (3 years) fails at about 80 MPa, and failure stress is considerably lower in younger tendons, falling to about 50 MPa at 6 months.

Fig. 9.10 also illustrates the tendon properties in the early development of the rat. The curve for the tendon from the 3-month-old animal proceeds directly from the linear region into yield region 2, and failure occurs at about 16% total extension and at a stress of about 35 MPa. The 1.7-month tendon does not quite achieve the 1.5 GPa initial stiffness seen in the older animals, and in this sample the yield region appears as a very long plateau to failure at high strain and low stress that occurs at a strain of about 0.18. This high strain to failure arises from the combination of a roughly 0.07 toe-region strain and a long yield region 2. The data for the 2-week-old animal came from J. Diamant et al.'s 1972 study that showed only the initial portions of the stress-strain curve, so the failure stress and strain are not known for this sample, although it seems clear that failure strain will be considerably larger and failure stress lower than that seen in tendons from older animals. The initial linear region had a stiffness that was about half that of the 1.7-month sample.

There are several structural factors of these tendons that can explain the observed changes in the tendon properties with age. First, there are significant changes in the collagen fibers' crimp morphology. In addition, there are changes in the composite organization of the collagen fibrils and their proteoglycan matrix, as well as changes in the extent of covalent cross-linking within the collagen fibrils. As shown in table 9.1, the crimp angle, θ, for rat tail tendon declines dramatically from the term fetus to the 2-week juvenile, and then declines gradually over the remaining life-span of the animal. These changes in crimp angle are largely responsible for the dramatic reduction in the toe region with age. In fact, simple trigonometry allows us to predict the tensile strain, ε, needed to rotate a collagen-fiber segment of fixed length from its starting crimp angle, θ, to become aligned with the applied tension, as $\varepsilon = 1 - \sec\theta$. Values for this calculated strain of the toe region are given in table 9.1. Note that for very young tendons, up to and including the 2-month-old tissue, the measured strain required to straighten the crimps is considerably larger than the strain predicted from the magnitude of the crimp angle, and this indicates that extension of these very young tendons results in both the rotation of the crimped tendon segments and also the straining of the composite structure that makes up the collagen fibrils. However, for 5-month and older tendons the observed and predicted strains to straighten crimps match closely, suggesting that the toe region is explained by fiber reorientation alone.

We do not know the reason for the discrepancy in the young tendons, but there are two plausible explanations. First, as shown in the two columns to the far right in table 9.1, we know that collagen-fibril length changes with age, and thus the composite structure that exists in collagen fibers may be less stiff than in fibers in more mature tendons. The data in these columns indicate that collagen-fibril aspect ratio is quite low in young tendons (about 500 in a 5-day-old animal). Aspect ratio rises dramatically with time, increasing to about 4500 at day 22, and by an additional order of magnitude at 6 months. Recall from our discussion of composite-materials design in chapter 8 that in a discontinuous fiber composite the aspect ratio of the reinforcing fibers plays a key role in determining the reinforcement efficiency and hence the stiffness of a composite. It is apparent, therefore, that changes in collagen-fibril length will contribute to the early rise in the initial stiffness of the youngest tendons. In addition, changes in volume fraction of the reinforcing fiber and in matrix stiffness may also affect the properties of these youngest tendons. We will discuss some of these possibilities shortly.

The changes seen in the shape of the stress-strain curves beyond the toe region in fig. 9.10 are likely to arise primarily from changes in collagen-fibril cross-linking that occur with age, as discussed above. It is also possible that some of the changes may be due to the quantity and the quality of the proteoglycan matrix that holds the collagen fibrils together in collagen fibers, or perhaps to interactions between fibers in fascicles and between fascicles in whole tendons. It is interesting that by 3 months the slope in the linear region is similar to that in the older tendons at a level of about 1.5 GPa, but at this age there really is not a yield region 1, and the tendon proceeds directly to yield region 2 and failure. However, with continued development yield region 1 is extended, producing tendons that fail at stresses that approach 80 MPa in samples from the very oldest animals. This rise in tensile strength most certainly is associated with the continued development of covalent cross-links, and it may also reflect the observed increase in fibril aspect ratio.

It is important to note that rat tail tendon is likely not the best model for the mechanical behavior of tendons in general. These tail tendons are quite large structures relative to the size of the animal, but in normal use they do not carry particularly large tensile loads. That is, they likely do not experience high stresses because rats do not use their tails in any significant way in their locomotion, other than perhaps switching them from side to side to help maintain balance. Other tendons in rat limbs and in the limbs of other animals experience stresses well above the yield regions that were established for rat tail tendon, and it will become clear that many tendons have material properties that are superior to those shown here for these tail tendons.

There have been a large number of mechanical studies on the kinds of tendons that play key roles in the mechanics of animal locomotion, and some interesting and important differences are observed. Fig. 9.11 shows stress-strain curves, redrawn from a study by Robert Shadwick (1990), for digital flexor and digital extensor tendons obtained from "mature" pigs, which were 4–5 months old and weighed 80–100 kg. The comparison between these curves and those for the rat is quite revealing. Note first that both pig tendons have a toe-region strain of less than 0.015, whereas the half-year rat tail tendon has a toe-region strain of about 0.04. That is, these pig leg tendons enter their linear elastic behavior quickly. In addition, their stress-strain curves remain linear to quite high strain, and they do not enter an extended yield region leading to failure until they reach about 80% of the tendon's failure stress. It is also important to note that there are large differences between the properties of the flexor and extensor tendons, with the flexor tendon having greater stiffness and strength by almost a factor of two (E_{flexor} = 1.66 GPa, $\sigma_{max,flexor}$ = 80–90 MPa; $E_{extensor}$ = 0.76 GPa, $\sigma_{max,extensor}$ = 40–50 MPa). The reasons for these differences likely arise from the fact that the functions of these tendons are quite different, and this can be visualized by looking at the anatomical diagram in this figure. The diagram shows a very simplified morphology for the distal end of the leg of a horse, which was used here because the anatomy of the horse's foot is much simpler than that of a pig. Both animals run and stand on the tips of their toes, but the horse's leg contains a single digit, which is formed from the evolutionary "fusion" of the three central digits. The pig's foot has four digits, where the central two digits carry the majority of the ground-contact force in locomotion. In this diagram, the two black bands represent the two digital tendons that were tested in this study. The deep digital flexor tendon is shown as the thick black band that runs along the right side of the column of digital bones. That is, it runs along the posterior face of the limb. The digital extensor tendon is shown as the thin black band on the anterior face of the digital bones.

The digital extensor tendon is relatively thin because its function is primarily in the extension of the forelimb (i.e., forward rotation of the digit) when the foot is off the ground. Thus this tendon carries relatively small loads. On the other hand, the deep digital flexor tendon, along with other flexor tendons that run parallel to it, resists the very large loads that occur during the ground-contact phases of locomotion. Tension created by flexor muscles, located well above the top of this diagram, pull the distal limb segments toward the right in flexion, and this movement contributes to thrust during locomotion. The locomotor mechanisms in terrestrial locomotion are complex because in gaits with aerial phases (i.e., trotting and galloping) this muscle/tendon system must resist the very large ground reaction forces that develop during landing phases, and it does this in large part by the stretching of these tendons in a process that stores a large amount of elastic energy. Then, during the thrust/takeoff phase, these tendons recoil and release the stored energy, and this contributes to the vertical and forward forces that power locomotion. Clearly, tendon size is a major factor in flexor-tendon design;

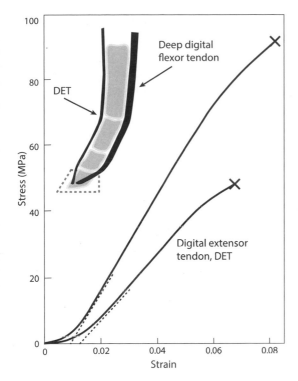

Figure 9.11. Stress-strain curves for digital flexor and digital extensor tendons from the pig limb. Here we see properties for tendons from young animals (ca. 6 months) that function at high stress in locomotion, and their properties are very similar to those of the oldest rats. The anatomical drawing shows the equivalent tendons in the distal horse's leg (used here because the morphology is simpler than in the pig). Note the greater thickness of the flexor tendon compared with the extensor tendon.

thicker tendons resist larger forces and store more elastic energy. But these tendons also appear to have better properties.

This gets us to the crux of the matter. What do we mean by better properties? It should be clear by now that it is not possible to provide single numbers for each of the properties that we can measure for a tendon or ligament. There is error in all measurements, but the variation we are talking about arises from two biologically relevant processes. Firstly, tendon development has a very large influence on material properties, as we saw above with rat tail tendon. In addition, it is possible that the properties of the various tendons in an animal skeleton will vary with the biomechanical function of each tendon. To investigate this possibility we must first establish a suite of material properties that we can measure, and then we should consider which properties are of particular importance to a particular tendon's function. We will start by comparing properties obtained from tests in which tendons are loaded at constant strain rate to failure, as we have seen in fig. 9.11.

It is typically quite difficult to obtain accurate stiffness and strength values for tendons and ligaments, largely because of their structural organization as tensile materials. That is, their basic structural organization is a rope-like design with bundles of parallel filaments. In ropes, however, the filaments are twisted around each other to create a very tightly packed structure, but in tendons the fascicles, fibers, and filaments are essentially parallel to the long axis of the tendon. Thus, when clamping tendons for testing with a simple mechanical clamp, the tendon will spread out and is often grossly distorted and mechanically compromised. This issue of gripping tendon ends in combination with the relatively low aspect ratio (length/diameter) of tendons makes it

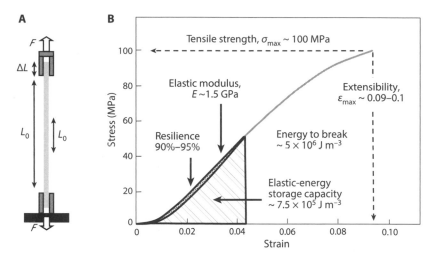

Figure 9.12. Testing tendon samples. **Panel A**, showing a typical tendon sample gripped between two clamps, illustrates some of the problems associated with the testing of tendon samples, as discussed in the text. *F*, force; L_0, initial length; ΔL, change in length. **Panel B** shows a tendon stress-strain curve that illustrates what is generally accepted as the behavior of mature tendons, along with typical values for the properties of these tendons.

difficult to get accurate values for stiffness and for strength. The origins of these problems are illustrated in fig. 9.12, which shows a typical tendon sample gripped between two clamps, which would then be attached to a testing machine. This sample has an aspect ratio of 30, and about 10% of sample length is used in gripping each end. In a horse tendon the sample would be about 1 cm in diameter, making the gripped section of the tendon sample about 3 cm long. Two problems arise with gripping this kind of sample. First, because the samples are quite thick, the stress distribution within the sample inside the grip will not be uniform. That is, the stress and strain of the tendon do not go from maximal just outside the grip to zero just inside the grip. Rather, the outer surface of the tendon that makes contact with the grip conforms to this pattern, but the central regions of the tendon inside the grips will carry less stress and strain than the outer surfaces. Thus, it is difficult to establish a true initial length, L_0, for the tendon sample from the separation of the grips, as indicated by the long arrow in this diagram, and it is not appropriate to measure tendon strain by tracking these grips. Indeed, Bennett and colleagues (1986) quantified this source of error and found that the effective length of a tendon of this size can be as much as 3–5 cm longer than the spacing between grips, so errors can be very large. The best way to deal with this problem is to determine the sample's strain with measurements taken over a small region at the center of the test sample, as indicated by the double-headed arrow on the right side of the test sample. This can be done by placing markers on the tendon's surface near the center of the tendon and following deformation with video systems, or by attaching a displacement transducer at the center of the tendon. In addition, clamping a tendon with sufficient force to keep it from slipping out of the end grips will dramatically flatten and distort the tendon's shape. This will cause large stress concentrations at the clamp that are likely to initiate tendon failure at stress levels that are significantly lower than the true strength of the material. To prevent this, the stress-strain curves in fig. 9.11 were obtained by "freeze clamping" the tendons, using grips that were cooled with liquid CO_2, and this prevented premature failure at the grips. Tendon strain was measured at the center of the tendon using the methods described above.

A study of the mechanical properties of limb tendons from wild turkeys by Andrew Matson and his colleagues (2012) used a similar protocol for testing tendons that is based on an unusual aspect of turkey tendons; namely, that portions of the tendons of the hind limb are ossified. This means that they become bone-like, and as such provide rigid ends that can be clamped without gross distortion of the structure of the intervening tendon segments. The animals used in this study had been used in locomotor studies and had undergone treadmill training, so it is likely that the tendons would be in a similar state to those of free-living animals. Mechanical testing of three digital flexor tendons and one extensor tendon revealed similar results to those shown for the pig extensor and flexor tendons in fig. 9.11. That is, the extensor tendon had a modest stiffness (1.07 GPa) and strength (68 MPa), similar but somewhat higher than those of the pig extensor tendon, whereas two of the three flexor tendons had significantly higher stiffness (1.39 and 1.48 GPa) and strength (112 and 107 MPa), similar to those of the pig flexor tendon. Interestingly, the third flexor tendon had properties that were similar to those of the extensor tendon. So, the distinction between flexor and extensor limb tendons seems to hold, although there may be some exceptions.

An important question to ask is whether these apparent differences in material properties are predetermined by the animal's developmental program, or if they are adjusted by mechanisms that respond to the mechanical-loading regimes experienced by the individual tendons. Answers to this type of question come from experiments where animals undergo training regimes to see if tendon size and tendon properties are increased by exercise, and one of the first such studies was carried out by Savio Woo and his colleagues (1980, 1981). This group first tested the properties of digital extensor tendons from 1-year-old miniature Yucatan swine, five of which were subjected to 1 year of a modest exercise program, and three served as nonexercised controls. The exercised animals showed significant increases in both extensor tendon size and, more importantly, tendon mechanical properties. That is, tendon stiffness increased from about 0.70 GPa in the control to 0.95 GPa in the exercised animals, and tensile strength increased from about 40 MPa in the control to about 50 MPa in the exercised animals. Thus, it seems clear that exercise can improve the properties of tendons.

Tests on the larger flexor tendons from these animals showed quite different results. First, the control flexor tendons were both stiffer (E_{flex} = 1.14 GPa versus E_{ext} = 0.7 GPa) and stronger than the control extensor tendons ($\sigma_{max,flex}$ = 96 versus $\sigma_{max,ext}$ = 40), which is essentially in agreement with the results shown in fig. 9.11. But the exercise regime did not cause any significant increase in the flexor tendons' properties. This suggests that the flexor-tendon system may be preprogrammed for greater strength and stiffness to deal with the high-level dynamic-loading regimes that are required during locomotion, and they therefore do not "need" to respond to the exercise regime. A recent review of this area of tendon biology by Marcio Bezerra and colleagues (2012) found only eight studies dealing with these issues, including the two papers by Woo discussed above, and no clear picture has emerged. However, in the two most recent studies carried out on rats, where Achilles tendons and digital flexor tendons were followed, there appeared to be no significant effects of exercise on the tendons. Thus, it seems likely that programmed development plays a major role for tendons that function in elastic-energy-storage mechanisms in terrestrial locomotion.

The importance of limb extensor tendons for elastic-energy-storage systems in locomotion requires that tendon collagen must have properties that are consistent with this function, and this is indeed the case. Fig. 9.12B shows a generalized stress-strain curve for this type of tendon, which we know has an appropriate combination of stiffness, strength, and extensibility, as shown. That is, typical stiffness is about 1.5 GPa, strength is about 100 MPa, and extensibility is of the order of 0.09 to 0.10 (9%–10% extension). In addition, we can estimate the energy required to break the tendon as the area under the stress-strain curve to failure, giving a value of about 5 MJ m^{-3}. Spring tendons must also have high resilience so that energy absorbed during stretch is recovered in elastic recoil, with little loss through internal friction. Tendon resilience has been measured in load-unload cycles to determine the dynamic mechanical properties at frequencies that are appropriate for the time scale of locomotor movements. Experiments of this sort have been carried out with both sinusoidal and linear load cycles at frequencies of two to four per second, and the hatched loop at the lower end of the stress-strain curve in this figure shows the typical form of this behavior. The stress amplitude of this load cycle is matched to the typical in vivo force-generating capacity of the flexor muscles attached to these elastic limb tendons, as will be discussed shortly. The hysteresis loop shown here is typical of spring tendons, and values for resilience fall in the range of 0.9 to 0.95 (90%–95% energy recovery). With this information we can estimate the elastic-energy storage capacity of tendons, which in this case is based on a strain beyond the toe region of 0.035, which is a reasonable estimate of the strain achieved when maximal isometric muscle stress is applied to the muscle's tendon. The value of 0.75 MJ m^{-3} shown is a conservative estimate of the spring capacity of limb flexor tendons, as in running gaits the ground-contact forces developed during landing phases will drive muscles into negative-work (eccentric) contractions, which will drive peak muscle stress to well above isometric levels. This will result in a significantly higher level of elastic-energy storage in the tendons.

These values for tendon material properties are summarized in table 9.2, where they are shown as relatively broad ranges of values in which the upper value is derived from the spring tendons described above. The lower end of the range of stiffness, strength, resilience, etc., is based on tests of tendons, like the pig extensor tendon, that do not experience the negative-work episodes that are characteristic of limb flexor tendons. Thus, the table attempts to summarize the material properties of tendons to show the full range of properties seen in vertebrate animals. It might be useful to comment on one additional aspect of the mechanical properties of tendons; namely, the relationship of tendon properties to those of other structural materials. The comparison presented in fig. 9.1G between tendon and other high-performance polymeric fibers suggested that tendon collagen was a relatively low-performance material. However, in one particular property collagen turns out to be very high performance indeed. The property is elastic-energy storage capacity, and when expressed as specific-energy storage capacity ($\sim$800 J kg^{-1}) it turns out that collagen is quite exceptional. The comparison, as documented by J. E. Gordon, in his book *The Science of Structures and Materials* (1988),

Table 9.2. Typical Values for the Mechanical Properties of Vertebrate Tendons

Modulus of elasticity	0.78–1.6 GPa
Tensile strength	40–110 MPa
Extensibility	0.06–0.09
Resilience	0.70–0.95
Hysteresis	0.05–0.20
Energy to break	1.5–5 MJ m^{-3}
Density	1120 kg m^{-3}
Specific energy to break	1400–4500 J kg^{-1}
Elastic-energy storage capacity (@3.5% extension beyond "toe region")	0.4–0.75 MJ m^{-3}
Specific-elastic-energy storage capacity	350–800 J kg^{-1}

is with high-tensile spring steel, which was used for the springs in mechanical clocks and watches. Estimates of the useful specific elastic-energy storage capacity of spring steel are about 120 J kg^{-1}, which is about one-sixth that of our conservative estimate for tendon collagen.

7. THE STRUCTURAL DESIGN OF TENDONS AND THEIR FATIGUE LIFE-TIME. This takes us to a brief consideration of the several possible functions for tendons in animal skeletons. The spring tendons we have just considered in some detail represent one class of tendons because their function as elastic-storage devices imposes key structural-design features upon them. These features allow cursorial terrestrial animals to place low-weight, high-capacity elastic-energy-storage structures (i.e., flexor tendons) at the distal end of long limbs to increase running performance and reduce energetic costs of locomotion. The way this is achieved is through the production of high-stiffness and high-strength tendons with the capacity to be repeatedly loaded to high stress levels so that they can store and release large amounts of elastic energy in repetitive locomotor movements. This means that the spring tendons of limb flexor muscles must function at relatively high strains, and as a result they will have quite low safety factors. The term "safety factor" is not really a material property; rather, it is a property of a structural design. That is, it's a property of a tendon within a locomotor limb. The safety factor, *SF*, in this case will be defined as the tensile strength of a tendon, which is assumed to be 100 MPa, divided by the maximal in vivo stress developed during the normal maximal activities of the animal:

$$SF_{\text{tendon}} = \frac{\sigma_{\text{max,tendon}}}{\sigma_{\text{max,in vivo}}}. \tag{9.1}$$

In the case of limb flexor tendons this maximal activity may be high-speed running (galloping) or maximal jumping, etc.

It is difficult to measure tendon stress and strain directly in maximal activities, but it is relatively easy to calculate maximal tendon stress levels by measuring the force-generating capacity of the muscles that the tendons are connected to. This is because the force applied to a tendon is determined by the force generated by the muscle attached to it. Robert Ker and his colleagues (1988) estimated the tendon safety factor

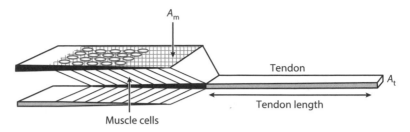

Figure 9.13. The organization of muscle and tendon in a pennate muscle system. The key feature of this pattern of muscle organization is that it allows a small mass of very short muscle cells to be organized into a compact structure that can apply very large forces onto tendons of insertion, as typically found in the distal limb segments of terrestrial animals. Tendons found in pennate muscle systems typically function at very low safety factors. A_m, muscle force-generating area; A_t, tendon cross-sectional area.

for a number of muscle-tendon systems in a variety of mammalian limbs, and they discovered that there are two relatively distinct classes of tendons. One of these is the spring tendons, such as the pig and turkey digital flexor tendon discussed above, and these tendons are characterized by being attached to muscles that contain short muscle cells, but with very large muscle cross-sectional area. This type of muscle is created by an interesting muscle architecture called pennate muscle-fiber design, as shown in fig. 9.13. In this diagram we see the organization of a bipennate muscle. A key feature of pennate muscle design is that the individual muscle cells do not run parallel to the axis of overall force production, which is parallel to the long axis of the tendon. Rather, they are arranged in V-shaped arrays, where the axis of force production of individual muscle cells is oriented at a small angle relative to the direction of force production through the tendon of insertion. In this bipennate muscle there are two layers of these short, angled muscle cells. Pennate muscles also occur in unipennate and multipennate structures, but the general properties of these muscles are the same. In the muscle shown here, the muscle-cell orientation is offset by about 22° relative to the direction of the tendon. Thus, the force generated within the muscle cells will be reduced, as a component of the force will act in a lateral direction. However, the component of the muscle force applied to the tendon is proportional to the cosine of the pennation angle, and at angles of this magnitude the tendon force is only slightly reduced ($\cos 22° = 0.93$) from the full muscle force.

The importance of this pennate design is that it makes it possible to organize a large number of short muscle cells into a compact design that has a very large muscle cross-sectional area, A_m (the hatched area in this diagram shows the effective cross-sectional area of one of these cell layers), and hence muscles with this type of design can produce a very large forces with a small mass of muscle, particularly in multipennate designs where there can be many such layers of very short muscle cells. In other muscles that do not function in elastic-energy storage, the most common muscle architecture has long muscle cells running parallel to the long axis of the tendon, and thus the force-generating area of a parallel muscle, A_m, is simply its cross-sectional area in a plane normal to the long axis of the muscle cells. The maximal in vivo tendon stress, or the "stress in life" as used by Ker et al., can be estimated with the following equation that relates tendon stress, σ_t, to muscle stress, σ_m, and the ratio of muscle cross-sectional area, A_m, to tendon cross-sectional area, A_t, as:

$$\sigma_{t,\max} = \sigma_{m,\max}\left[\frac{A_m}{A_t}\right].$$

(9.2)

The analysis assumes that muscle stress is equal to the maximal isometric tension of muscle, $\sigma_{m,\max}$, which is generally considered to be about 0.3 MPa. The muscle and tendon cross-sectional areas required in this kind of analysis can be measured easily by dissection.

The result of measurements for a range of limb tendons from a variety of animals is that the safety factor for spring tendons typically falls in the range of 1.5–2, whereas the safety factor for the majority of tendons, which do not serve as locomotor springs but simply act as relatively inextensible, but flexible, links that transfer muscle force and displacement to distant locations in a skeletal system, typically are in the range of 8 ± 4. This interesting dichotomy suggests a different design plan for the nonspring tendons. Ker et al. proposed that nonspring tendons may be optimized for minimizing the weight of the combined muscle + tendon system, and their analysis predicted that this would occur if tendon stress were maintained at about 10 MPa. Their logic goes as follows. Most tendons function to transmit the shortening displacements of contracting muscles, and as such they would work best if they did not stretch at all in the process of transmitting muscle displacements. The material that a tendon is made from is capable of stretching about 10% before it breaks, and if muscles are attached by long, thin, low-safety-factor tendons to their points of action, then a system with long tendons might see much of the muscle displacement taken up in the stretching of the tendon, reducing the magnitude of the limb rotation. To compensate for tendon stretch, the muscles would need to be longer so that they could contract over larger distances. However, if the tendons are thicker, and hence have much higher safety factors, the tendons would stretch less when they transmit the force and displacement of their muscles, and the total weight of muscle + tendon could be reduced. The analysis indicated that the optimum tendon stress for minimizing muscle + tendon weight would occur if the tendons carried only 10 MPa of stress when the muscles were maximally activated (i.e., safety factor = 10), and this is very close to the safety factors they observed for the nonspring tendons. The net result of this analysis is that different classes of tendons may require different material properties, assuming that all of the tendons show equivalent levels of tensile stiffness. Only the spring tendons require the highest levels of tensile strength because they are actually required to withstand stress transients that approach the tensile strength of the materials that they are made from. The extremely high stresses that develop in spring tendons arise in large part from the fact that when muscles are stretched while they are active, a process called negative work, they can develop stress levels that may be as much as 80% higher than the maximal isometric stress of 0.3 MPa. Because of this it appears that the in vivo tendon stress in spring tendons can approach the 100 MPa failure stress observed in single extensions to failure in mechanical tests, which suggests that spring-tendon safety factors can approach values of 1. It is important to note, however, that these extreme events are rare, and for the most part the spring tendons function at much more modest stress levels, keeping their safety factors to the range indicated above, 1.5–2. However, it is clear that there may be differences in material properties between the two tendon classes, low-safety-factor spring tendons and the other tendons that function exclusively in positive-work contractions at high safety factor. Indeed, the data in fig. 9.11 for pig digital extensor and flexor tendons conform to this possibility.

These observations lead us to the final class of material properties that we will discuss for tendon and ligament; namely, their behavior under long-term static loading or under repeated load-unload cycles. These kinds of loading regimes lead to the process of fatigue, and the quantification of fatigue is usually achieved by estimating the relationship between the intensity of loading, as indicated by the magnitude of the applied stress or strain, and the time or load-cycle number to failure. Robert Ker and his colleagues (2000) have carried out an extensive series of fatigue experiments, and their results are very interesting. They investigated the two basic fatigue processes, those that occur with loading at constant stress (i.e., creep failure), and those that occur with long-term cyclic loading. We will call the fatigue properties measured in these tests the static fatigue lifetime and the dynamic fatigue lifetime. In both cases, the outcome of fatigue testing is the time to failure, and this is measured at different stress levels so that fatigue lifetimes can be related to the in vivo conditions found within the animal's normal behavior.

Ker et al. investigated static and dynamic fatigue in wallaby tail tendons, a high-safety-factor tendon that has a calculated in vivo stress, $\sigma_{\text{in vivo}}$, of 13.5 MPa and hence has a safety factor of about 8. In static-fatigue tests they increased initial strain rapidly until the set stress level was reached, and then the test system was set to maintain the initial load and to record the increase in strain over time. Initially, strain increased linearly with time, but with a low slope, but after times that varied with the stress level, the slope of the strain curve increased exponentially, ending in the rupture of the tendon. The failure time gave a value for fatigue lifetime at the stress applied in the test. When data from experiments at different stress levels are combined, the log of fatigue lifetime (in seconds) versus the experimental stress level produced a linear relationship that established the creep-fatigue behavior of this tendon, as shown in fig. 9.14A, where it is seen that with static loading, fatigue lifetime at 20 MPa is about 3×10^5 s, falling to about 10^3 s at 80 MPa. Extrapolation of the fatigue line to the calculated "stress-in-life," as shown by the dashed line in this figure, indicates that the fatigue lifetime of this tendon under static loading from its attached muscle would be about 5×10^5 s, which is about 5 days. Let's consider the implications of this result, because wallabies live much longer than 5 days. The tendon's lifetime in the live animal will be determined by the accumulation of fatigue damage over a large number of locomotor movements that would occur when the tail muscles are generating their maximal isometric tension. These events occur only during brief periods of ground contact during the animal's locomotor cycle, and wallabies do not hop all the time. So the fatigue of the tail tendon in a live animal would proceed over a much longer time scale than that seen in the experimental tests, where the stress was applied continuously. In addition, the tendons in living animals, where the blood supply remains intact, would be able to sustain cellular processes that repair the mechanical defects that develop during periods of intense activity. Indeed, tendon rupture in the absence of a catastrophic injury is quite rare, and it is typically associated with periods of intense activity in which the rate of fatigue damage exceeds the relatively limited capacity in tendons for repair.

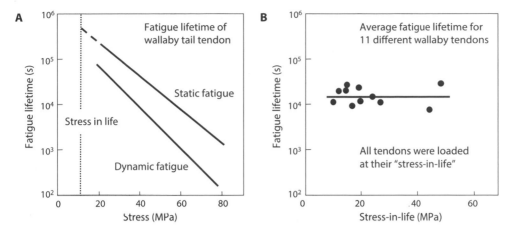

Figure 9.14. Fatigue-lifetime tests in wallaby tail tendon. **Panel A** shows the general pattern of static- and dynamic-fatigue behavior of wallaby tail tendon. The "stress-in-life" line indicates the estimated stress applied to the tendons when the tail muscles are activated and produce their maximal isometric tension. **Panel B** shows the results of fatigue tests on a variety of wallaby tendons, including high-safety-factor tendons such as the tail tendon, as well as low-safety-factor tendons such as the digital flexor tendons of the hind limb. The data plotted show that the fatigue lifetimes for all the tendons when they are loaded at the estimated maximal isometric tension are essentially identical.

It is also important to note that there was considerable variation in the fatigue behavior of tendons from the six different animals used in this study. The mass of the animals ranged from 12.6 to 25 kg, but ages were not provided. It is likely, however, that age variation explains the variation in fatigue lifetime between animals. The use of wallaby tendons is very convenient because their tails are large, and they contain many testable tendons. In fact, one can isolate up to 60 tendons from each animal, making it possible to characterize the fatigue behavior at different stress levels with tendons from a single animal, and the data from a single animal showed a very strong linear correlation between stress and lifetime.

There were two other interesting observations made in the analysis of static fatigue in the wallaby tail tendons. First, fatigue lifetime was significantly affected by the length of the test sample, with samples longer than about 80 mm having approximately one-quarter the lifetime of 20 mm samples. This effect was seen over the full range of test stress, and it suggests that there is a structural feature in the tendon with a length in the range of 20–80 mm that is involved in the fatigue process. This length scale is somewhat above the length of collagen fibrils, which have been suggested to have a length range of 1–10 mm, and somewhat below what we might expect for collagen fibers, which may run the full length of the tendon, and tendons range from 200 to 500 mm in the animal. The other observation is that static-fatigue behavior is very temperature dependent. Most experiments were run at room temperature, at about 20°C, but fatigue lifetime drops rapidly as temperature is elevated above this level, being reduced by more than an order of magnitude at 38°C. Both of these results will become useful when we discuss the nanostructural basis of tendon mechanics in the next section.

Fig. 9.14A also shows a line labeled "dynamic fatigue," which presents the general form of the fatigue lifetime of wallaby tail tendons under dynamic-loading conditions. In this case the tendons were tested in a machine capable of applying sinusoidal strains to samples with programmed stress amplitudes at frequencies that varied from 1 to 50 Hz. In most conventional dynamic-fatigue experiments a test specimen would be strained on load cycles that went from zero stress to a maximal stress, and the maximal strain of each load cycle would be tracked as a measure of dynamic fatigue. In the experiments with wallaby tail tendons the load cycling was slightly modified to maintain the sample in the linear region of its stress-strain curve. That is, the minimum stress applied was 11.5 MPa, a level that would ensure that the tendon crimps were fully straightened at all times in the experiment, and each sinusoidal load cycle was between this stress and the stated maximal stress for the experiment. The line in this figure for dynamic fatigue is an average for all the test samples, which were from two animals, and in this case there were no differences observed in the properties of the samples from the two animals. What we see is that the general characteristics of the dynamic-fatigue data follow the trend seen for static fatigue, but the dynamic fatigue lifetimes are significantly shorter, by about 20%, and the discrepancy grows somewhat as the peak stress increases.

These interesting results tell us a great deal about the fatigue behavior of wallaby tail tendons, but since this tendon is one of the high-safety-factor tendons, it would be useful to know if the low-safety-factor spring tendons show similar behaviors. Ker and his colleagues (2000) measured the static fatigue lifetime for 11 different tendons from the wallaby, and this provides us with some very interesting results. These 11 tendons included the tail tendon and several extensor tendons that had "stress-in-life" values in the range of 12–15 MPa, as well as digital flexor tendons and the plantaris and gastrocnemius tendons, which all function as spring tendons. There were two key changes to their experimental protocol. First, they tested the tendons at 37°C so that the measured fatigue lifetimes would apply directly to tendon behavior in living animals, and in addition they tested the tendons at the estimated "stress in life" of the tendons, as defined in eq. (9.2). The key result of this study is shown in fig. 9.14B, where we see that the static fatigue lifetimes for all 11 of these tendons, each tested at its "stress in life," are essentially the same, at a value of about 15,000 s, or 4.2 hours. This very striking result clearly demonstrates that tendons with widely different functions in an animal are apparently adapted structurally to produce fatigue behavior that is matched to the stress levels over which individual tendons function in vivo. This leads us to the key question, what are the structural origins of these differences in fatigue resistance?

8. THE NANOMECHANICS OF TENDONS AND LIGAMENTS. In the past 10–15 years numerous attempts have been made to translate the macroscopic properties of tendons, described in the last two sections, to the nanoscale structures described earlier. The approaches have ranged from the use of synchrotron X-ray diffraction to resolve the strains at the whole-tendon level into the strain of collagen fibrils and strain of the collagen molecule's triple-helical-rod domain, to the direct mechanical testing of isolated collagen fibrils and isolated collagen molecules, to molecular modeling of collagen fibrils and molecules. Although there are some uncertainties that remain, we can now provide a reasonable description of the molecular origins of tendon mechanics.

Synchrotron X-ray diffraction studies provide information on the strain of collagen molecules and collagen fibrils as a function of the macroscopic straining of intact tendon samples. The important feature of synchrotron X-ray systems is that their extremely high intensity beams allow the collection of X-ray diffraction patterns over a small number of seconds, which means that they can be used to track the changes in tendon structure with strain in real time in mechanical tests of hydrated tendon samples. These studies indicate that the strain of tendon fibrils is considerably less than the macroscopic tendon strain, and the strain of individual collagen molecules within a fibril is smaller still. Information on collagen-fibril strain comes from the analysis of low-angle diffraction patterns that result from changes the 67 nm D-period repeat that is created by the overlapping array of tropocollagen molecules in the collagen fibril. Peter Fratzl and his colleagues (1997) estimated that the collagen-fibril D-period strain, ε_D, was about 40% of the whole-tendon strain, ε_T, over the linear range of the stress-strain curve. This result indicates that collagen-fibril stiffness is roughly 2.5 times that of the whole-tendon stiffness, and thus that collagen fibrils should have a stiffness of about 3–4.5 GPa. Interestingly, they used tail tendons from 6-week-old rats in their experiments, but they report that the tendons failed at a stress of 120 MPa and at strains that would suggest that the linear modulus of the tendons was in the normal, 1.5 GPa range. These properties are dramatically different from those shown in fig. 9.10 for developing rat tail tendons at similar ages, where we see failure at a stress of about 20 MPa and at strains of the order of 18%, and this discrepancy may arise from the fact that they were testing small fiber bundles or a single fascicle with a length between grips of about 12 mm. Thus, the sample length was likely only about 2–4 times the length of fibrils in the tendon (see table 9.1), and this may have allowed their samples to behave like more mature tendons.

Collagen fibril strain arises from two kinds of mechanism. One involves the stretching of tropocollagen molecules within the fibril crystal lattice, and the second involves the sliding of tropocollagen molecules relative to one another and the stretching of the cross-linked telopeptides within the gap zone. X-ray diffraction can also be used to track changes in the diffraction pattern that arise from the helical pitch of the collagen polypeptide chains, or the rise per amino-acid residue of the individual polyproline-II helices within the triple-helical rod of the collagen molecule. This information should provide insight into the stiffness of individual collagen molecules, as they exist in the semicrystalline lattice of the collagen fibril. Nemetschek and his colleagues (1978) used wide-angle X-ray diffraction to follow collagen-molecule strain in 2-year-old rat tail tendons, and they observed that the rise per residue in the unstretched triple helix was 0.286 nm. When 5% strain was applied to the sample this spacing increased by 0.5% to 0.2875 nm. This suggests that the collagen-molecule strain, ε_C, is about one-tenth that of whole tendon, but the X-ray exposure required to capture collagen-molecule strain lasted 300 s, and the samples were held at constant strain during exposures. Thus, stress relaxation would cause fibril strain to decrease with time. Fortunately, they also provided information about the relaxation of the D-period strain as well as the stress in their test samples, both of which were consistent, showing a relaxation of the fibril strain, ε_D, from 2.7% to 1.6% during the exposure over 300 s. Taking this relaxation into account, it is likely that the wide-angle diffraction pattern was obtained at an average D-period strain of about 1.6%. This suggests that the collagen-molecule strain, ε_C, is about one-third the D-period strain, or that the collagen-molecule stiffness is about three times greater than the collagen-fibril stiffness. Thus, the collagen-molecule stiffness, E_C, should be about 10 GPa.

Other methods for measuring the properties of collagen fibrils and molecules have been used to verify the predictions based on the X-ray diffraction studies described above, with varying success. Steven Eppel and his colleagues (2006) tested collagen fibrils that were isolated from the connective tissues of sea cucumbers in MEMS (microelectromechanical systems) devices to measure their material properties directly, and they observed highly variable stiffness values, with modulus ranging from 0.4 to 1.6 GPa. However, they also showed that some of their "single-fibril" test samples actually contained multiple fibrils, so it is not surprising that there is variability. It is possible that the upper end of this range does represent the true stiffness of individual collagen fibrils, but, as we will see shortly, the collagenous tissues of echinoderms have some very unusual properties. A 1.6 GPa fibril stiffness seems much too low for vertebrate tendons, where the whole-tendon stiffness is about 1.5 GPa, but it may be reasonable for fibril stiffness in echinoderm tissues.

Rene Svensson and his colleagues (2010, 2012) used atomic-force microscopy (AFM) to test single collagen fibrils from human patellar tendon, and they report that the stiffness of these fibrils is approximately 3 GPa, which is in reasonable agreement with the estimates from the X-ray diffraction studies described above. Chapter 8 of Fratzl's 2008 book *Collagen, Structure and Mechanics* has a good description of molecular-modeling approaches to determining the mechanical properties of collagen molecules, as well as comparisons with other data for collagen-fibril mechanics. For our purposes, however, we will make the assumption that a reasonable value for the stiffness of whole tendon is 1.5 GPa, and that the stiffness of collagen fibrils is about 3 GPa. With these values in mind we can address a final question about tendon collagen, how we can incorporate these stiffness levels with the structural organization of the discontinuous-fiber-composite structure of collagen fibers in tendons.

To answer this question we must return to the theory of discontinuous fiber composites, which was introduced in chapter 8. The two key equations for this theory, eqs. (8.11) and (8.12), allow the calculation of stiffness in a discontinuous-fiber composite, and are repeated below:

$$\eta_{ref,\theta} = 1 - \frac{\tanh(ns)}{ns},$$ (9.3)

$$E_{c,0} = \eta_{ref,0}E_f V_f + E_m(1 - V_f).$$ (9.4)

Eq. (9.3) gives the reinforcement coefficient, η_{ref}, which incorporates input on the fiber volume fraction, the ratio of fiber stiffness to matrix stiffness in the variable n, and the fibril aspect ratio in the variable s. Eq. (9.4) shows how this coefficient is used in the calculation of the composite modulus, using the rule of mixtures for the fiber and matrix components. Fig. 9.15 shows the results of calculations with these equations for a tendon composite made from collagen fibrils with an assumed modulus of 3 GPa and an assumed fiber volume fraction of 0.55, which with eq. (9.4) predicts a collagen-fiber stiffness of 1.65 GPa at high fibril aspect ratios.

The fibril modulus used in this calculation is based on the discussion above, and it is likely a good estimate for most tendons. The fiber volume fraction is basically a guess,

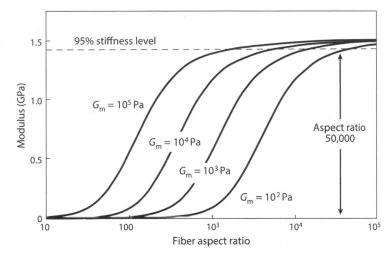

Figure 9.15. Calculated properties of the discontinuous-composite structure of a tendon collagen fiber. The individual curves show the predicted shear modulus, G_m, of a collagen-fibril/proteoglycan-matrix composite as a function of the collagen-fibril aspect ratio. The four curves show data for different levels of proteoglycan-matrix stiffness. The assumed aspect ratio for collagen fibrils in mature tendon is indicated as 50,000, and at this aspect ratio the predicted collagen-fiber stiffness is essentially unaffected by dramatic differences in matrix stiffness.

but it is based on two arguments. First, it gives us a value for the stiffness of an individual collagen fiber of 1.65 GPa, which seems reasonable, and second, it likely accounts quite well for the actual fraction of the fiber's cross-section that is occupied by collagen fibrils. Electron micrographs of collagen fibers, as shown diagrammatically in fig. 9.6, give us a view of the packing density of collagen fibrils within the matrix "space," and estimates of the area fraction of the collagen fibrils in these images fall at around 0.7–0.8. We could assume that a fibril volume fraction of 0.55 allows us to account for the shrinkage of the highly hydrated matrix space during sample preparation for electron microscopy, but we should also consider the contribution of other compartments within the whole tendon cross-section. Tendons contain multiple collagen fibers encased in sheaths to form fascicles, which are then bundled together to form the whole tendon. The fluid-filled spaces between collagen fibers in the fascicle are likely not load bearing, and the fascicle sheath and spaces between fascicles contribute to the cross-sectional area of the whole tendon, but likely do not contribute significantly to tendon stiffness. If we assume that collagen fibers occupy 91% of tendon cross-sectional area, the tendon modulus becomes 1.5 GPa, which is the value we have decided to adopt for the modulus of tendons in general. It is important to keep in mind that this analysis is not meant to be exact, and calculations based on these equations likely only provide reasonable estimates of composite properties. Nonetheless, with the information we currently have, the model appears to be giving us reasonable answers. So let's look at what additional information we can get from the model.

The four lines in fig. 9.15 present the results of calculations using four different values for the shear modulus, G_m, ranging from a high value of 10^5 Pa to a low value of

100 Pa. The upper end of this range of shear modulus was chosen because a number of investigators have suggested that 10^5 Pa is a reasonable estimate for the stiffness of the tendon matrix, based on estimates of the stiffness of the proteoglycan matrix in cartilage. This is a plausible estimate, and it is a good starting point. The vertical dashed arrow to the right indicates the location of fiber aspect ratio AR = 50,000, which was at the upper end of the range of aspect ratios in the two studies that have attempted to measure it. Given what we have seen for the fatigue properties of the low-safety-factor limb tendons, it seems likely that some high-performance tendons may have fiber aspect ratios at this level or above.

Look first at the line for $G_m = 10^5$ Pa. Note that at high fiber aspect ratio the composite modulus for the whole tendon is at its maximum value, 1.5 GPa, and that it would require that fiber AR drop to about 1000 for the modulus to fall to 95% of the maximal stiffness. At lower values of G_m the curves are shifted to the right, and higher aspect ratios are required to bring composite modulus to the 95% level, but even when the shear modulus has been reduced by three orders of magnitude, to 100 Pa, the tendon-composite stiffness remains only slightly below the 95% stiffness level at AR = 50,000. This insensitivity of the tendon-composite stiffness to changes in the matrix shear modulus likely explains why the several studies described earlier in this chapter failed to show any dramatic changes in tendon mechanical properties when proteoglycans were digested with enzymes or eliminated in knockout mouse strains. The modeling in fig. 9.15 suggests that digesting 99% of load-carrying proteoglycans, and hence lowering G_m to 10^3 Pa, would not significantly lower composite stiffness if fiber aspect ratios are at the level 10,000 or above. Alternately, for the knockout mouse strains, even if all decorin proteins were missing, a small amount of other proteoglycans (e.g., biglycan) could maintain the structural integrity of the tendons. That is, the role of proteoglycans in the tendon composite remains intact as long as there is a reasonable fraction of mechanically competent proteoglycans to stabilize tendon structure.

This brings us to a final thought about the composite design of the tendons. To this point we have considered composite models where the fiber and the matrix are characterized by stiffness values alone. The behavior of tendons in static-fatigue tests described above reveals a characteristic feature of tendons under constant high stress; namely, that they creep continuously, typically at low rates at short times, but that creep rate increases as tendons approach their failure point. Time-dependent creep behavior requires that some component in the tendon hierarchy must exhibit viscoelasticity, and the matrix that exists between collagen fibrils is the ideal candidate for this component. We already know from the discussion of viscoelasticity in chapter 3 that cartilage, a composite of collagen fibers and a proteoglycan matrix, creeps over more than four decades of log time, and that stiffness falls by more than an order of magnitude over this period (see fig. 3.6A). If this behavior is characteristic of the proteoglycan matrix in the collagen-fiber composite, then we already have a model for the static-fatigue behavior of tendons. That is, matrix creep allows time-dependent strain of tendons, which over time will create defects that lead to collagen-fiber failure, which ultimately leads to tendon rupture. In addition, the time-temperature superposition principle of polymer viscoelasticity (chapter 3) provides us with a simple explanation for the effect of temperature that Robert Ker saw on the fatigue lifetime of tendons. Higher temperature lowers the viscosity of the interfibrillar matrix, which allows more rapid creep of tendon samples. With these ideas in hand, we can begin to think about possible explanations for the range of fatigue behaviors seen for the high-safety-factor and low-safety-factor tendons illustrated in fig. 9.14B. That is, what kind of biomechanical adaptations allow some tendons to achieve better high-stress-fatigue resistance than others? If tendon failure is initiated at the level of the fibril-matrix composite within the collagen fiber, then we know enough to think about mechanisms for fatigue control. That is, increased tendon fatigue lifetime could be controlled by changes in the creep rate of the matrix proteoglycan, by changes in the fibril aspect ratio, or by both. Increased matrix concentration will increase the dynamic stiffness of the matrix, even if the matrix is an un-cross-linked, entangled network. Thus, this will reduce matrix creep rate at the shear-stress levels that occur at the fiber-matrix interface. Alternatively, or in combination, increased fibril aspect ratio will reduce the magnitude of the shear stress at the fiber-matrix interface, and through this reduce the creep rate. It remains for someone to test these hypotheses.

To this point we have assumed that tendon properties, including failure, are determined at the level of the collagen-fibril-decorin composite within collagen fibers. However, studies of tendon mechanics that focus on interactions between fibers within tendon fascicles by Hazel Screen and her colleagues (2004, 2009, 2015) suggest that processes at this level may be more important than we have assumed to this point. These papers track the time-dependent stress-strain and stress-relaxation responses of isolated tendon fascicles to various levels of initial strain. The isolated fascicles from 4–6-month-old rat tail tendons were mounted in a test chamber with a sample length of about 25 mm, which was designed to fit onto the stage of a confocal microscope that allowed the inspection of the internal structure of an intact fascicle. The fascicles were stained with the vital dye acridine orange, which binds selectively to DNA and RNA in cell nuclei, and this allowed these authors to track the strain within an individual collagen fiber by tracking the position of the cells that were synthesizing collagen for incorporation into the fibers. In addition, they could track the strain between adjacent collagen fibers by following the movement of cells on adjacent fibers. Their observations are quite revealing because they indicate that there can be significant sliding between adjacent collagen fibers within an intact fascicle. Specifically, they observed that at fascicle strain below about 0.03, the fiber strain, both within and between fibers, was about 25% of the fascicle strain, and since the specimens were from relatively young animals, this 0.03 specimen-strain level coincided with the initial portion of the non-linear toe region of the stress-strain curve. Above 0.03 specimen strain the within-fiber strain increased slowly, tracking the actual strain of the collagen-fibril-matrix composite that we have been focusing on to this point. However, the between-fiber strain rose about an order of magnitude more rapidly at strains above 0.03, indicating that fiber sliding is a major component of whole-fascicle deformation. It is interesting to note that the sample lengths, at 25 mm, are only slightly longer than what we have assumed as the upper end of the range of collagen-fibril length (ca. 10 mm) in collagen fibers, and this length is below the 80 mm length at which Ker et al. (2000) found a significant decline in wallaby tail tendon fatigue lifetime. Thus, the test specimens may well have been dominated by the within-fiber composite structure, and fiber sliding in longer structures, such as the full-length tendon, may be even more important in the failure process of tendons in vivo.

Screen et al. also carried out macroscopic stress-strain curves to failure for a number of additional 25 mm fascicle specimens in a conventional materials-testing system, and they observed a linear modulus of about 700 MPa and a failure stress of about 45 MPa at a strain of about 0.14. These properties, overall, seem quite similar to those described in fig. 9.10 for rat tail tendons at age 6 months to 1 year, but with one important difference. The linear modulus is almost exactly one-half that reported by many other authors as 1.5 GPa. This suggests that the methods used to isolate and prepare the individual fascicles have allowed the fascicles to swell to a level that roughly doubled their cross-sectional area. Indeed, a recent paper from this lab compared whole-tendon properties with those of individual isolated fascicles from a horse leg tendon, and they saw precisely that: isolated fascicles had twice the cross-sectional area of fascicles observed in the intact tendon, and hence had half the stiffness and strength of the intact tendon. So perhaps we should be cautious in applying these observations rigorously to the overall picture of tendon structure-property relationships. However, they do point to a need for more research into the discontinuous-composite structure of collagen fibers in fascicles, and for that matter of fascicles within whole tendons. It appears that our assumption that collagen fibers and collagen fascicles run the full length of the tendon may not be correct, and it is indeed likely that the key mechanisms leading to tendon fatigue and traumatic rupture lie in these larger-scale levels of structural organization in whole tendon.

9. ECHINODERM LIGAMENTS AND MUTABLE CONNECTIVE TISSUES. Up to this point we have dealt with the structural organization and material properties of tendons and ligaments exclusively through studies of tendons. It is important to note that in the vast majority of animals where ligaments have been investigated, their basic structural design is essentially identical to that of tendons. Clearly, there will be differences, but they are relatively minor. This is because tendons and ligaments have very similar functional requirements. That is, a tendon is a structure that connects an active contractile component (a muscle) to a passive and rigid structural element (e.g., a bone), and this requires a stiff, strong, and flexible fiber-based material that can transfer muscle displacements to articulated structural elements in the skeleton, thus creating postural movements and locomotion. In addition, tendon fibers are crimped so that their stress-strain curves show a low stiffness "toe region" that allows a small amount of initial low-stiffness stretch that can be adjusted to suit the specific functional requirements of a particular tendon. Ligaments function in animal skeletons to maintain interconnections between bones in a way that allows the rotation of the bones at joints, but at the same time prevents the disarticulation of those joints. In this case, the properties required of ligaments are very similar to those of tendons, where flexibility, stiffness, and strength are crucial.

In echinoderms, however, collagenous tissues show some quite remarkable differences, at the level of the individual collagen fibers, that impart radically different mechanical properties, and we will concentrate on one particular kind of tissue, the sea-urchin spine ligament. This ligament will provide an introduction to the molecular mechanisms that allow animals in this phylum to have collagenous tissues whose mechanical properties are under the control of their nervous system. That is, echinoderm structural tissues have the ability to change the stiffness, strength, and extensibility of collagenous tissues over quite broad ranges on relatively short time scales. This unusual capability arises from interesting molecular mechanisms that apparently function at the level of the collagen-fibril/proteoglycan-matrix composite present in collagen fibers.

Sea-urchin spines function as struts for support and locomotion on the oral surface of these animals, and they function largely as defensive devices on the aboral surface, although they can also be used to wedge these animals into cracks in rock surfaces in the wave-swept intertidal environment. Each spine is able to rotate on a ball-and-socket joint at the spine's base, and this rotation is controlled by a ring of muscle and a ring-shaped collagen ligament, which are arranged in parallel with each other, as shown in fig. 9.16A. This diagram shows the base of a single spine and its ball-and-socket joint with the rigid "test," or body wall, of the animal. Both the test and the spines are rigid structures made from blocks of porous, crystalline calcium carbonate, and the spines are moved under the active control of the nervous system. That is, they can be actively rotated by the contraction of appropriate sectors of the muscle ring. However, it is important to note that the spine muscles and the spine ligament are arranged in parallel, so that contraction of a spine muscle sector on one side of the spine will cause the spine to rotate toward the side of muscle shortening, and this rotation will extend the muscle and the ligament on the opposite side of the ball-and-socket joint.

The spine in the figure has been pulled to the right, to an angle of about 20°, by the contraction of the muscle block on the right side of the spine. In this process both the muscle and the ligament on the left side are extended, and accordingly they are shown as long, thin strips of tissue. The muscle block on the right is shown as a shorter, thicker strip of tissue because of its contraction. In addition, the segment of ligament adjacent to the contracting muscle will shorten. Note that the ligament strips are shown with lines running parallel to their long axis to indicate the orientation of collagen fibers

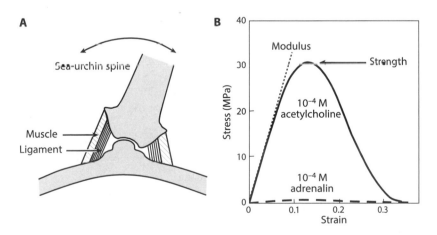

Figure 9.16. Sea-urchin spine ligaments. **Panel A** shows a diagram of the sea-urchin spine and its articulation with the body wall of the animal. The muscle and ligament that control spine rotation and stabilization are indicated. **Panel B** shows the form of stress-strain data from spine-ligament samples tested in constant-strain-rate tensile tests. The *solid line* indicates the behavior of the ligament in its stiff state, and the *dashed line* indicates the behavior of the ligament in its compliant state.

that run from the spine at the top to the spine socket on the body wall at the bottom. If the echinoderm spine ligament had properties similar to those of the tendons and ligaments in vertebrates, as described above, these ligaments would impose quite severe restrictions on the movement of the spines. Interestingly, the collagen fibers in the spine ligaments are not crimped; rather, they are made of bundles of straight collagen fibers, and they do not appear to limit the reorientation of the spines, as they should if they were as stiff as normal tendons and ligaments. So the properties of the collagen fibers in this ligament are likely quite different from those in "normal" collagenous tissues. In addition, they do not become crimped in places where they run adjacent to contracted spine muscles. Thus, it seems that spine ligaments can have structural properties that are quite exceptional.

Anyone who has snorkeled on a reef in the tropics has likely found areas where the black sea urchin, *Diadema*, is seen in large numbers, and probably learned to avoid them because they have long defensive spines that can easily penetrate the skin and cause painful wounds. However, if you observe these sea urchins closely without disturbing them, you will note that the animal typically moves its spines slowly at small amplitudes, but if you gently poke at a spine, the animal will reorient multiple spines toward your finger and then lock the spines in place. This behavior is the basis of the name given to the spine ligament, the "catch ligament," because it has low stiffness in its "relaxed" state, but this stiffness can be rapidly increased in the "catch" state, which will cause the spine to be locked at a specific orientation. This behavior requires active control of muscle as well as control of the material properties of the ligament. It turns out that the muscles in the spine-ligament complex are quite conventional, but the ligaments are anything but conventional.

Studies by Hidaka and Takahashi (1983) on the mechanical state of isolated spine-ligament samples noted that changes in the ionic composition of test solutions, including changes in pH, calcium ions, and potassium ions, could dramatically affect the stiffness and strength of the spine ligaments. That is, mechanical tests revealed that stiffness and strength could vary by up to two orders of magnitude with changes in inorganic ions. This suggests that purely biophysical processes involving the interaction of inorganic ions with the collagen-fibril–proteoglycan-matrix composite within collagen fibers might control material properties. It turns out, however, that changes in material properties are created by biological processes, involving the coordinated action of stiffening and relaxing proteins that are released by neurosecretory processes, as well as by the action of muscle cells and elastic fibers that exist within the spine ligament, in spaces between the collagen fibers. Fig. 9.16B shows the results of tensile tests by Hidaka and Takahashi in which spine-ligament samples were strained at various strain rates to determine the effective stiffness and strength of the ligament tissue. The initial linear rise in stress allows the calculation of the strain-rate-dependent modulus of the ligament, and the peak stress provides an indication of the strength of the material. The stress-strain curves shown in the figure were obtained at a low strain rate of about 0.003 s^{-1}, and at this strain rate the tests would have taken about 100 s to run. Tests were also run at strain rates of 0.03 and 0.3 s^{-1}, which would have taken about 10 and 1 s respectively. The samples exhibited linear behavior for the initial portion of the test, but yielded at strains of about 0.1, and failed at strains of the order of 0.3. The figure shows curves for two conditions, which represent the extremes of material behavior seen for

Table 9.3. Stiffness and Strength Values for the Spine Ligament from a Sea Urchin

Strain Rate	Modulus in 10^{-4} M Adrenalin (MPa)	Strength in 10^{-4} M Adrenalin (MPa)
0.003 s^{-1}	2	0.1
0.03 s^{-1}	90	7
0.3 s^{-1}	220	18

Strain Rate	Modulus in 10^{-4} M Ach (MPa)	Strength in 10^{-4} M Ach (MPa)
0.003 s^{-1}	240	20
0.03 s^{-1}	260	24
0.3 s^{-1}	320	28

Adrenalin induces a relaxed state in the tissue, and acetylcholine (Ach) induces a "catch" state. In the catch state the tissue behaves like an elastic material, with high stiffness and strength at both high and low strain rates. With adrenalin, however, the tissue shows a dramatic strain-rate dependence, which arises from the tissue's viscoelastic behavior.

the catch-ligament material in general. The solid line indicates the behavior of a ligament that has been immersed in seawater with 10^{-4} M acetylcholine, which is thought to induce the release of compounds that create the "catch" condition of the ligament. The dashed line indicates its behavior in seawater with 10^{-4} M adrenaline, which is thought to cause the release of compounds that create the softer, compliant condition of the ligament. It is important to note that these two organic compounds function in animals as neurotransmitters and/or as hormones, and at these concentrations they should have no direct effect on the biophysical interactions between collagen fibrils and their proteoglycan matrix. Rather, they apparently act on the nerve cells within the isolated ligament, causing them to release compounds that stiffen or soften the ligament by modulating interactions between the collagen fibrils and their matrix. Calculated values for the stiffness and strength are given in table 9.3 for the three strain rates used in these experiments, and the test results clearly document the viscoelastic behavior of this material. At low strain rates, the modulus and strength of adrenaline-treated tissues are 2 and 0.1 MPa respectively, values that are roughly two orders of magnitude lower than those of acetylcholine-treated tissues. However, the adrenaline-treated tissues are exceptionally strain-rate dependent, and at the highest strain rate the stiffness and strength rise by roughly 100-fold, to levels that are about half those of the acetylcholine-treated tissue. The acetylcholine-treated tissues are stiff and strong at all strain rates, with modulus at about 300 MPa and strength at about 25 MPa. It is also important to note that these properties are significantly different from the properties of the conventional ligaments and tendons seen in vertebrates (1.5 GPa and 100 MPa respectively), over broad ranges of strain rate. Thus, normal tendons and ligaments are roughly 1000 times stiffer and stronger than "relaxed" spine ligaments tested at modest strain rates, and about 5 times stiffer and stronger than spine ligaments in the "catch" condition.

John Trotter and Tom Koob (1989) were able to isolate collagen fibrils from catch ligaments, and they reported that the average fibril aspect ratio (length/diameter) was about 2500, or about an order of magnitude lower than our estimate for mature mammalian tendons. In addition they observed that the individual fibrils are tapered, which is very likely the case for collagen fibrils in mammalian tendons and ligaments. Indeed, this taper likely explains the observation that electron micrograph cross-sections of tendons typically show a broad range of fibril diameters, which could easily be explained by the presence of tapered fibrils. Another important feature of echinoderm ligament fibers is that it is possible to isolate individual fibrils with relatively gentle chemical treatments. Collagen fibrils of similar size have been isolated from vertebrate tissues, but only in fetal tissues. The unusual mechanical properties of catch ligaments, plus the ease of isolating collagen fibrils, strongly suggest that the structural organization of the collagen composite in echinoderm tissues may be quite different from that in other collagens.

The first clues to the biochemical nature of the control process came from experiments in which various body-fluid and tissue extracts were added to experimental test samples taken from echinoderm structural tissues, and over the past two decades experiments have demonstrated that it is possible to isolate specific proteins that provide insights into the origins of these behaviors. However, much remains to be learned. What we know is largely based on studies of mechanisms that control the stiffness of body-wall connective tissues in sea cucumbers, another group within the echinoderms. In 1999, Koob et al. isolated the first stiffening and relaxing proteins that are released from the cells in the sea-cucumber dermis when the tissue layers are subjected to freeze-thaw cycles. This process breaks open cells in the dermal tissues to release proteins from membrane-bound vesicles in nerve endings that have been seen in electron micrographs of sea-cucumber dermis and of sea-urchin catch ligaments to be in close proximity to collagen fibers. Extracts from the sea-cucumber inner dermis contained a protein that increased tissue stiffness in the absence of calcium (Ca^{2+}) ions, and extracts from the outer dermis contained a relaxing factor that decreased stiffness in the presence of Ca^{2+} ions. These factors are inactivated by protease digestion, suggesting that the mechanical state of the catch ligament is controlled by the release of protein from nerve-cell endings in the tissues, but it may also be influenced by Ca^{2+} ion concentrations in the tissues.

Several groups have discovered additional protein factors in sea-cucumber dermis that are involved in the control of mechanical properties of echinoderm connective tissues. These proteins include stiparin, a glycoprotein that causes isolated echinoderm collagen fibrils to aggregate, and a stiparin inhibitor that specifically inhibits stiparin's action of aggregating collagen fibrils. In addition, a tissue stiffener called tensilin has been cloned and sequenced, but its exact function and mechanism remain to be determined. Tensilin shows homology to tissue inhibitors of metalloproteinase, and it may, therefore, function in the control of a metalloproteinase that functions in the regulation of material properties. Yamada and his colleagues (2010) have isolated an additional stiffening protein that appears to be involved in shifting these tissues into their stiffest state, and that paper provides references to the full range of these studies. The current consensus is that these tissues can exist in three mechanical states. The release of the relaxing factor puts them into a low-stiffness "compliant state," and the release of tensilin raises stiffness to an intermediate level, which is called the "standard state." The newly isolated stiffener identified by Yamada raises stiffness to the "stiff state."

A more recent study by Ana Ribeiro and her colleagues (2011) documented the ultrastructural changes seen in a sea-urchin ligament between the compliant, standard, and stiff states. The most striking structural difference was the dramatic reduction in interfibrillar distance in the three states, with mean interfibrillar distance in the compliant and standard states being about 40 nm, falling to about 20 nm in the stiff state. This means that the transition to the stiff state occurs with a large reduction in the matrix volume fraction, a change that would certainly tend to increase stiffness in a fiber-reinforced-composite structure. But the roughly 100-fold changes in stiffness and strength seen in these systems must certainly arise from processes such as the introduction of cross-links within the proteoglycan matrix in the collagen-fiber composite, or between the matrix and the collagen fibrils.

Some insights can be gained by considering how discontinuous-fiber-composite models can deal with the properties of catch ligaments, and fig. 9.17 shows the results of calculations for this system. The axes of this figure are identical to those of fig. 9.15, which modeled the properties of the tendon composite, and the dotted line in the figure shows the model behavior for a vertebrate tendon, with parameters as described above and a matrix stiffness $G_m = 10^4$ Pa. The solid line labeled "urchin ligament" provides a model for the acetylcholine-treated ligament, which is assumed to be equivalent to the ligament in the stiff state. Collagen fibrils were given a stiffness of 3 GPa, just like vertebrate tendon fibrils, and an aspect ratio of 2500, which is approximately the value seen in echinoderm tissues and is 20-fold lower than the value assumed for mature vertebrate tendons. The volume fraction of collagen fibrils in the collagen fibers was set at $V_f = 0.55$, and to account for the presence of nerves, muscles, and elastic fibers in the spaces between collagen fibers we assume that the fibers occupy about 60% of

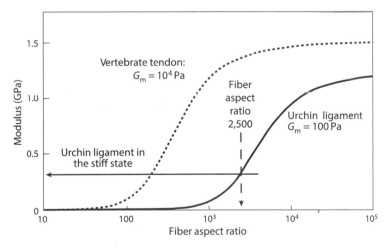

Figure 9.17. Calculated properties of sea-urchin spine ligament. The results of calculations for the properties of the discontinuous-composite structure of the sea-urchin spine ligament, along with a calculation for the fibers in a typical vertebrate tendon. The urchin-ligament calculation is based on a fiber aspect ratio of 2500, and the shear modulus, G_m, of the urchin-ligament matrix was set at 100 Pa to bring the stiffness of the composite to a value of 0.3 GPa, which is the observed stiffness of the urchin spine ligament it its stiff state.

whole-ligament cross-sectional area. With these structural parameters and a matrix shear modulus of $G_m = 100$ Pa, the model predicts the ligament's stiffness as 300 MPa, which is the level shown for the acetylcholine-treated tissue in table 9.3. Thus, a composite model based on a soft elastic matrix seems appropriate for the catch ligament in the stiff state, and to put the stiffness of the assumed matrix into perspective, an elastic material with a stiffness of 100 Pa is one ten-thousandth the stiffness of a typical rubber band. One important conclusion from this analysis of the stiff state is that its behavior seems consistent with a fixed matrix modulus over broad time scales, and this suggests that the matrix contains a dilute polymer network that is cross-linked.

The strongly viscoelastic behavior of the adrenaline-treated tissue will require significant modification to explain the effect of strain rate on the ligament's behavior in the compliant state. Specifically, the matrix must exhibit a strong viscoelastic response in which the effective matrix stiffness at high strain rate approaches the value used for the tissue in the stiff state, but at low strain rate the matrix stiffness must fall to levels of 1 Pa or lower. That is, the matrix in the compliant state must behave like the un-cross-linked and entangled network of polymer chains shown in fig. 7.2C to illustrate the viscoelastic behavior of Plexiglas at temperatures that are well above its glass transition. Here we see that the stiffness of Plexiglas is very low because the polymer chains are highly mobile. In this un-cross-linked system we see a transition from the elastic plateau zone, where molecular entanglements create temporary cross-links, into the viscous flow zone, where stiffness falls rapidly as temperature increases or as dynamic-loading frequency falls. Indeed, the dynamic stiffness of Plexiglas over this temperature range shows about a 100-fold drop in stiffness over a 100-fold decrease in strain rate (i.e., fall in test frequency). This behavior of Plexiglas is very similar to the strain-rate-dependent fall in matrix stiffness required to explain the behavior of the urchin ligament in the compliant state. There is, however, one key difference between Plexiglas at high temperature and the urchin ligament matrix in seawater. The Plexiglas is pure polymer, while the matrix proteoglycan is dissolved in a solvent similar to seawater. Thus, the ligament matrix can be 1000-fold less stiff than Plexiglas at 150°C, because it is dissolved in a low-molecular-weight solvent. In summary, our analysis suggests that in the stiff state the sea-urchin ligament functions as a discontinuous-fiber composite of collagen fibrils in a cross-linked soft elastic gel matrix with a shear stiffness of about 100 Pa, which is made from hydrated cross-linked proteoglycans. The transition to the compliant state occurs with the loss of cross-links, and this transforms the elastic matrix into a largely un-cross-linked, but entangled, system where stiffness varies dramatically with strain rate. It will be interesting to see how future studies deal with these ideas.

Much remains to be done to document the mechanisms of mutable connective tissues, but we are getting closer to an understanding of the basic principles of this system. That said, there are other issues that arise from the changes that occur when a sea urchin transforms its catch ligament from its compliant state, where it can reorient the spine, into the stiff state, and then back to the compliant state. Look again at fig. 9.16. The stretched ligament on the left side must have been in the compliant state, and was then stretched when the spine was deflected to the right by the contraction of the muscles on the right side of the spine. Assuming that the animal locked this spine into "catch" when it was moved to the right, we can think about the processes that are required to return the spine to the vertical position in its compliant state. First, the animal will need to return the ligaments to the compliant state, and then contraction of muscles on the left side can reorient the spine to the vertical position. This process will extend the "contracted" ligament on the right side and compress the previously stretched ligament on the left side. Recall that we do not see crimped collagen fibers in catch ligaments, suggesting that the shortening of the ligaments involves some sort of sliding mechanism that allows straight collagen fibrils to slide together to shorten the collagen fibers that make up the ligament without crimping. Similarly, the short, thick band of ligament on the right side must have shortened by a similar mechanism of sliding collagen fibrils in straight collagen fibers during the initial deflection of the spine. This sliding of collagen fibers is driven in part by a passive elastic mechanism and in part by active muscular contraction.

In 1981, Smith and his colleagues documented the ultrastructure of the spine-ligament system, and they observed the presence of two additional components of these catch ligaments. First, there are fine filaments between collagen fibers in the ligament that run parallel to the long axis of the fibers. These have been identified as "microfibrils," a term that is applied to a relatively poorly understood class of structural material that exhibits long-range, low-stiffness elasticity that is similar to that of a rubber band. The microfibrils are made from a protein called fibrillin, which has a filamentous structure that appears in electron micrographs like beads on a string. Microfibrils have been observed throughout the metazoans, starting with their presence in the soft, elastic bell of the hydromedusa *Polyorchis*, and extending to the arterial tissues of arthropods and cephalopods. They are also found in the elastic tissues of vertebrates. One particular system, the zonular filaments of the vertebrate eye that support the lens, appears to be constructed entirely from fibrillin microfibrils. Mechanical tests of these filaments by David Wright and his colleagues (1999) indicate that they are indeed low-stiffness elastic structures that can be strained reversibly to quite large extensions. Think of these as nanoscale rubber bands that contribute a passive elastic-recoil mechanism to shorten catch ligaments that are in their compliant state.

In addition, Smith and colleagues observed that there were very thin muscle cells running between the collagen fibers. Like the microfibrils, these cells are apparently stretched passively when a catch ligament is extended in its compliant state, and when they are activated by motor neurons, these cells will shorten actively and hence contribute to the shortening of the compliant ligament that is adjacent to the muscles that are contracting to deflect the spine. So, we see that the catch ligament is a much more complex structural organ than a conventional tendon or ligament in other animals, and it will be interesting in the future to discover the full details of the interaction of the active and passive components of this remarkable structural system.

This brings us to the end of our discussion of collagen-based tensile materials, where we have seen how the self-assembling hierarchical structure that we call collagen creates an exceptional structural-fiber system that plays a dominant role in the tissue mechanics of the full spectrum of animal phyla. Our next case study will consider the mechanical design of a system of tensile fibers that are assembled in processes that resemble those used by the polymer industry in the spinning of high-performance fibers such as nylon and Kevlar. That is, we will consider the design and assembly of the silks that spiders use for the frame of their orb webs and for their safety lines. As fig. 9.1G suggests, these silks have properties that in some ways are superior to those of the best tensile fibers made by the polymer industry.

Chapter 10

The Structural Design
of Spider Silks

1. THE FUNCTIONAL DIVERSITY OF SPIDER SILKS. Walking through a garden, field, or forest edge in the late summer or fall you are likely to come across a large orb web in a bush or tree, with a large spider sitting at its center. The web is constructed from two types of tensile fiber that are superbly adapted to their function in this prey-capture device. Fig.10.1 shows the basic organization of the web and also provides typical stress-strain curves for the two types of silk that are used in the web's construction. The silk that makes up the polygonal frame, the radial fibers of the web, and the spider's dragline is all synthesized in a pair of major ampullate (MaA) glands and is therefore called MaA silk. The silk that makes up the sticky capture spiral of the web is synthesized by two glands, the flagelliform (FL) glands, which produce the core threads, and the aggregate (AG) glands, which produce the glue compounds that coat the threads of the capture spiral. This silk is often called the "viscid silk," but we will call it the FL silk.

Stress-strain curves for these two silks produced by the spider *Araneus diadematus* are plotted in this graph, and the values shown are reasonably typical for the silks produced by other orb-weaving spiders. It is clear that both materials are quite extensible, but the two silks show dramatically different stress-strain curves. MaA silk starts out with a high initial stiffness, at about 10 GPa, and then at a strain of about 0.02 the material goes through what looks like a "yield" process, where stiffness falls dramatically, but this yield does not lead to a failure process. Rather, the material enters a second linear zone and continues to extend to failure at extensions of 20%–30%. The FL silk, on the other hand, shows a "J-shaped" curve, with low initial stiffness, followed by a dramatic rise in stiffness leading to failure at extension of about 300%. Because both materials fail at quite large strains it is appropriate that we use true stress and true strain to quantify their material properties, and so those are shown in this graph. Both of these silks deform at essentially constant volume, so the relationships introduced in chapter 2 (eqs. [2.7] and [2.8]) were used to compute these true-stress-strain curves. Note also that there is an engineering strain axis along the top of the graph, and that at high strains there are quite significant differences in the magnitudes of these two strains.

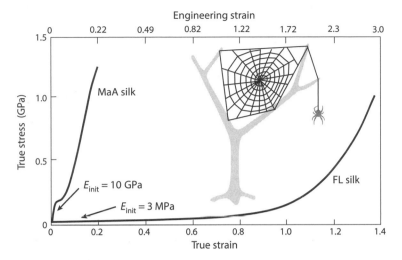

Figure 10.1. Material properties of orb-weaver spider silk. The material properties of the MaA-gland silk that forms the spider's safety line and the web frame and radii, and those of the FL silk that forms the sticky spiral of the orb web. E_{init}, initial modulus.

The initial modulus of MaA silk is about 10 GPa, its tensile strength is about 1.25 GPa, and its extensibility or true strain to failure is about 0.20. This material is in a totally different league to collagen-based fiber systems, like those we saw previously in fig. 9.1G, but it is somewhat less impressive than the synthetic superfibers Kevlar and Spectra. That is, MaA silk's initial modulus is about 8% of Kevlar's, and its strength is about 40% of Kevlar's. These properties are compensated by the fact that the silk has roughly a tenfold greater failure strain, and this means that its "toughness," or the energy to break, calculated as the area under the stress-strain curve to failure, is in the range of 160 MJ m^{-3}. Thus, MaA silk's toughness is higher than Kevlar's by a factor of about three. These properties are compared with a number of natural and synthetic materials in table 10.1, where it should be clear that spider MaA silk is quite an exceptional material.

The true-stress-strain curve for the FL silk is dramatically different from that of the MaA silk because this fiber's properties are quite different from those of the high-stiffness fibers discussed to this point. Indeed, FL silk's properties would place it amongst the rubberlike proteins that will be presented in section V, pliant materials, because its initial stiffness is three to four orders of magnitude lower than the initial stiffness of other tensile fibers we have discussed. However, as will soon be clear, this silk protein has an amino-acid sequence organization that is quite similar to that of the MaA silk discussed above, and it is actually the combination of the FL silk protein with the aggregate gland secretions that make it behave like a rubber. It is clear from fig. 10.1 that there are major differences in properties between the MaA silk and the FL silk, with the FL silk having an initial modulus about 3000-fold less than the MaA silk and an extensibility that is roughly 10 times greater. Interestingly, the true tensile strength of FL silk of about 1 GPa is nearly as high as the strength of MaA silk. This is particularly impressive when compared with the strength of natural rubber, one of the strongest commercial rubbers, which we saw in fig. 2.4 had a true-stress tensile strength of the order of only 100 MPa.

Table 10.1. The Mechanical Properties of Spider Silks

Material	Initial Modulus, E_{init} (GPa)	Breaking Stress, σ_{max} (GPa)	True Breaking Stress, $\sigma_{max,true}$ (GPa)	Breaking Strain, ε_{max}	True Breaking Strain, $\varepsilon_{max,true}$	Breaking Energy (MJ m^{-3})	Hysteresis (%)
Araneus diadematus MaA silk	9	1.1	1.40	0.27	0.24	160	65
A. diadematus FL silk	0.003	0.27	1.00	2.70	1.31	150	65
Caerostris darwini MaA silk	11.5	1.4	1.85	0.39	0.33	270	
Tendon collagen	1.5	0.1	0.11	0.12	0.11	7.5	7
Elastin at 100% relative humidity	0.001	0.002	0.005	1.50	0.92	2	10
Natural rubber	0.001	0.05	0.48	8.50	2.25	100	15
Nylon fiber	7	0.95	1.12	0.18	0.17	80	
Kevlar 49 fiber	130	3.6	3.70	0.03	0.03	50	
Carbon fiber	300	4	4.04	0.01	0.01	25	
High-tensile steel	200	1.5	1.51	0.01	0.01	6	

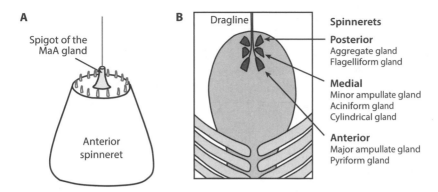

Figure 10.2. Silk-producing structures of the orb-weaving spider. **Panel A** is a diagram of an anterior spinneret, shown with a single MaA silk thread emerging from the central spigot, and a ring of small spigots from multiple pyriform glands. **Panel B** shows the ventral surface of a spider's abdomen, with three pairs of spinnerets, and it lists the silk-producing glands of each spinneret.

Spider silks are produced by three pairs of spinnerets that are located on the posterior ventral surface of the spider's abdomen. Fig. 10.2A shows the external structure of an anterior spinneret, which contains the major ampullate gland that produces the MaA silk. Here we see a single MaA silk thread emerging from the large spigot at the center of the spinneret, and in this diagram the silk thread is shown in reasonable proportion to the dimensions of the spinneret and spigot. In addition to the single MaA silk thread produced by an anterior spinneret, another type of silk is produced by multiple pyriform glands within the anterior spinneret and released through the ring of small spigots that surrounds the MaA spigot. These spigots produce very thin fibers and glue, which the spider uses to form the attachment discs that it applies onto the trailing dragline it releases as it moves through its environment. That is, every so often as a spider moves, it will drop its anterior spinnerets to the surface and wiggle them back and forth over a recently released segment of dragline, and through this process it will firmly attach the dragline so that if the spider falls or decides to lower itself in its environment, it will always have a firmly attached safety line to regulate these movements. These attachment discs are also used by the spider to attach the guy lines and frame of the web to leaves and branches.

Orb-weaving spiders produce an additional five types of silk. Each of these silks is produced in a different gland, and they are spun from gland spigots that are located on the three pairs of spinnerets, as shown in fig. 10.2B. This figure shows a ventral view of the abdomen of a spider that is dangling from an MaA-silk dragline. The dragline is a paired fiber, and a single silk-gland spigot is located on each of the two anterior spinnerets. It is important to note that the dragline silk strands in this diagram are not presented accurately to scale because the typical width of a mature spider's abdomen would be of the order of 5 mm, and the diameter of individual dragline fibers would be of the order of 2–3 μm. That is, their actual widths differ by over 1000-fold, and thus the silk diameters are grossly exaggerated in this diagram. The MaA silk gland, where the silk proteins are synthesized and transformed into silk fibers, is deep within the spinneret and abdomen, and the MaA gland in each spinneret has a single spigot through which the spun silk is released. The medial and posterior spinnerets produce the remaining silks. The medial spinnerets contain three gland types, the minor ampullate gland, which produces the accessory silk, or MiA silk, that is used in the construction of the capture spiral; the aciniform gland, which produces multiple strands of the swathing silk that is used to immobilize prey captured in the web; and the cylindrical gland, which produces the silk that forms the outer layer of the spider's egg case. The posterior spinnerets contain the flagelliform gland and the aggregate gland, which together produce the FL silk, as mentioned above.

We will start our analysis of silk design by considering the amino-acid-sequence designs of the proteins that form the dragline silks of spiders. Before we look at these sequences, however, it will be useful to consider the general pattern of amino-acid-sequence designs that are known for several well-studied kinds of silk. Two general patterns are commonly found; one is based on repeating sequence motifs that include repeats of the dipeptide glycine-alanine (GA). This kind of repeat is well represented in the silk fibroins produced by the silkworm *Bombyx mori*, which is the moth that forms the basis for the world's silk industry. The most abundant repeat motif in *Bombyx* silk is GAGAGS, where the S represents the amino acid serine. Glycine alternating with alanine and serine, organized in continuous arrays, has a high probability of directing the formation of β-sheet crystals, which are known to stabilize and reinforce the molecular networks in animal silks. In addition, other moth species make silks that contain alanine in long stretches, and these polyalanine sequence blocks can also form stable β-sheet crystals.

The general features of β-sheet crystals are illustrated in fig. 10.3. Fig. 10.3A shows the basic structure of a β-sheet crystal formed from polyalanine sequence blocks. The crystal structure is based on layers that are formed when individual peptide chains run next to each other in appropriate conformations, which turn out to be nearly extended peptide chains. Looking at the top surface of this crystal you will see that it contains four peptide chains that run adjacently to each other in a zigzag pattern that is often described as a pleated sheet. The bold arrows at the end of the chains in this top sheet indicate the amino-to-carboxyl-terminal direction of the polypeptide chains, and it should be clear that this is an antiparallel β-sheet structure, in which the direction of the polypeptide chain alternates from one chain to the next. β-sheet crystals can also form between parallel polypeptide chains, but it is well established that the packing of the polypeptide chains in parallel β-sheet crystals is not as tight, and hence they are not as stiff or strong as antiparallel β-sheet crystals.

Note also the dotted lines that run between carbonyl groups (C=O) and amino groups (NH) on adjacent chains. These indicate the presence of hydrogen bonds between the chains that hold the structure laterally into the pleated sheet, and this zigzag-chain conformation allows every amide group in the protein to form two hydrogen bonds, one with each of the chains on either side of it. The other important feature of the pleated-sheet structure is that the zigzag-chain conformation orients the amino acids' side groups vertically, either up or down from the plane of the pleated sheet, depending on the direction of the zigzag. This arrangement of side groups can be seen on the side face of the crystal in fig. 10.3A, where we see four zigzag chains stacked on each other. Since this is a polyalanine crystal, the side group is a methyl

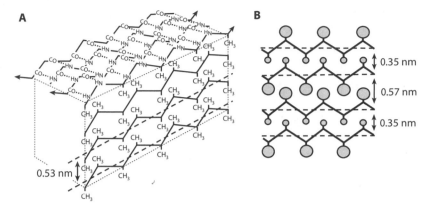

Figure 10.3. Features of β-sheet-crystals. **Panel A** shows the structure of an antiparallel β-sheet crystal made from polyalanine. The top surface is the hydrogen-bonded pleated sheet that forms when nearly extended antiparallel polypeptide chains align. In this diagram four sheets are stacked, bringing the alanine methyl side chains together to form hydrophobic bonds, which stabilize this stack. **Panel B** shows the arrangement of stacked β sheets in the silk crystals found in silkworm silk with $(GAGAGS)_n$ amino-acid repeat motifs. The intersheet spacing differs in alternate layers, with the alanine methyl groups on stacked peptide chains interacting with each other to form the thick layer, and glycine hydrogens associating to form the thin layer.

group ($-CH_3$), and these stacked sheets assemble in a way that brings every side group into a bonding interaction with a side group from the sheet above or below it. In the case of polyalanine chains, the nonpolar methyl side groups will form hydrophobic interactions with each other, which will involve the expulsion of water and the formation of tightly packed sheets in the β-sheet crystal. The spacing of the pleated sheets in the polyalanine crystal is illustrated in fig. 10.3A, which shows that the alanine methyl group creates an average intersheet spacing of 0.53 nm. The net result is that in these β-sheet crystals, adjacent chains are strongly attached through the numerous hydrogen bonds in the plane of the sheet, and the stacked sheets are linked together largely through hydrophobic interactions, which exclude all water from the crystal interior. This anhydrous environment further strengthens the hydrogen bonds between internal amide groups and produces a very stiff and very strong crystal structure that connects the protein chains in silk together into a robust polymer network. With this information in hand, we can look at the evolution of amino-acid-sequence designs in spiders. Similar β-sheet-crystal structures can form with $(GA)_n$ repeats as well, as is the case with silkworm silk as mentioned above. The hydrogen-bonded pleated sheets formed with $(GA)_n$ sequences stack like the polyalanine crystals shown in fig. 10.3A, but they are spaced differently because the sheets are organized so that all of the alanine methyl groups are on one side of the sheet and all of the glycine hydrogens are on the other. This results in a crystal structure in which the intersheet spacing alternates, as shown in fig. 10.3B, with the spacing on the glycine side being 0.35 nm and the spacing on the alanine side being 0.57 nm. In the case of silkworm silk, where serine often replaces alanine in the sequence, the intersheet spacing on the alanine/serine side is about 0.57 nm even though the serine side group (CH_2-OH) is somewhat larger than the alanine methyl group.

With this general information on the structure of silk crystals, we will consider first the evolution of the silk fibroins that exist in spiders, and we will then look at the protein-sequence structure of the various silks produced by orb-weaving spiders. Gatesy and his colleagues (2001) sequenced dragline-silk genes from a number of spider species that spanned the broad diversity of spiders (order Araneae), and table 10.2 provides examples of the amino-acid sequence motifs that were found. Starting with the most primitive species and progressing down the list to the more advanced araneoid species, which include the very successful and diverse orb-weaving spiders, we see some interesting trends. *Eugarus* is a member of the Mygalomorphae, primitive spiders that include the tarantula and funnel weavers. Its sequence is notable primarily because it contains a large fraction of the amino acid alanine (A), which we know can play a major role in the formation of β-sheet crystals. In *Eugarus* these presumptive crystal sequence blocks are not particularly well organized into regular structured arrays to direct crystal formation. However, there are certainly many alanines present. In addition, serine (S) is interspersed between the alanine residues, but again in a relatively unstructured manner. If we assume that these alanine and serine blocks could form β-sheet crystals, as shown in fig. 10.3, then we can estimate the potential for crystal formation and express it as a percentage of the total sequence available. This calculation is given in the third column, labeled "% crystal." At 57% crystal potential, the fibroin produced by *Eugarus* should be able to form β-sheet crystals, but perhaps in a haphazard way. This number should not be taken as quantifying the actual crystal content of the fiber; rather, it is intended to indicate an upper limit for crystal content.

Next down the list is *Plectreurys*, a member of a relatively rare group of spiders that were abundant in the Jurassic period, and its silk fibroins show some interesting long runs of $(GA)_n$ and poly(A) repeats. One block of sequence has 62% crystal-forming potential, and the other has 27%, so there appears to be strong potential for the formation of β-sheet crystals. However, they also have a fairly haphazard sequence organization between the crystal-forming blocks. *Dolomedes*, a more advanced genus of fishing spiders, produces a silk fibroin with a repeat sequence that shows a compact and quite regular sequence organization. That is, they seem to have evolved proteins with multiple copies of these small sequence elements that encode both GA and poly(A) crystal-forming motifs, separated by somewhat larger blocks of sequence that are rich in glycine, often in a tripeptide motif based on the sequence (GGX). In this case, the crystal-forming sequence makes up 36% of the total sequence.

The orb weavers *Nephila*, *Araneus*, and *Argiope* all show sequence motifs that are very similar to that of *Dolomedes*. They have short sequence motifs that are repeated many times, but they come in two variations, which are called MaASp-1 and MaASp-2. MaASp stands for major ampullate spidroin, where the term spidroin is a variation of the general term, fibroin, that is used to describe proteins that have evolved to form fibers. A spidroin is a fibroin produced by a spider. MaASp-1 and MaASp-2 contain β-sheet-crystal-forming motifs that include poly(A) repeats, as well as GA motifs and some serines. Typically these are about 10 amino acids long, and this likely limits the length of the β-sheet crystal to about 4–6 nm. Also, they all contain somewhat longer,

Table 10.2. Major Ampullate Fibroin Sequence Motifs from a Variety of Spider Species that Track the Evolution of Dragline Silks

Spider Genus	Sequence Motifs in the Evolution of Dragline Silks	% Crystal
Eugarus	**ASQIAASVASAVASSASAAAAAASSSAAAAAGASSAAG** **AASSSS**TTTTTTST**S**TSSS**AAAAAAAAAAAAS**ASGASSASAA ASASAAASAFSSALISDLLGIGVFGNTFGSIGSASAASSI ASAAAQAALSGLGLSYLASAGASAVSASVAGVGVGAG AYAYAYAIANAFASILANTGLLSVSSAASVASSVASAIAT SVSSSSAAA	57
Plectreurys		
CDNA-1	**GAGAGAGAGAGAGAGAGSGAA**TAVST**SSSSGSGAG**	62
CDNA-2	**AGAGSGAGSGAGAGSGAGAGAGAGGAGA**GFGSG LGLGYGVGLSSAQAQAQAQ**AAA**QAQAXAAQ AQAYAAAQAQAQAQAQAQ**AAAAAAAAAAAA**TIAGLG YGRQGQGTD**SSASS**VSTSTSVSSSATGPDTGYPVGYYG AGQAE**AAASAAAAAAAASAA**EAA	27
Dolomedes	GGAGSGQGGYGGQGGLGGYGQ**GAGAGAAAAAAA**	36
Nephila		
MaASp-1	GGAGQGGYGGLGXQGAGRGGQ**GAGAAAAA**	28
MaASp-2	GPGQQGPGYGPGQQGPGGYGPGQQGPSGPG**SAAAA** **AAA**	21
Araneus		
MaASp-2	GGYGPGSGQQGPGQQGPGQQGPGQQGPYGP**GASAA**	23
MaASp-2	**AAAA** GPGYYGPGSQGPSGPGGYGPGGPG**SSAAAAAAAAS**	31
Argiope		
MaASp-1	GGQGGQGGYGGLGSQGAGQGGYGQG**GAAAAAAAA**	26
MaASp-2	GPGYGPGAGQQGPGSQGPGSGGQQGPGGQGPYGPS **AAAAAAAA**	21

Based on data from Gatesy et al. (2001).

The letters represent the individual amino acids, the important ones here being: A, alanine; G, glycine; S, serine. These three amino acids, when organized in blocks, are known to direct the formation of β-sheet crystals in the silks produced by a variety of insects and spiders. The percentage crystal is based on the ratio of known crystal-forming amino acids, which are shown in bold, to the total number of amino acids in the repeat motif.

glycine-rich repeats, with lengths that range from 20 to 35 amino acids. In general, the repeat motifs of the orb weavers have crystal-forming potentials in the range of 20%–30%, so that the crystal domains are clearly not the dominant feature of these spidroins. As we shall see, this is a relatively unusual feature of the major ampullate spidroins, and it gives these silks a set of quite unique properties. It is important to note that there are two variations of these glycine-rich repeats. MaASp-1 has many glycine tripeptides, such as GGX, where X is often tyrosine (Y) or glutamine (Q). MaASp-2 is rich in proline, which is often found in pentapeptide motifs of the form GPGXX, such as GPGQQ. The key difference between these two spidroins is the presence of proline (P) in MaASp-2. Proline is not actually an amino acid; it is an imino acid because its side chain forms a ring with its amino group. This makes proline a much more rigid unit in a polypeptide chain, and it imposes significant restrictions on the conformations that are available to proline-rich peptides. We saw an example of this in collagen, where the presence of glycine in every third position enabled the formation of the quite extended triple-helix structure of tropocollagen. In other glycine- and proline-rich proteins, such as the rubberlike protein elastin, less regular combinations of glycine and proline enable the peptide backbone to be highly flexible and to behave as a random coil. Thus, elastin exhibits rubberlike elasticity, as described in chapter 6. We will see soon that under some conditions dragline fibers from some spiders show rubberlike elasticity when they are wet.

Table 10.3 shows sequence motifs that are found in the full set of spidroins that are produced by orb weavers. That is, it provides data for the six types of fiber produced by the silk glands illustrated in fig. 10.2B. The top row of data gives the repeat structures found in MaASp-1 and MaASp-2, as shown previously in table 10.2. This table shows data for the spider *Argiope*. The second row shows data for minor ampullate silk fibroin from two spider species. Its organization is similar to that of the MaASps, but the crystal-forming potential of the MiASp is considerably larger at 60%–65%, and β-sheet crystals likely dominate the structure of this silk.

Clearly there can be considerable variation in crystal-forming potential of spidroins because when we look at the third row, for the flagelliform fibroin FLSp-1, we see an extremely low value for the crystal content in the FL silk, at values that range from 4% in *N. clavipes* to 20% in *A. trifasciata*. This protein shows some similarities with those of MaASp-2 because they contain many of the glycine- and proline-rich motifs in the form of GPGGX and GPGGX$_n$, where the subscript "n" indicates that there may be additional, variable, amino acids. For *N. clavipes* FL fibroin these two motifs are repeated 26 and 41 times, and for *A. trifasciata* FL fibroin there are similar but considerably fewer GP-rich motifs. In addition to the GP-rich motifs, there are somewhat fewer GGX sequence blocks, similar to those seen in MaASp-1. Finally, there are sequence blocks with a more complex organization that have been called spacers, which are indicated with bold letters. Interestingly, computer analysis of the secondary-structure-forming potential of these sequences suggests that the "spacers" in both species have high potential for forming β-sheet crystals. The underlined letters indicate the amino acids in the spacer that have a 90% chance of being present in β-sheet-crystal conformations. Thus, the spacers are very likely the sites in FL spidroins where β-sheet crystals form to create cross-links between neighboring chains, and these cross-links create a polymer network that exhibits rubberlike elasticity. When all these sequence blocks are added together, the repeat motif of the *N. clavipes* FLSp-1 protein is about 410 amino acids long.

The tubuliform gland produces TuSp-1, a spidroin that contains at least 10 long (about 180 amino acids) sequence repeats, as shown in table 10.3, and contains a number of interesting motifs that are rich in alanine and serine. These amino acids are organized into several quite long runs of poly(A) and multiple alanines interspersed with serines, with S+A (shown in bold in the table) accounting for 41% of the amino acids in this particular repeat block. Amino-acid analysis of the tubuliform-gland protein indicates that S+A accounts for about 50% of the protein's composition. Thus, this protein likely has strong β-sheet-crystal-forming potential. The other amino-acid sequence motifs in this protein are quite different from those seen in the MaA, MiA, and FL spidroins.

The aciniform gland produces AcSp-1, which contains 14 near-identical repeats of a 200-amino-acid sequence. This sequence shows essentially no similarity to the other fibroins. The G+A+S content of this peptide is about 50%, but there are only quite small blocks of sequence that look like the crystal motifs observed in the MaA and MiA fibroins. Glycine, G, is quite abundant, but much of it is associated with proline, P, and with GPGXX motifs that are found in the noncrystalline domains of MaA and FL silks. Thus, it seems that the G+A+S content does not provide a useful indicator of potential for forming β-sheet crystals. Clearly, much remains to be learned about the molecular design of aciniform silk fibers.

Finally, recent studies have revealed even more unusual sequence organization for the fibroin produced in the pyriform gland. The final row of table 10.3 shows a large block of sequence from *A. trifasciata* PySp-1, which will be taken as typical of this silk. The G+A+S content is about 40%, suggesting that crystal-forming potential is quite high. However, there are only a few small blocks where these amino acids are collected together as they are in the MaA and MiA spidroins. But there are two very interesting and novel sequence motifs that appear in this protein. One is a repeat motif with a proline in every other position, and this motif is shown bold and white on gray in the table. The other repeat is based on a QQSSVA motif, and it is shown in bold, white on gray and underlined. It is not clear if these motifs play a role in β-sheet-crystal formation in this silk, or if they function in adhesive interactions between the pyriform fiber and the glue or the substrate. Clearly, there is a great deal to learn about the silks produced by the pyriform gland.

The sequence data discussed to this point have focused on repeat patterns and on the motifs that create β-sheet crystals and noncrystalline domains. It is also important that we consider the size and sequence organization of the full protein. Much of the sequence data in tables 10.2 and 10.3 shows motifs, and many of the motifs are repeated many times. In most cases we do not know the full size or sequence of these proteins, but it is clear that these are very large proteins. Most sequence data is derived from complementary DNA (cDNA) clones, which are created from mRNAs isolated from cells in the secretory portions of the various silk glands. In the process of characterizing the mRNAs it is common to carry out Northern blots on total RNA isolated from the individual glands, and these are then probed with restriction fragments of isolated cDNA clones to identify individual gene products. This process allows the determination of the approximate size of the mRNA that encodes the protein being studied, so it will provide information about the size of the various proteins. Typically, the mRNAs isolated from the various silk-gland sources have sizes in the range of 8–15 kb, which means that the mRNA is of the order of 8000–15,000 nucleotide bases long. mRNAs contain varying amounts of untranslated sequence at both their C-terminal and N-terminal ends, and, allowing for this, it is likely that the expressed portion of the mRNAs is in the range of 7.5 to 14.5 kb. Since each amino acid is encoded in a triplet codon, this means that the full size of the proteins encoded by the mRNAs would be in

Table 10.3. Amino-Acid Sequence Designs for the Silk Proteins Made by Orb-Weaving Spiders

Spidroin Type	Typical Sequence Motif	% Crystal or % G+A+S
MaASp		
MaASp-1	GGQGGQGGYGGLGSQGAGQGGYGQGG**AAAAAAAA**	26
MaASp-2	GPGYGPGAGQQGPGSQGPGGQGPYGPGGQPS**AAAAAAAA**	21
MiASp-1		
Nephila clavipes	GGAGGYGR**GAGAGAGAAAAGAGA**GGYGGQGGYG**AGAGAGAGAAAAGAGA**	65
Araneus diadematus	GGYGGQ**GAGAGA**GGYGQGYGQGGYG**AGAGAGAAAAXGAGAGAA**	61
Flagelliform: FLSp-1		
N. clavipes	$(GPGGX_n)_{26}$ $(GPGGX)_{41}$ $(GGX)_7$ **TIIEDLDITIDGADGPITISEELTIS**GAGGS	4
Argiope trifasciata	$(GPGGX_n)_4$ $(GGX)_6$ $(GPGGX_n)_3$ $(GPGGX_n)_4$ **EGPVTVDVDVTVGPEGVGG**	20
Tubuliform: TuSp-1		
Argiope argentata	TTTTTSTAGSQAASQSASSAAAASQASASS**FARASSASLAASSSFSSAFSS**ANSLSALGNVGYQLGNVANNL GIGNAAGLGNALSQAVSSVGVG**ASSSTYANAVSNAVGQFLAGQGILNAANAGSLASSFASALSASAASVAS** SAAAQAASQSQAAAASAFSRAASQSAQSAARSGAQSSS	~40
Aciniform: AcSp-1		
A. trifasciata	**GSAGPQGGFGATGGASAG**LISRVANALANTSTLRTVLRTGVSQQIASSVVQRAAQSLASTLGVDGNNLAR FAVQAVSRLPAGSDTSAYAQAFSSALPNAGVLNASNIDTLGSRVLSALLNGVSSAAQLGINVDSGSVQSDI SSSSFLSTSSSSASYSCASASSTSGAGYTGPSGPSTGPSGYPGPLGGGAPFGGSGFG	~50

Table 10.3 (continued)

Pyriform: PySp-1	
A. trifasciata	~40

APLPAPAPRPRPRPAPRPAPVYAPAPVVSQIQAFTSNRGSTQQNSFAQQSSVAQQSSVAQQSSVA
QQSSTAQQSSVAQQSAVAQSQQSSVAAASSGGSSASQAQATVSSQAPVYFNSQILTNNLASSLQSLSGLNYV
SSGQLSSSDVASIVAEAVSQSLGVSQGSVQNIISQRLNGIGPGASPSSVSAAIANAVSSAVQGSASAAPGQEQ
SIAQSMSSAISSAFQQIISQRTAVAPAPST LPSPLPAPRPRPAPLQPGPVYAPALAYAPAPVYAPAPVVSQFHS
SASTEATAEQNSFVQTSLAQSQ

MaASp, MiASp, and flagelliform data are from Gatesy et al. (2001). Tubuliform data are from Tian et al. (2006), aciniform from Hayashi et al. (2004), and pyriform from Perry et al. (2010). The percentage crystal is based on the ratio of known crystal-forming amino acids to the total number of amino acids in the repeat motif for MaASp, MiASp, and FLSp-1. For TuSp-1, AcSp-1, and PySp-1 the percentage shown is determined as the percentage G+A+S in the total amino-acid composition of the peptides shown in the table. Each letter in a sequence motif indicates a different amino acid, but the important ones are: A, alanine; G, glycine; P, proline; S, serine; X, variable amino acid.

Bold underlining denotes crystal-forming sequences. Bold without underlining in the flagelliform row denotes spacers, in the tubuliform row denotes runs primarily of poly(A) interspersed with S, in the aciniform row denotes small blocks that look like the crystal motifs in MaA and MiA but which may not have crystal-forming potential. In the pyriform row, bold white on gray indicates a repeat motif with a P in every other position; bold underlined white on gray indicates another repeat motif. See text for details.

the range of 2500 to 4800 amino acids in length. These are indeed very large proteins, and the sequence information described above and in table 10.3 for the aciniform-gland silk indicates that the repeating structural domain of the protein is 2800 amino acids long.

In the case of the MaA and MiA spidroins, the known repeat motifs are very much shorter than the full-length proteins. For example, the repeat motifs in the MaA spidroins are 30–45 amino acids long. The mRNAs for the two MaA spidroins produced by *Araneus diadematus* are about 7.5 and 8.5 kb respectively, and this means that these proteins contain 66 and 70 repeats of motifs similar to those shown in table 10.2. The repeat sequence for the *N. clavipes* FL spidroin, as shown in table 10.3, is about 410 amino acids long, and the mRNA for this FL spidroin is reported to have a molecular size of about 15 kb. This suggests that it encodes a protein that is about 4800 amino acids long, and the expressed protein should therefore contain about 12 of the FL spidroin repeats. In fact, the full protein contains 11 repeats, plus an additional 126 amino acids in an amino-terminal peptide and 87 amino acids in a carboxyl-terminal peptide. We will hear more about the function of these terminal peptides when we look at the mechanisms that the spider uses to direct the assembly of these enormous proteins into structural fibers.

2. THE MECHANICAL PROPERTIES OF SPIDER SILKS. Fig. 10.1 provides a good starting point for the discussion of spider silk's material properties, but there is much more to discover. At this point we know that orb-weaving spiders produce six kinds of structural fiber, which are used by the spiders for various functions. Unfortunately, we do not have a complete understanding of all silk types, but, with the exception of the pyriform silk, we do at least have some mechanical-test data for the other silks. Fig. 10.4A shows the results of tensile tests on the MaA, MiA, Tu, and Ac silks produced by *Argiope argentata*. These data came from a study by Todd Blackledge and

Figure 10.4. (*right*) The mechanical properties of spider MaA silk. **Panel A** shows typical mechanical properties for four of the five rigid silks produced by orb-weaving spiders. The gland origins of each silk are indicated in the diagram. The curves are based on test data from the spider *Argiope argentata*. **Panel B** shows stress-strain data that indicate the range of material properties that can be seen in the MaA silks of *Argiope trifasciata*. Forcibly drawn silk is typically stiffer, stronger, and less extensible than naturally spun silk produced by freely moving animals. **Panel C** shows typical true-stress-strain curves and engineering-stress-strain curves for MaA silks from *Argiope* and from *Caerostris*. Eng., engineering. **Panel D** shows the load-cycle behavior of the MaA silk from *Araneus sericatus*. The first cycle (*solid line*) shows a low resilience of about 35%. Note that the second-cycle behavior (*dotted line*) is quite different from the first cycle behavior. **Panel E** shows the results of continuous dynamic testing on the MaA silk from *Argiope argentata*. This test procedure involves slow, constant-strain-rate extension of a test sample that is simultaneously loaded dynamically with small-amplitude 20 Hz sinusoidal oscillations to determine the dynamic stiffness (E') and mechanical hysteresis ($\tan\delta$) of the material as it is deformed to large strain.

Cheryl Hayashi (2006), and it presents a good survey of these silks. Looking first at the curve for major ampullate (MaA) silk, we see properties that are similar to those for the MaA silk of *Araneus diadematus* in fig. 10.1. However, the *Argiope* silk appears to be a bit stiffer, stronger, and less extensible than the *Araneus* silk. As we will see later, spiders have remarkable control over the properties of the silks that they produce, and this curve for the MaA silk of *Argiope* was likely produced with silk that was forcibly drawn from a tethered animal. This is a common procedure for obtaining samples of spider

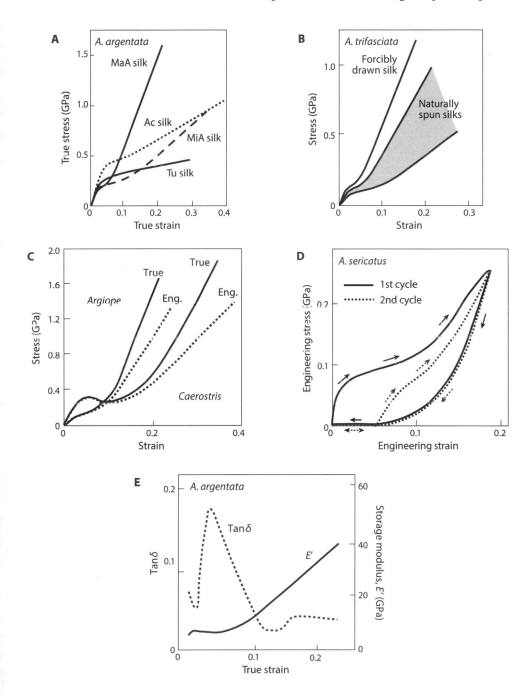

silks, but spiders usually resist having silk drawn from their spinnerets. The result of this resistance is that the silk fibers are formed under greater tension, which increases the alignment of fibroin molecules and the crystals that form during the spinning process. Thus, forcibly drawn silks are typically stiffer, stronger, and less extensible than naturally spun silks produced by unrestrained animals moving in their environment. Fig. 10.4B illustrates this important feature of the silks produced by spiders. Tensile-test data were obtained with silk samples produced either by forcible silking or by collection of samples that were spun by spiders freely moving in their environment, and the differences are clearly visible. With this image in mind, it seems clear that the range of properties seen in the silks produced by the four different glands is not really very different from the range of properties seen for the MaA silks. It is interesting, therefore, to consider why there are so many different kinds of silk produced. Certainly, the different glands produce silks of different diameters, and properties other than their stress-strain behavior may be important to the different functions of these silks.

Fortunately, we know quite a lot about the properties of the MaA silks produced by orb-weaving spiders, and, as shown in figs. 10.1 and 10.4A, that MaA silks all have quite similar properties. However, some seem to have "better" properties than others, and fig. 10.4C shows a nice comparison between forcibly silked MaA silk from *Argiope* and a naturally spun silk fiber taken from the web of *Caerostris darwini*. The data in this panel are plotted both in true stress/strain and in engineering stress/strain to make the differences more apparent. The *Caerostris* silk is very strong, with a true strength at about 1.9 GPa, but what is most impressive is the combination of high strength and extensibility. The engineering strain at failure is almost 0.4, and it is this extensibility that raises the true strength to this high level, and it also makes the energy to break, or toughness, = 270 MJ m^{-3}, which is significantly higher than that for other MaA silks. The stress-strain curves for *Caerostris* are drawn to show the average strength and extensibility of naturally spun fibers taken from this spider's web. However, tests conducted on some naturally spun and as well as on forcibly silked samples show extensibilities that are considerably greater, with breaking energies that are even higher. Thus, *Caerostris* currently holds the record for producing the toughest spider silk.

The high toughness observed for MaA silks is due in large part to these materials being highly viscoelastic. That is, much of their toughness arises from the fact that much of the strain energy imparted to the threads when they are stretched is dissipated as heat, and it is therefore not available to be released through elastic recoil at flaws to power the growth of cracks that ultimately initiate the fracture process. Stress-strain loops document this viscoelasticity very clearly, as shown by Mark Denny's work on the mechanical design of the web silks of *Araneus sericatus* (Denny, 1976). Fig. 10.4D shows results for stress-strain tests at relatively high strain rates, with loops taking about 4 s, and they reveal very interesting behaviors. The graph shows a test in which the sample was subjected to two load/unload cycles. The first cycle is indicated by the solid line, and the second cycle is indicated by the dotted line; the second cycle followed immediately after the first. There is a clear difference between the shapes of the first and second cycles. The first cycle took the sample to about 75%–80% of the expected failure stress, and it is clear that the recovery arm of the loop follows a very different path than the extension arm. The fiber's resilience for the first cycle was about 0.35 (35% energy recovery), which means that 65% of the energy input to strain the material is dissipated as heat and is therefore not available for crack growth. Interestingly, the sample's behavior is dramatically changed for the second cycle. The sample did not actually recover its initial length during the first cycle, so the sample went slack at low strains, and the second-cycle loading did not begin until strain reached about 0.05, and continued cycling would maintain the second-cycle behavior. However, if cycling were stopped after the second cycle, and the sample were heated or just allowed to recover at zero load for a long time, the material would return to its original state, and the sample would show the first-cycle behavior in subsequent tests. There are clearly some interesting structural changes occurring in this material.

Another way to document the viscoelastic behavior of MaA silks is to combine small-amplitude dynamic testing with large-amplitude stress-strain testing, as was done by Todd Blackledge and Cheryl Hayashi in their study of the silks produced by *Argiope argentata* (2006). They carried out constant-strain-rate (1% s^{-1}) extension tests to failure with small-amplitude 20 Hz dynamic testing to determine storage modulus (E') and damping (tan δ). This procedure is called continuous dynamic analysis because it allows us to quantify the contribution of viscous and elastic processes at different stages during the extension of a sample to its final rupture. A typical result is shown in fig. 10.4E. At strains below about 0.02 the dynamic storage modulus (E') is at about 10 GPa, which is similar to the initial slope of the stress-strain curve, as shown in fig. 10.4A, and the damping (tan δ) is low, at about 0.07. That is, at low strain the silk behaves like a rigid elastic material. Specifically, eq. (3.22) in chapter 3 allows us to calculate the resilience in this initial zone at a value of about 90%. However, at strains of about 0.02 the material enters the yield zone, and properties change dramatically. From strains of 0.02 to 0.05 the dynamic stiffness remains constant, but damping rises to about 0.17. This transition indicates the disruption of elastic structures within the material and the start of frictional processes associated with the onset of molecular rearrangements that we will discuss in the next section. As extension proceeds past the yield zone and into the linear rise to failure, we see that the material's damping falls quickly and dynamic stiffness rises gradually to the failure point. The fall in damping indicates the strain-induced alignment of the protein chains in the silk fiber and the formation of new elastic structures. Indeed, at strains of about 0.14 the material reaches very low damping with rising stiffness, which together indicate the formation of new extended elastic structures at the molecular level. Interestingly, the dynamic stiffness, E', reaches very high levels, up to 40 GPa, which is about four times higher than the slope of the stress-strain curve in fig. 10.4A for the same material. Clearly, there are some intriguing things happening with these dynamic behaviors, and it suggests that we might expect that failure testing of MaA silks at high strain rates will provide some interesting results.

The dynamic-mechanical-test results described above clearly indicate that at high strains MaA silks should show evidence of viscoelastic behavior, and a study by Yang and his colleagues (2005) on the effect of temperature on the mechanical properties of MaA silk from *Nephila edulis* clearly documents this. Low-strain-rate tests were carried out over a broad range of temperatures, and dramatic effects on material properties were observed. The logic to this experiment comes from the time-temperature superposition principle described in chapter 3 in the discussion of polymer viscoelasticity. Briefly, lowering temperature on low-strain-rate tests of polymeric materials is

Table 10.4. The Effect of Temperature on the Mechanical Properties of Forcibly Silked Dragline from *Nephila edulis*

Temperature (°C)	Breaking Stress, $\sigma_{max,eng}$ (GPa)	Breaking Strain, $\varepsilon_{max,eng}$	Initial Modulus, E_{init} (GPa)
−60	1.93	0.45	14.4
−40	1.75	0.39	14.8
−20	1.37	0.35	13.7
0	1.35	0.38	11.8
16	0.9	0.24	12.7
30	0.95	0.24	11.2
50	0.85	0.22	12.4
70	0.72	0.17	14.5

Data from Yang et al. (2005).
Note that tensile strength and extensibility increase dramatically at low temperature.

equivalent to increasing the strain rate of tests carried out at higher temperatures. This is exactly what was shown in these experiments, and the results are summarized in table 10.4. Values for tensile strength, extensibility, and initial modulus for the material at normal environmental temperatures are shown in bold, and we see that the "normal" behavior, as seen in most tests of silk materials, is a tensile strength of about 1 GPa, a breaking strain of about 0.24, and an initial modulus of about 12 GPa. These values are similar to those for forcibly silked samples obtained from other species. The key data in the table are the values at reduced temperature, where we see dramatic increases in the tensile strength and extensibility. At −60°C the tensile strength reaches 1.9 GPa, essentially twice that seen at normal ambient temperatures, and the failure strain reaches 0.45, again nearly twice that at normal temperatures. Initial modulus was not strongly affected over this temperature range. Taken together, these data strongly support the notion that spider MaA silk is a viscoelastic material. This means that its properties should be affected by strain rate, and impact testing of MaA silk confirms this conclusion.

My colleagues and I (Gosline et al., 1999) measured the behavior of MaA silk at strain rates that are in the range of those that might occur when a flying prey item impacts a radial strand in an orb web, as indicated in fig. 10.5A. Insects can fly at speeds in the range of 10 m s^{-1}, and for impacts at these speeds the strain rate imposed on the silk strand will be of the order of 100 s^{-1}, depending on the length of the silk segment that is impacted. At this strain rate, a failure strain of 0.3 would be reached in about 3 ms. A pendulum impact apparatus loaded the silk samples at constant velocity and at right angles to the fiber axis, as illustrated in fig. 10.5A, and it is important to note that the strain rate imposed on the fiber in this test depends of the angle of deflection reached in the fiber. At the earliest stages of the transverse impact an increment of lateral deflection will impose a very small strain on the silk, but as the deflection angle increases, the same lateral deflection will cause a larger strain. Thus, the initial portions of the stress-strain curve are generated at a lower strain rate than the later stages of the test, and average strain rates are appropriate. The average strain rates in figs. 10.5B and

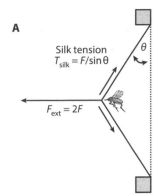

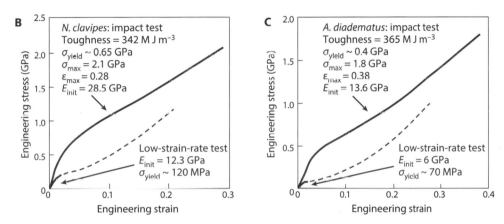

Figure 10.5. Impact test on MaA silk. **Panel A** shows the typical loading regime that would occur when a flying insect hits a radial supporting thread in an orb web. This transverse-loading regime was used in high-strain-rate impact tests of MaA silk samples, shown in panels B and C. F_{ext}, total applied force; F, force in each half of the thread; θ, angle of deflection. **Panel B** shows the results of an impact test on an MaA silk sample obtained from a forcibly silked *Nephila clavipes* spider. The properties indicated at the top left are from the impact test, and the properties at the lower right are from a low-strain-rate test of the initial properties of the same sample tested prior to the impact test. The dashed line indicates the likely form of the properties if the sample had been tested to failure at low strain rate. E_{init}, initial modulus; ε_{max}, breaking strain; σ_{max}, breaking stress; σ_{yield}, yield stress. **Panel C** shows the results of an impact on an MaA silk sample obtained from a freely walking *Araneus diadematus* spider, presented in the same format as panel B.

10.5C were about 100 s^{-1}; the impacting arm was traveling at about 4 m s^{-1}, and the test duration was about 2 ms. The test apparatus used in more recent tests was designed so that a micrometer could be used to extend the test sample at low strain rate through the initial linear region and into the yield zone just prior to the impact test. This made it possible to measure the low-strain-rate initial modulus and yield stress, and these values provided a reference for determining the enhancement of properties that could be attributed to the dynamic strain rate. Fig. 10.5B shows results for an impact test on a sample of *N. clavipes* MaA silk that was forcibly drawn from a tethered animal. The full-length solid line shows the results for the impact test, and the short solid line near the graph origin shows the quasi-static properties that were used to determine the initial modulus and yield stress at low strain rate. The dotted line that continues from this line is an indication of the properties that this silk sample would have shown if it were tested to failure at low strain rate. Fig. 10.5C shows similar impact test data for a dragline produced by a freely walking *Araneus diadematus*. Again, the short solid line shows a portion of the quasi-static stress-strain curve, and the dotted line gives an estimate of its behavior at low strain rate. For the two panels, the values in the upper left corner are the results of the impact test, and the values at the lower right corner are from the low-strain-rate test. The results are quite clear: high strain rate at normal ambient temperature has a large effect on mechanical properties. It is important to note that this graph is plotted in engineering stress/strain units. The graphs show that initial stiffness, strength, extensibility, and yield stress all increase, and thus there is a large increase in energy to break, or toughness, of more than twofold. Toughness values are in the range of 350 MJ m^{-3}, which is quite a bit higher than the value reported for *Caerostris* MaA silk, above. If these dynamic tensile-strength values are converted to true stress, the two samples would have tensile strengths in the range of 2.6 GPa, which is nearing the strength of the synthetic-polymer superfibers. Returning to our initial comparison of spider silks with high-performance industrial fibers in table 10.1, the dynamic toughness seen for MaA silks, at about 350 MJ m^{-3}, makes them 7 times tougher than Kevlar, 14 times tougher than carbon fiber, and 60 times tougher than high-tensile steel. Evolution has produced a fiber that is truly exceptional in its ability to function in a prey-capture device.

3. THE NETWORK STRUCTURE OF MAJOR AMPULLATE SILKS. Fig. 10.6A shows diagrams that represent the nanometer-scale molecular structures that exist in the load-bearing regions of MaA silk fibers. These model structures are based on the design of the sequence repeats observed for MaASp-1 and MaASp-2, and on the biophysical characterization of spider silks. For reference, the dimensions of the β-sheet-crystal structures are based on the polyalanine blocks, which typically contain about 10 amino acids. This would make the crystals about 5 nm long. The crystals are shown to contain four stacked β sheets, which is plausible but not really a known parameter. These are reasonable guesses for the dimensions of the crystals. When the spidroins assemble into these crystal-cross-linked polymer networks and are drawn from the spider's MaA gland, the silk is stretched to varying degrees. The imposed stretch elongates the network chains into extended conformations, and tension in these chains reorients the β-sheet crystal toward the long axis of the fiber. This is the state that is represented in these two diagrams.

The network chains that interconnect the crystals come in two varieties, as we saw in tables 10.2 and 10.3. MaASp-1 proteins have about 20–25 residues of the GGX motif that form the network chains that run between the β-sheet crystals, and the MaASp-2 proteins have similar lengths but these are dominated by proline-containing motifs that are typically 5 amino acids long, in sequences like GPGXX. Biophysical analysis of the network structure of MaA silks provides us with some interesting insights

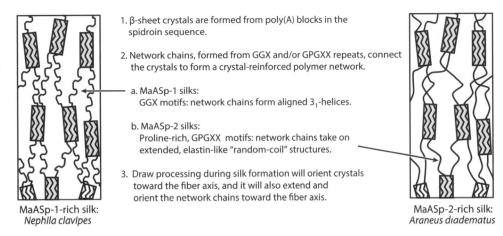

1. β-sheet crystals are formed from poly(A) blocks in the spidroin sequence.

2. Network chains, formed from GGX and/or GPGXX repeats, connect the crystals to form a crystal-reinforced polymer network.

  a. MaASp-1 silks:
     GGX motifs: network chains form aligned 3_1-helices.

  b. MaASp-2 silks:
     Proline-rich, GPGXX motifs: network chains take on extended, elastin-like "random-coil" structures.

3. Draw processing during silk formation will orient crystals toward the fiber axis, and it will also extend and orient the network chains toward the fiber axis.

MaASp-1-rich silk:
  Nephila clavipes

MaASp-2-rich silk:
  Araneus diadematus

Figure 10.6. Draw processing MaA silk. The structural organization of the network formed with MaASp-1- and MaASp-2-based fibers in different MaA silks. The *rectangular blocks* represent the polyalanine-based β-sheet crystals, and the *lines* interconnecting the crystals represent the extended network chains that interconnect the crystals. G, glycine; P, proline; X, variable amino acid.

into the origins of the material properties that are presented above (2. The Mechanical Properties of Spider Silks). It is now generally agreed that the GGX sequences seen in MaASp-1 spidroins lead to the formation of a helical structure, called the 3_1 helix, likely facilitated by the stretching of the maturing silk fiber within the spinneret and after its emergence from the gland spigot. This helix is characterized by a repeat structure in which one turn contains three amino acids; indeed the name, 3_1 helix, is derived directly from this feature. This helix is also called a 3_{10} helix because the three amino acids that form a full helical turn are joined together by a hydrogen bond between the carbonyl oxygen in the first amino acid to the amino nitrogen of the third amino acid, and the ring formed by this hydrogen bond contains 10 atoms.

The 3_1 helix is somewhat different from the α helix and the collagen helix, which we have discussed previously. The collagen helix is a highly extended structure, with a single turn of an individual polypeptide helix having a length of about 9 nm. The α helix makes a single turn with a length of 5.5 nm. This difference in helical pitch has an important effect on the mechanical properties of materials that are constructed from proteins that contain these helical structures. That is, collagen molecules are very stiff, in part owing to the extended shape of the helix formed by each of the three peptide chains that together make up the triple-helical collagen molecule, and in part because of the tight twisting of the three peptide chains to form the triple helix. Materials that are based on α-helical proteins, like hair keratins, can undergo large-scale extensions, owing to the opening and unfolding of their constituent α helices. The 3_1 helix is similar to the α helix, with a pitch of about 6 nm, so it can extend a considerable distance before forces are carried entirely by the backbone of the polypeptide chain. This ability of the 3_1 helix to extend when hydrogen bonds are broken plays a key role in the extensibility and the toughness of MaA silks. That is, the energetic cost of breaking the hydrogen bonds in these helices contributes to the energy required to extend the silks in their postyield extensions to failure.

The GPGXX motifs function in a similar manner, but they do not involve the formation of stable helical structures. Rather, these motifs likely form "random" network chains similar to those seen in the rubberlike protein elastin, where they exhibit entropic elasticity when they are hydrated. In silks, however, these chains function in webs and in draglines in air, where they typically experience humidity levels low enough to remove the much of the bound water. This allows the chains to form stable hydrogen bonds between the peptide groups in the protein backbone, which transforms the dynamically mobile random chains into a rigid polymer-glass state. This transformation occurs as the silk fiber is drawn from the spinneret into the air, where it dries. In the draw process that occurs during fiber formation, the "rubberlike" random chains are extended to varying degrees depending on the amount of tension that the spider applies to the thread as it is pulled from the gland spigot. Because the structures formed by the GPGXX motifs are less regular than those seen in silks dominated by GGX sequences, there may be somewhat different levels of hydrogen-bond formation in silks expressing different levels of MaASp-1 and MaASp-2, but as we will see in the next section, this does not seem to affect the ability of orb-weaving spiders to control fiber formation and to produce silks that have very similar mechanical properties.

There are, however, significant differences that can be seen in the mechanical properties of different MaA silks, but they are particularly apparent in the fully hydrated state. Savage and Gosline (2008) investigated the properties of MaA silks from *Araneus diadematus* and *Nephila clavipes*, spiders whose MaA silks show a large difference in MaASp-1 and MaASp-2 content, and they documented differences in mechanical properties and in the nature of the elastic mechanism in hydrated silk samples. One very interesting feature of MaA silks is their ability to be transformed from a rigid state when held in air, at humidity levels below about 75%, to a soft state when hydrated at high humidity or when immersed in water. MaA silks undergo a dramatic transformation, called "supercontraction," upon immersion in water, in which the silk absorbs water, contracts, and becomes a very much softer material. This shift in properties with hydration is illustrated in fig. 10.7, which shows force-extension data for MaA silks from *Araneus* and *Nephila*, with the dotted and dashed lines indicating the properties of the silks when immersed in water, and the solid lines showing the force-extension curves for dry samples of the silks. The tests with hydrated samples start at a contracted length, expressed as a fraction of the initial dry length, which is 0.5 for *Araneus* and 0.7 for *Nephila* silks. The use of force-extension rather than stress-strain data eliminates the effect of changes in sample dimensions with the absorption of water, and this allows the curves from the hydrated state to merge nicely with the curves from the dry state. Clearly there are dramatic changes in properties that occur with supercontraction. The silk's initial stiffness falls by two to three orders of magnitude, from a dry stiffness of about 10 GPa to a wet stiffness of 10 MPa for *A. diadematus* and a wet stiffness of 100 MPa for *N. clavipes* MaA silks. Thus, supercontracted and hydrated *Nephila* silk has an initial modulus that is about tenfold greater than *Araneus* silk. Thermoelastic testing to determine the molecular origins of the elastic mechanism for these hydrated silks indicates that the GPGXX-rich segments in *Araneus* silk exhibit rubberlike elasticity when hydrated, and the GGX-rich segments in *Nephila* exhibit a large component

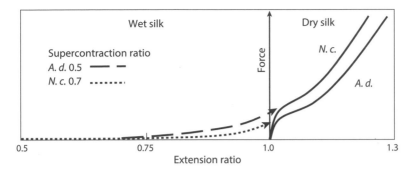

Figure 10.7. Effect of hydration on properties of MaA silk. Typical force-extension data for supercontracted MaA silk from *Araneus diadematus* (*A. d.; dashed line*) and from *Nephila clavipes* (*N. c.; dotted line*) on the left, for samples that are immersed in water. The *solid lines* that continue to the right show the force-extension behavior for these two silks in their dry state. The extension ratios shown in this diagram are based on the resting length of the dry silk as obtained from the spider. The extension ratios for the wet silks indicate the degree of contraction that occurs when the dry silk is wetted, called the supercontraction ratio.

of bond-energy elasticity when the hydrated silk is stretched by only a few percent from its fully contracted state. These differences in the elastic mechanisms of supercontracted silks highlight the differences in the structural stability of the hydrogen-bonded structures that form in the GGX-rich and the GPGXX-rich fibroins. Specifically, the 3_1 helices that form with MaASp-1 GGX repeats likely remain largely intact in supercontracted *Nephila* dragline silk.

Yi Liu and colleagues (2008) continued this comparison of MaA silks with different proline levels by investigating the properties of MaA silks produced by spiders with widely varying expression levels of MaASp-1 and MaASp-2, which they documented by measuring the proline content of the MaA silks produced by eight species of araneoid spiders. Proline content varied from a high 14.3%, in *Araneus*, a spider that expressed only MaASp-2 proteins, to a low of 0.6% in a spider that expressed almost exclusively MaASp-1 protein. These authors carried out stress-strain tests on samples of forcibly silked dry samples that were formed at a draw rate that produced silks that all failed at extensions of about 24%, which were identified as the "native" silk. They also determined the supercontraction ratio of samples immersed in water and determined the mechanical properties of samples that had been dried at their contracted dimensions. They observed a strong correlation between the "native" silk's capacity to supercontract in water and proline content; the higher the proline content (MaASp-2 content) the greater the supercontraction ratio, which ranged from 0.5 for *A. diadematus* to 0.9 for *Cyrtophora citricola*. In addition, the "native" silk's initial stiffness, E_{init}, showed a strong negative correlation with proline content, with E_{init} for *C. citricola* silk at 18 GPa and E_{init} for *A. diadematus* silk at 10 GPa, and values for other silks falling at intermediate levels in proportion to their proline levels. So the stress-strain properties of this broad range of MaA silks are strongly influenced by the relative expression of MaASp-1- and MaASp-2-type genes.

What is really interesting is that when Liu et al. varied the draw rate for silk collection over a range that spanned from 0.1 to 20 cm s^{-1}, they were able to produce "native" silks with a full range of properties, as shown in fig. 10.4B, for all of the species. That is, if spiders can control the conditions of silk formation in the silk gland, then they can all produce an essentially identical range of material properties, regardless of the gene composition. Clearly, the control of silk formation in the spinneret is a crucial factor in the production of functionally competent dragline silks. With these final insights into the molecular mechanics of the MaA silks, it is clear that we need to understand the nature of the biochemical and biophysical mechanisms that are employed in the silk glands and spinneret in the transformation of proteins synthesized in cells to the production of the finished silk fiber.

4. SILK FORMATION IN THE GLAND/SPINNERET COMPLEX. How do spiders make these remarkable fibers? The answer lies in the biophysical processes that take place within the silk gland where high-molecular-weight proteins, with sequence designs as described above, are transformed into high-performance tensile fibers of remarkable strength and toughness. Two review articles (Vollrath and Knight, 2001; Eisoldt, et al., 2012) provide many of the details of the process that will be described here briefly.

Fig. 10.8 shows a simplified structure for the major ampullate silk gland and associated spinneret seen in the orb weavers, such as *Araneus* and *Nephila*. The scale bar indicates the approximate dimensions of this structure in a mature female spider, but it will be useful to note that some of the components in this diagram have not been drawn accurately to this scale. For example, for clarity, the S duct and terminal duct are drawn with much larger diameters than specified by the scale bar. The S-duct lumen near the funnel has a diameter of about 100 μm, and as it approaches the valve its diameter is about 10 μm. The diameter of the terminal duct is about 15 μm, in spite of the fact that the scale bar would suggest that the terminal duct lumen is of the order of 1 mm wide.

The basic organization of the gland is that there are four functional zones. Zone 1 is the region of the gland where the majority of the proteins for the silk are synthesized and stored. This zone includes a long, thin tail that reaches well into the abdomen and can be several centimeters in length if uncoiled, and is the site of spidroin synthesis. It also includes the larger sac at the end of the tail where the silk secretion is stored, typically at concentrations that approach 50% protein. The line that shows the shape of the tail and sac is the inner surface of the gland lumen, and this line therefore represents the apical face of a single layer of large cells that is about 100 μm thick. These cells synthesize and release the spidroins into the gland lumen. Zone 2 is the downstream end of the storage sac, and it is lined with cells that synthesize components that form a surface "skin" for the silk fiber. The funnel is an area lined with a thickened cuticle, which anchors the thinner cuticle that lines the wall of the S duct. The S duct has three arms, which fold back and forth before proceeding through the valve into the terminal duct and the spigot, where the silk thread is released into the atmosphere. The S duct plays a key role in the manipulation of the spidroins from a concentrated "liquid" solution into liquid-crystalline arrays of proteins that become extended and aligned parallel to the duct axis, and ultimately are transformed into the solid state as highly oriented protein fibers. Knight and Vollrath's (1999) microscopic analysis of a spider that was preserved

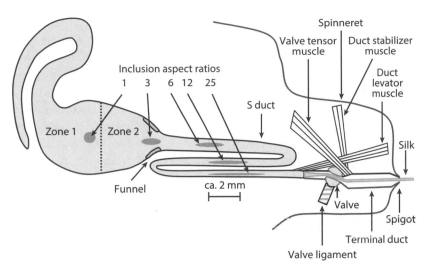

Figure 10.8. Silk formation in the gland spinneret complex. A diagram that illustrates the structure and organization of the silk gland and associated spinning-duct structures that are responsible for the formation of MaA silk, as described in the text.

while actively spinning silk revealed some important information about the structural transformations that occur within the silk secretion as it passes from the storage zone, through the S duct to the valve region, and out through the spigot. Polarized-light microscopy of the gland contents revealed that the highly concentrated solution of spidroins in the gland are organized into liquid-crystalline structures that change as the gland fluid moves from the storage sac, down through the S duct. These authors were also able to follow small fluid inclusions in the gland contents, with dimensions on the scale of a few micrometers in diameter. These droplets were spherical in the center of the storage sac, but they became ellipsoidal as they approached the funnel, with an aspect ratio (length/diameter) of about 3. Then, as they proceeded down the tapering S duct, they elongated, and the aspect ratio of the droplets increased to about 80 as the fluid approached the valve. Fig. 10.8 illustrates this process with the elliptical drawings placed strategically along the length of the S duct. The round, gray circle in the center of the storage sac represents the undistorted shape at the start of the alignment process. The ellipse located at the funnel has an aspect ratio of 3 and an area equal to that of the starting circle. The three thinner ellipses within the S duct illustrate the extended shapes of inclusions with aspect ratios of 6, 12, and 25. This transformation represents the effect of extensional flow in extending the molecular structures in the silk dope as it proceeds down the S duct. The equal area of the transformed elliptical shapes is used to represent the fact that the volume of the inclusions remains constant as they move down the duct. This elongation of droplets arises from the taper of the S-duct diameter over its length, which follows a hyperbolic decay profile, and it is believed that the extensional flow that causes droplet elongation in the S duct will also cause the elongation of the very long and flexible spidroin molecules, which start out as compact globules, organized into micelle-like structures. These structures and their constituent spidroins are elongated and aligned as they proceed down the tapering S duct. Thus, as the spidroin "solution" approaches the valve, the aligned spidroin molecules begin to interact laterally to form the β-sheet crystals that will ultimately transform this structured fluid into a crystal-cross-linked polymeric fiber.

As the fluid approaches the end of the third arm of the S duct, the spidroin molecules become sufficiently interconnected through the development of β-sheet crystals that the fluid takes on the qualities of a solid, and the silk dope pulls away from the walls of the duct several millimeters upstream from the valve. This is described as the draw-down taper, and it is important to note that it is not caused by pressure from the bulb forcing the gland "fluid" down the tapering duct. Spiders form their silk by pulling it out from their bodies, usually by moving away from a point of attachment, and the draw-down process occurs because the solidifying silk thread develops sufficient tension that it is physically pulled away from the duct walls. This process is shown explicitly in fig. 10.8 by the change in the color of the thread as it pulls away from the duct walls just upstream from the valves. Also, note that the thread thins rapidly as it progresses beyond the draw-down point, as once the developing silk fiber pulls away from the duct surface the external tension acts directly to stretch the fiber and further align the spidroin molecules. This tension completes the assembly of the β-sheet crystals and elongates and aligns the GGX and GPGXX network chains in the direction of the thread. Thus, the final structure, and hence properties, of the silk thread are created by the magnitude of the external tension and resulting strain that are transferred from the outside world into the terminal duct of the spinneret. Indeed, the range of material properties seen for dragline silks, as shown in fig. 10.4B, is determined largely by this external tension.

There are a number of chemical changes that accompany the process of moving the spidroin solution through this silk-forming system. First, there are changes in pH and ion concentrations that occur in the silk dope as it moves down the duct, and these are known to influence the interactions between the fibroin molecules in ways that facilitate the biophysical processes described above. The interactions here are ones that arise from the fundamental design of the spidroin proteins themselves. Recall our description of the protein structure; it is made up of a large number of the amino-acid-sequence repeats that form the crystal and network-chain components of the silk's structure, and it is capped with N-terminal and C-terminal domains that to this point have been ignored. The May 13, 2010, issue of *Nature* contained two consecutive papers on the structure and function of these two terminal domains that opened up the issue of how these terminal sequence blocks influence the assembly of spider MaA silk (Askarieh et al., 2010; Hagn et al., 2010). The short answer is that the terminal domains assemble into globular structures constructed from α-helical bundles that aggregate, bringing the C- and N-terminal domains of MaA fibroins together into preorganized dimers and ultimately tetramers and larger aggregates, which facilitate the molecular transformations described above. A review of these processes by Eisoldt and colleagues in 2012 provides an excellent overview of the recent research in this area.

The diagrams in fig. 10.9 illustrate some of the processes involved in the formation of silk fiber in the S duct. This shows a model for the full spidroin molecule at the top, with the zigzag central region representing the repeated crystal/network chain motifs, with the square and the circle at the ends representing the N- and C-terminal domains of the

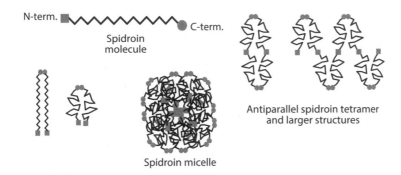

Figure 10.9. The structural organization of spidroin molecules. Spidroin molecules include the central domain that contains multiple repeats of the β-sheet-crystal and network-chain domains, as well as the N-terminal (N-term.) and C-terminal (C-term.) domains that function in the assembly of the silk network structure during its passage through the silk gland.

protein, respectively. The event that starts the assembly of the spidroins occurs early in the gland system and is likely completed before the proteins reach the storage sac. The gland contents in the upstream regions of the gland exist at a pH of about 7, and with relatively high concentrations of sodium and chloride ions. Under these conditions the spidroin molecules form dimers by linking together their C-terminal domains, but the N-terminal domains remain free. The C-terminal domain comprises a compact bundle of five short α-helical segments. It has a hydrophobic face that contains a single cysteine residue. This domain is constructed so that two C-terminal domains will join together, initially through hydrophobic bonding, but when brought together the two cysteine side chains, each of which has a free thiol (-SH) group, form a disulfide bond that permanently links the two spidroins together at their C-terminal ends. This dimer formation conceals the hydrophobic surfaces of the C-terminal domain, and leaves the C-terminus with a very hydrophilic end that interacts strongly with the aqueous phase of the gland fluid.

The N-terminal domain and the structural sequences of the spidroin are both quite hydrophobic, and this polarization of the spidroin dimer likely causes the formation of micelle-like structures, as illustrated in fig. 10.9. At the top left the spidroin dimer is drawn in its fully extended form, which makes it easy to identify the terminal domains and central structural domain. However, the drawings below, which show a much less ordered structure, provide a better picture of the dimer as it probably exists. The long peptide chain is flexible, and it is likely to crumple into a folded and perhaps disordered structure. The spidroin micelle drawing to the right of the dimers illustrates a possible arrangement of spidroin dimers that would allow them to be stored at such high concentrations (i.e., 50% protein) but still retain fluid properties that allow the spidroin to flow through the spinning system.

As these stable dimers in their micelle-like storage form move into the S duct, they enter a region where the cells lining the duct alter both the pH and the ion content of the fluid, and this triggers a reorganization of the dimers into tetramers or similar larger structures, which create polarized precursors that will facilitate the assembly of the silk network structure as the fluid passes through the S duct. Experiments with isolated fibroins indicate that the N-terminal domain, which is also constructed from a bundle of five short α-helical segments, remains free in the storage sac because of the high pH and ionic content mentioned above. However, as the silk fluid moves into the S duct, the cells that line the duct actively lower the pH to below pH 6.4, reduce sodium- and chloride-ion concentrations, and increase potassium- and phosphate-ion concentrations in a way that triggers a structural transformation in the N-terminal domains, causing them to form dimers with neighboring N-terminal domains. A key feature of these dimers is that they form antiparallel structures. This means that it is highly likely that pairs of dimers or multiple dimers will bind to create antiparallel tetramers or larger antiparallel structures, as shown on the right of the figure.

The result of this transition is the creation of a macromolecular precursor in which the crystalline components are preorganized into an antiparallel arrangement. The significance of this arrangement is that the β-sheet crystals that will form as the silk matures will be predisposed to form as antiparallel crystals, which is the most stable and hence the strongest form of β-sheet crystals. It is not clear where exactly this process is completed and what happens to the micellar organization seen in the storage sac, but it is clear that the extensional flow that occurs in the tapering S duct will stretch out the somewhat collapsed spidroin chains, bringing overlapping antiparallel sequence blocks into alignment, where they can begin the process of forming the β-sheet crystals that link the spidroin molecules into the rigid structure that is responsible for the exceptional properties of the MaA silks.

We already know that the spider has considerable control over many of the parameters that determine the properties of the silk, and we also need to consider the active processes that are available to the spider to control the final stages of the silk-production process. Silk properties are largely controlled by manipulation of tension in the silk during the final stages of silk formation. Starting at the draw-down taper, seen in fig. 10.8 just upstream from the valves, we note that the silk fiber has pulled away from the S-duct wall, and it then passes through the valves into the terminal duct, which is much expanded from the 10 μm diameter of the S duct leading to the valves. The Knight and Vollrath study (1999) was based on an experiment with a large *Nephila edulis*, and the silk thread in the terminal duct was about 8 μm in diameter, well below the 15 μm diameter of the terminal duct. This chamber functions as a space to collect and remove the large amount of water that is released during the draw-down process. The terminal duct is lined by cells that are designed to actively transport water from the duct into the spider's tissues that surround it, so that it is conserved by the animal. This kind of process is well known in insects, where active-transport mechanisms move ions from fluid-filled compartments into spaces on the other side of the cells that form the epithelial sheets to create osmotic gradients that drive the diffusion of water out of the duct lumen and across the epithelium. Similar mechanisms drive the water recovery in the spider silk gland.

The final stages of silk-fiber formation occur as the silk thread passes through the valve and terminal duct. In this region, draw forces imposed upon the thread by the movement of the spider away from the silk's point of attachment will stretch the fiber to continue the draw-induced crystallization that was initiated in the S duct, and to extend the network chains running between the crystals. The extent of this transformation is determined by the amount of tension that the spider applies to the silk as it is pulled from the spigot, and we need to consider how the spider controls this tension. The valves provide an obvious possibility, although there are others. Ronald Wilson investigated the structural organization of the spinneret (Wilson, 1969) and made some suggestions, but they remain to be confirmed experimentally. The spinneret contains muscles that control the movement of the spinneret as a whole. In addition, there are three other muscles that control the movement of the valves and the terminal duct, which are opposed by a ligament on the opposite side of the valve, as shown in fig. 10.8. The valve itself is quite complex, with several different cuticular layers that include an internal set of "lips" that can be inflated/deflated by the action of the associated muscles. In addition, the duct, which enters the spinneret at its center, bends toward the posterior wall of the spinneret at the location of the valve and then runs along the posterior wall to the spigot. So the maturing silk fiber does not necessarily have a straight path through the terminal duct to the outside environment.

There are a number of possibilities for the function of the terminal regions of the gland system in the control of silk formation. The lips of the valve are formed by an inner layer of lamellar cuticle, which is a ring of cuticle that can be inflated to expand and constrict the valve opening, or deflated to open the valve. One possibility for powering this process is that the contraction of the valve tensor muscle pulls the ventral side of the valve away from the duct lumen, drawing fluid from the lips outwards and opening the valve. Alternatively, the valve tensor muscle and the duct levator muscle are arranged in opposition, and thus the simultaneous contraction of these two muscles would squeeze the valve in a way that could force fluid into the lips, causing the valve to squeeze down on the silk or close the duct. In addition, because the duct levator muscle spans the full length of the terminal duct, the simultaneous contraction of these muscles would shorten the spinneret, and in this process would bend the terminal duct, creating curved surfaces against which the silk fiber would have to slide under considerable tension as the silk is drawn from the spigot. Indeed, Wilson shows pictures of isolated terminal ducts with a strong S-shaped curvature. Silk pulled through the curved structure would create frictional forces as it slid along the duct wall, which would add to the frictional forces created at the valve. The result of these frictional forces would be the heating of the frictional surfaces, but perhaps spreading the frictional forces over the full length of the terminal-duct complex would limit the rise in temperature when silk is formed.

As discussed above, spiders can control the properties of their silks, presumably through the control of tension developed in the final stages of silk production, and, as shown in fig. 10.4B, forcibly silked threads are stiffer, stronger, and less extensible than naturally spun silk produced by freely moving animals. Christine Ortlepp (Ortlepp and Gosline, 2004) measured the silking force produced when tethered spiders were forcibly silked, and she observed that the spiders can apply very large forces indeed. A well-fed and rested spider can be forcibly silked at a speed of 1 cm s^{-1} for several hours, and during that process can produce about 100 m of silk. At times in this process the spider will apply a silking force that is in the range of three to four times the body weight of the spider. Adult spiders produce silk fibers that break in low-strain-rate tensile tests at forces that are about five spider body weights, and this means that at times during forced silking the draw tensions can reach levels that are up to 80% of the tensile strength of the silk. It is the high levels of tension seen at high draw speeds that creates the high stiffness and strength of forcibly silked fibers shown in fig. 10.4B. The spiders, however, do not maintain these high tensions for more than about 10 s at a time, presumably because the frictional heat generated in the terminal duct raises temperatures to uncomfortable levels, and the spider reduces tension to levels that correspond to 20% of breaking tension or less for much of the time at this silking speed. Forcibly silked spiders will produce silk at rates as low as 1 mm s^{-1}, and as high as 30 cm s^{-1}, but at the highest speeds they can only be silked for very short periods of time, and the 30 cm s^{-1} draw rate likely sets the maximal rate of forced silking because of the high heat levels that develop in the terminal duct of the spinneret.

5. THE FUNCTIONAL DESIGN OF SPIDER DRAGLINES. Spiders use their MaA silks for two major functions: for draglines that are assumed to function as safety lines in falls, and structural lines in the construction of orb webs, which includes guy lines that attach webs to the branches of trees and shrubs that the spiders use to support their webs, the web frame, and the radial fibers that support the sticky capture spiral. What is not clear from the previous discussion of silk formation is whether spiders can vary the structural properties of their dragline silks. We saw in fig. 10.4B that spiders can vary their silk's material properties over a fairly wide range, presumably by altering the tensions that they apply in the final stages of fiber formation. It is not clear, however, if this range of properties is sufficient to allow spiders to construct all the things that they use their MaA silks for. Webs typically have four or five guy lines to support the entire web, and these must support the total forces applied to the web, including environmental forces such as wind and branch movement, as well as forces required to support the spider's weight and the impact forces generated in prey capture. Mark Denny (1976) documented the number of MaA silk pairs that existed in various locations of the web and dragline of *Araneus sericatus*, and he observed that guy threads typically contained four to five MaA pairs, frame threads contained two to four MaA pairs, and radii contained a single MaA pair. Thus, adjustment in the cross-sectional area of supporting threads in a web is used by the spider to compensate for the load distributions that occur in the orb web.

Draglines, however, are always found as a single MaA silk pair, and it would be interesting to know if spiders can adjust the size, as well as the properties, of a single MaA pair. Garrido and colleagues (2002) clearly demonstrated that spiders are not only capable of changing MaA-silk material properties, but also capable of altering the cross-sectional area of their MaA fibers. This study involved collecting naturally spun draglines from spiders walking on horizontal and vertical surfaces. This difference may seem negligible, but a fall with a preexisting length of safety line that is attached at the level from which the fall occurs generates an impact energy that is half that produced by a fall from a distance above the attachment point equal to the length of the preexisting safety line (as we will see in detail below). So dragline failure is much more likely in a fall from above the dragline attachment. The researchers collected a number of vertical draglines (produced by spiders climbing a vertical surface) and horizontal draglines from spiders, and they measured the stress-strain properties of the threads, which showed a very interesting difference in material properties. The draglines spun by vertically climbing spiders always had the same material properties; that is, their stress-strain curves all overlapped tightly, regardless of the spider's size. The draglines spun by spiders walking on a horizontal surface, however, showed a broad range of properties, as indicated by the variable range in fig. 10.4B. They then selected individual stress-strain curves for draglines, produced by a single 1 g adult spider, that had identical stress-strain curves, one from a vertical dragline and one from a horizontal dragline. When they converted the stress-strain curves into force-strain curves, they discovered that the vertical dragline required almost exactly twice the force to break as the horizontal dragline. That is, its cross-sectional area was twice that of the silk thread produced by the same spider when walking on a horizontal surface. Thus, even when a spider is creating silk fibers with identical material properties, they are capable of adjusting the cross-sectional area of the silk strands to adjust the breaking force of the silk, at least by a factor of two. This ability almost certainly can be attributed to the activities of the valve and valve muscles in the terminal duct of the spinneret.

Interestingly, the vertical dragline broke at a force that was about three times greater than the weight of the spider, and this measurement forms the basis of a safety factor that we can use to evaluate the effectiveness of the design of spider draglines to function as safety lines. This safety factor of 3 is actually a static safety factor, because the applied load, as determined by the spider's weight, is essentially the load required to support a motionless spider. That is, the static safety factor, (S_{BW}), is:

$$S_{BW} = F_{max}/Mg, \tag{10.1}$$

where F_{max} is the silk thread's breaking force, M is spider mass, and g is the acceleration of gravity. But the dynamic-loading forces that occur in a fall will always be greater than the spider's weight in static loading, and we need to understand the energetics of dynamic loading in safety lines. Christine Ortlepp (Ortlepp and Gosline, 2008) analyzed the scaling of the safety factor in the mechanics of spider draglines, and produced the following analysis.

Understanding the mechanics of dynamic loading in a safety line requires understanding the ability of the safety line to absorb the kinetic energy of motion that a falling object gains as it falls, and assuming a fall with negligible loss due to aerodynamic drag, the kinetic energy gained in the fall will equal the loss of gravitational potential energy, Mgh, where h is the height of the fall. The height of the fall includes two stages. Initially, free fall will proceed until the existing safety line is straightened to its resting length, x_0, and the fall will continue as the safety line is stretched by Δx to its failure point, as shown in fig. 10.10A. A perfect system, with a dynamic safety factor of 1, would require that the energy needed to break the safety line is just equal to the gravitational energy released in the fall. Energy to break is measured as the area under a force-extension curve to failure, W_{break}, which for a material with a linear force-extension curve to failure is:

$$W_{break} = \frac{1}{2}F_{max}\Delta x_{max}. \tag{10.2}$$

Where Δx_{max} is the extension of the thread beyond its resting length x_0. W_{break} must be just equal to the gravitational energy released in the fall, W_g, and this is calculated as:

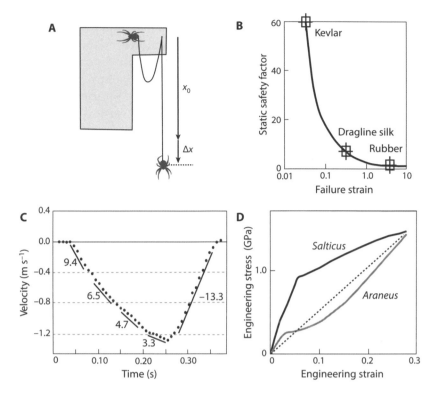

Figure 10.10. Functional design of spider draglines. **Panel A** shows the scenario considered for a spider falling from the height at which a preexisting length of safety line is
attached to the substrate. The energetics of this fall are described in the text. x_0, resting
length of safety line; Δx, length of stretch in safety line. **Panel B** shows the results of the
analysis of safety-line performance for the drop illustrated in panel A. The plot is based
on eq. (10.5), which gives the relationship between the failure strain of the safety line
and its static safety factor. For rigid safety lines, such as Kevlar ropes with failure strains
of 0.03, the breaking load for the rope must be 60 times greater than the weight of
the falling body. For spider silks with failure strains of about 0.30, the breaking load
for the safety line must be eight times the weight of the falling spider. **Panel C** shows
the time course of a vertical drop by a juvenile spider escaping from its web when a
predatory insect approached. The data show the velocity of descent, which involves
the production of new silk as the spider falls below its safety-line attachment at the
hub of the web. Numbers to the side of line segments indicate the approximate acceleration at different points in the fall. Note that at a time of about 0.25 s, the velocity
trajectory shifts rapidly as the spider applies its internal friction brakes to the silk that is
being drawn from its silk gland. The friction brakes cause decelerations of 1.6 to 2.4 g,
which determines the loads that are actually applied to a safety line during a typical
fall by a spider. **Panel D** shows stress-strain curves for MaA dragline silks from the orb
weaver *Araneus* and from the jumping spider *Salticus*. The r-shaped stress-strain curve
of the *Salticus* safety line indicates that its toughness is roughly 50% greater than the
Araneus safety line, in spite of their having the same strength and extensibility. Eng.,
engineering.

$$W_g = Mgx_0 + Mg\Delta x_{max}. \tag{10.3}$$

Thus, failure occurs when $W_{break} = W_g$, as:

$$\frac{1}{2}F_{max}\Delta x_{max} = Mg + Mg\Delta x_{max}. \tag{10.4}$$

By applying eq. (10.1) to eq. (10.4) and rearranging, it is possible to calculate the magnitude of the static safety factor that is required for a safety line made for a material with a failure strain ($\varepsilon_{max} = \Delta x_{max}/x_0$) to just survive a fall as:

$$S_{BW} = (2/\varepsilon_{max}) + 2. \tag{10.5}$$

Fig. 10.10B shows the basic form of this relationship and highlights some familiar examples for comparison with the behavior of spider MaA silk. Kevlar is very strong but it has a very small failure strain of about 0.03, and this means that a safety line made from Kevlar would need a static safety factor of about 60 to function safely. A spider dragline, with an extensibility of about 0.3, will require a static safety factor of about 9 to survive a free fall with a fixed length of silk. Stretchy rubber bands require even lower safety factors that approach the lower limit of 2. A similar relationship can be generated for calculating the static safety factor required for a spider climbing a vertical wall to survive a fall from a distance x_0 above the dragline's attachment point, as:

$$S_{BW} = (4/\varepsilon_{max}) + 2, \tag{10.6}$$

which predicts that a vertically climbing spider with a dragline extensibility of 0.3 would require a static safety factor for its dragline of 15. It is very interesting, therefore, that this analysis indicates that the spider silk produced by the vertically climbing spider described above had a static safety factor of only 3. This is not even close to the strength level required for a static safety line. This spider's dragline would certainly fail in a fall from its position above the height of its dragline's attachment. Indeed, it would fail in a fall from the height of the dragline's attachment.

Does this mean that spider draglines can't function as safety lines? The answer is that in some cases they cannot, but in reality spiders don't use fixed lengths of dragline that loop down as shown in fig. 10.10A. Rather, when they fall they spool out new silk, and then apply their internal friction brakes to slow their fall gradually. This prevents the abrupt rise in force that would occur when a fixed-length safety line becomes taught after the initial free fall. The effect of spooling out new silk is equivalent to making the dragline much more extensible than its material extensibility. In effect, spooling new silk makes the Δx_{max} in the equations above much larger than the silk's extension owing to the tensile material that the safety line is made from.

Christine Ortlepp (Ortlepp and Gosline, 2008) analyzed this situation by investigating how dragline silk strength varied with spider size, and she discovered some interesting things about spider growth. She reared spiders through their full life cycle, which in the case of *Araneus diadematus* involves seven developmental instars of gradually increasing size, with body-mass values that span from 4×10^{-7} to 6×10^{-4} kg, about a 1500-fold range of body mass. The tensile strength and extensibility of MaA silks obtained from these spiders remained constant with animal size, but interestingly the cross-sectional area of the draglines did not increase apace with the increase in body mass. That is, the cross-sectional area, A, of draglines varied with body mass, M, as:

$$A = 5.86 \times 10^{-9} M^{0.74}, \tag{10.7}$$

The important feature of this tight scaling relationship (which explains 90% of the variation in A) is the mass exponent of 0.74. Because this exponent is less than one, the dragline cross-sectional areas do not increase in line with body mass as these spiders grow, and this means that the breaking force and hence the static safety factor must fall as spiders get larger. Indeed, Ortlepp's analysis of breaking force versus body mass indicated that the breaking force, F_{max}, and safety factor, S_{BW}, scaled with body mass as:

$$F_{max} = 11.2 M^{0.79}, \text{ and } S_{BW} = 1.14 M^{-0.21}, \tag{10.8}$$

relationships that explain 96% of the variation in both F_{max} and S_{BW}. Again, the exponent of 0.79 for the breaking force, and the negative exponent for the scaling of the safety factor mean that the safety factor drops as animal mass increases. Thus, in the case of $A.$ $diadematus$ the range of safety factors starts at a high level of 20+ for 2 mg second instars to a low level of 5 in 1 g gravid females.

As mentioned above, the reason that spiders can make their safety lines smaller than needed to survive a free fall with a fixed length of fiber is that they are able to produce more silk as they fall. Investigation of the dynamics of this process provides additional insights into the silk-spinning process. Fig. 10.10C shows the results of high-speed-video analysis of a spider's fall. The animal in question was a small second instar $A.$ $diadematus$, with a body mass about 2.5 mg, that dropped suddenly from its position in the center of its web. This behavior of dropping from a web is elicited by the presence of foraging wasps or hover flies, and the trigger is the sound produced by the wing-beat frequency of the predator. The behavior can also be elicited by making a loud humming noise near to the web.

This behavior can be recorded by high-speed video, and the data in fig. 10.10C show a plot of fall velocity as a function of time. Note that the entire behavior lasts about 0.3 s, from the start of the fall to the stop at the bottom. The way to recognize the end of the fall is to look for the point in time when the velocity returns to zero. Actually the spider bounces up and down once it has stopped falling because there is elastic energy stored in the dragline from the stretch caused by the tension that develops as the spider puts its brakes on. The diagram traces the fall as a series of black dots, which indicate the velocity of the spider over each time interval of the video record, which was 8 ms. The numbers next to the dots indicate the calculated acceleration over several time intervals adjacent to the number. Thus, at the very start of the fall, the dots form a steep segment, and the calculated acceleration is 9.4 m s^{-2}. This means that the spider is in true free fall. It is accelerating at essentially 1 "g", and this means that it is spooling out new silk with very low draw force. As time proceeds, however, the acceleration declines, likely because this is a very small animal, and it is experiencing a rapidly rising drag force. Despite having a mass of only a few milligrams, the legs and body of the animal generate considerable drag, which causes the observed deceleration. If the animal continued in its starting state, where apparently it is applying a very small friction force onto its dragline, it is likely that the animal would reach terminal velocity at some point in time not long after the point where the velocity trace makes a sharp deviation.

This is the point where the spider decides to put on the brakes, and one can see that over a time interval bounded by three data points the velocity switches from a gradually rising downward velocity to a rapidly falling velocity. This is when the spider puts on its internal friction brake, and over the course of about 16 ms the acceleration of 3 m s^{-2} shifts to a deceleration of -13 m s^{-2}, which brings the spider to a halt in 0.12 s. Presumably, this rapid transition is created by the cocontraction of the opposing muscles in the spinneret that span the valve in the silk production line, and muscle activation over this short time interval provides a plausible explanation for the rapid onset of deceleration. During this period following application of the friction brake, we can estimate the actual force that is applied to the dragline. There are three forces to consider. The gravitational force, F_g, will be Mg and will always act to increase velocity. There are two forces that will reduce velocity: (1) an external aerodynamic drag force, F_d, which is proportional to the effective area of the animal's body and velocity squared, and (2) the tension in the silk created by the spider's internal friction system, F_f. At the moment when the spider turns on the brakes, there is a shift of acceleration from $+0.3$ g to -1.3 g, or a total shift of 1.6 g. Since force equals mass $\times$ acceleration, $F = Ma$, the braking force generated at the time that the animal turns on its brakes will be 1.6 body weights. As time proceeds, the deceleration remains constant at about -13.3 m s^{-2}, but the velocity falls. This means that the aerodynamic drag force will also fall, which means that the internal frictional force will rise to values that approach 2.3 g, or 2.3 body weights.

This kind of behavior has been observed for falls of both adult and juvenile spiders, and the results are remarkably consistent. The main difference is that in falls by adults, with higher body mass, the initial free fall occurs at accelerations that remain essentially constant at values near 9 m s^{-2} because drag forces are much less important for the larger animals. Regardless of the spider's size, in vertical free falls spiders consistently turn on their internal friction brakes at fall velocities of -1.0 to -1.4 m s^{-1}, values that are about four times greater than the velocities that spiders will allow us to forcibly draw silk from their tethered bodies. In addition, the friction forces that they apply, once they turn on their brakes, consistently cause decelerations in the range of 13–15 m s^{-2}, and this means that adult animals also will generate internal friction forces that approach 2.4 body weights.

Clearly, this spider has developed a very effective silk-production system that is matched to one of its key functions; namely, a dynamic safety-line system that can minimize the use of silk in the creation of the safety lines. It achieves this by making its draglines much thinner than they would need to be if they were required to function as fixed-length safety lines. With this information it is now possible to define a dynamic safety factor, one that quantifies how close the dragline strength is relative to the loading that it will experience in a fall. As mentioned previously, the static safety factor of the dragline from the upper range of body mass, a 1 g gravid female, was about 5 body weights, and the dragline made by this animal apparently experiences only 2.4 body weights of force in a fall, so the dynamic safety factor is about 2. In addition, we saw that spiders climbing vertically can double the cross-sectional area of their draglines, and hence this dynamic safety factor should also apply to falls from above the attachment point of the silk. Indeed, given that MaA silks are even stronger when deformed at high strain rate, it seems that the *Araneus* safety line may be somewhat overdesigned. Perhaps the size and strength of *Araneus* MaA silk fibers is more about building webs than making safety lines.

Our analysis of MaA silks has focused on orb-weaving spiders, where the majority of MaA silk is used in making and maintaining orb webs, but not all spiders make webs. Indeed, the jumping spiders (family Salticidae) are wandering spiders that make continuous draglines, but do not make prey-capture structures. Ortlepp also documented the biomechanics of safety lines produced by the jumping spider *Salticus scenicus* (Ortlepp and Gosline, 2008). Fig. 10.10D shows stress-strain curves for draglines produced by *Araneus* and *Salticus*, and it is clear that they have similar strengths and extensibilities, but there is also a clear difference in the shapes of the stress-strain curves. The *Araneus* curve quite closely follows the linear relationship assumed in the calculation of the static safety factor (as indicated by the dotted line), and hence the linear model gives an energy to break that is essentially identical to the area under the actual stress-strain curve. The r-shaped stress-strain curve of the *Salticus* dragline has a longer initial slope and a much higher yield stress, properties that likely arise from a higher β-sheet-crystal content in this silk. The result of this stress-strain curve is that, for a comparable strength and extensibility, the *Salticus* dragline requires about 45% more energy to break than that predicted by the linear model. At present nothing is known about the dynamic properties of the *Salticus* dragline material. However, Ortlepp's scaling studies clearly demonstrate that *Salticus* draglines are thinner than *Araneus* draglines at equivalent body weights (ca. 20 mg), with cross-sectional areas that are one-fifth of those of the *Araneus* draglines. This means that the static safety factor for adult *Salticus* is about 2. If jumping spiders show similar deceleration profiles in their jumps and falls, it is likely that the dynamic safety factor for *Salticus* draglines is very close to 1, but this remains to be confirmed. If it is correct, however, it means that this jumping spider has evolved a dragline-silk-forming system that is perfectly matched to the mechanics of the silk-tensioning systems within its spinneret.

This concludes our discussion of spider silks, and the two case studies presented here for tensile materials should provide a useful basis for understanding the design principles that govern the assembly, properties, and functions of tensile biomaterials. There are a number of additional tensile materials in living organisms, including two very important polysaccharide-based materials, cellulose and chitin. Cellulose is the primary reinforcing fiber in composite materials that form the cell walls of plants, and it is a truly remarkable material. The literature on this material is immense, and readers should consult books such as Julian Vincent's *Structural Biomaterials* (2012) for entry into this literature. Chitin is a similar, high-performance fiber that is used extensively in the phylum Arthropoda (insects, spiders, crustaceans), where it exists in the exoskeletal cuticles of these animals. We will consider arthropod cuticles in section IV, on rigid materials. In addition to these two polysaccharide-based fibers, animals make other protein-based tensile fibers. One very interesting example is byssal threads, which are produced by intertidal molluscs, primarily mussels, to attach these animals to rocks in the wave-swept intertidal zone of the marine environment. Herbert Waite has made major contributions to our understanding of the molecular and biomechanical design of byssal threads, and a paper by Harrington and Waite (2007) will provide interested readers with access to this literature. In addition, tensile fibers produced by hagfish when disturbed to reinforce their defensive slimes are made from intermediate filament proteins (Fudge et al., 2003), and this study provides important insights into the molecular design of intermediate filaments as tensile structures in living cells. A review by Herrmann et al. (2007) will provide access into the cellular nanomechanics of intermediate filaments. Intermediate filaments also play a dominant role in the molecular design of the rigid composite material called keratin, which is found in vertebrate structures such as hair, hoof, horn, and the epidermal layer of skin.

THE MECHANICAL DESIGN
OF RIGID MATERIALS

Chapter 11

The Structural Design of Bone

Chapter 8 introduced the basic concepts of composite-materials design and provided basic information on the properties of some artificial rigid composites. For example, fig. 8.1B showed calculated values for the stiffness of parallel and series composites made from Kevlar fibers and an epoxy matrix, with varying levels of fiber and matrix components. We then considered the roles of laminate organization and of discontinuous-fiber-composite structures. We will now apply these concepts to develop an understanding of the structural design of rigid biomaterials, and we will start by looking at the design of vertebrate bone. Probably the most useful general reference on the structure and mechanics of bone is the monograph by John Currey, *Bones, Structure and Mechanics* (2002), and I recommend this book to all readers interested in bone biomechanics. In addition, a recent review by Wang and Gupta (2011) provides excellent updates on the details of the nanoscale bone composite and fracture mechanisms.

Bone is a material that largely defines the skeletal systems of vertebrates, as it is present throughout their fossil record, which starts about 500 million years ago (Ma). The calcified hard tissues of the early Cambrian invertebrates, about 550 Ma, were based on calcium carbonate crystals of calcite or aragonite, and structural materials based on calcium carbonate are found in many invertebrate phyla today. Bone, however, is based on a composite structure that forms between the tensile fiber, collagen, and mineral crystals made from calcium phosphate. It is normally accepted that this mineral phase is found primarily in the form of crystalline hydroxyapatite, with a composition of $Ca_{10}(PO_4)_6(OH)_2$, although it may also exist in other, similar, crystal forms, and in amorphous forms as well. Ruben and Bennett (1987) present an interesting proposal for why early vertebrates made the transition from calcium carbonate materials to calcium phosphate materials to create their rigid skeletons. They base this proposal on the observation that living vertebrates show a remarkable ability to accumulate blood and tissue lactate following periods of intense activity, much more than the nonvertebrate chordate amphioxus and the majority of invertebrates. Lactate accumulation causes the acidification of tissue fluids, and these authors tested the hypothesis that this acid will cause much greater dissolution of calcium carbonate than of hydroxyapatite. They carried out a simple in vitro experiment by comparing the dissolution of calcite and hydroxyapatite crystals in fish serum at pH 7.8 (normal) and pH 7.1 (acidified), and they observed a 50% increase in the serum calcium concentration in the acidified calcite serum, and only about a 3% increase in the hydroxyapatite serum. In vivo experiments with live fish that were implanted with calcite and hydroxyapatite crystals followed the dissolution of the implanted minerals over 35 days, in fish that remained undisturbed for the full period as well as in fish that were forced to exercise intensely for 90 s every day. Both test conditions created roughly order-of-magnitude differences in the loss of mass in calcite and hydroxyapatite systems, with the exercised fish losing about eight times more calcite mass than hydroxyapatite mass. So it seems plausible that the presence of calcium phosphate crystals in the bone composite evolved to provide active early vertebrates with the ability to produce an acid-resistant, rigid skeletal material.

John Currey mentions in his book another, very interesting, possibility as an anecdote, which is based on a discussion that he had with another bone biomechanicist. The notion was that evolution produced calcium phosphate–based bone, as opposed to the calcium carbonate–based shells of invertebrate animals, because "apatite is very reluctant to form large crystals" (Currey, 2002, p. 108). The logic of this statement is that the mineral crystals in bone exist as nanoscale structures, and they are, therefore, able to form within individual collagen fibrils to create a nanocomposite in bone. Calcium carbonate, on the other hand, easily forms very large crystals. Indeed, mollusks and other invertebrates that produce mineralized materials must actively control crystal formation in shell production, and typically they do this by surrounding actively growing crystals with layers of protein that very effectively limit these crystals to the micrometer scale. We will see shortly that in bone the structural design of collagen fibrils provides a nanoscale internal space, the gap zone of the collagen fibril, where hydroxyapatite crystal nucleation first occurs. Later, when apatite crystals form on the fibril surface, they remain as platelets with lengths on the order of 50–100 nm. This leads us directly to the key question, how do collagen and hydroxyapatite combine structurally to produce a useful rigid skeletal material for highly active vertebrate animals?

1. THE STRUCTURAL HIERARCHY OF BONE. One way to begin thinking about this question is to consider how hydroxyapatite crystals interact with collagen fibers. Our thoughts about composites that were introduced in chapter 8 included the tacit assumption that the high stiffness and strength came from the stiff and strong tensile fibers, and the matrix functioned to tie the flexible, slender fibers into a single object to provide rigidity in compression and bending. A "composite" made of collagen fibers and mineral crystals seems somewhat backward, as the fiber is the low-stiffness component. However, the composite that emerges when tendon-like collagen-fiber arrays are impregnated with hydroxyapatite crystals becomes a structure in which the mineral component infiltrates the collagen fibrils and fills spaces between fibrils to form the high-stiffness elements in the composite.

The analysis of the structure and development of bone will start off where we ended in our discussion of the structural hierarchy and assembly of collagen fibers in tendons, presented in chapter 9. Bone formation begins with arrays of collagen fibrils that are infiltrated by and then encased in crystalline hydroxyapatite, and to some degree amorphous calcium phosphate minerals. This process is initiated by the nucleation of hydroxyapatite crystals at the nanometer scale within collagen fibrils. That is, studies of the earliest stages of calcification in bone, and also in the calcification of avian tendons, reveal that the first location for the formation of hydroxyapatite nanocrystals is within the gap zones that are present in collagen fibrils. Interestingly, Landis and Silver (2009) describe the presence of amino-acid clusters that line the surface of this internal space, which is about 1.5 nm wide and 40 nm long, that create focal areas of high positive and negative charge density and are thought to bind calcium (Ca^{2+}) and phosphate (PO_4^{3-}) ions at a number of sites in the gap zone. In addition, Sebastian Kalamajski and colleagues (2009) have discovered that asporin, a small leucine-rich protein similar to decorin but lacking a proteoglycan chain, competes with decorin for binding onto collagen fibrils in developing bone tissues. Asporin stimulates bone mineralization, likely because it binds to sites near the gap zone in mineralizing collagen fibrils, and it contains a run of 14 aspartic acid residues plus some additional glutamic acid residues, amino acids whose side chains carry a negative charge. These charged groups are thought to attract positively charged Ca^{2+} ions to the site of mineral nucleation in the gap zone.

Thus, local high concentrations of calcium and phosphate ions in the gap zone initiate crystal formation, which leads to the creation of thin platelet crystals of hydroxyapatite, with dimensions that initially fit into the gap zone. This process has been extensively studied in embryonic bone and in calcified turkey tendons. Peter Fratzl and his colleagues (1991) used X-ray diffraction techniques and observed that initial crystals are about 1 nm thick, which is about the thickness of the crystalline unit cell of hydroxyapatite; the crystals are about 30–40 nm long, and hence are somewhat shorter than the collagen gap zone. However, as mineral continues to be deposited the platelets thicken to about 3 nm and they essentially fill the gap zone in the collagen-fibril crystalline structure. Interestingly, the crystals become thicker than the width of the gap zone, which means that the growing crystals actually compress the collagen molecules adjacent to the gaps. Another important feature that is apparent from X-ray studies is that the c axis of the hydroxyapatite crystal is always oriented essentially parallel to the long axis of the collagen fibrils, a feature that likely reflects the specific orientation of the nucleation sites on the fibrils.

Landis and his colleagues (1993) used high-voltage electron-microscopic tomography and graphic image reconstruction to visualize other aspects of crystal growth in the early stages of mineral deposition in calcified tendon. They showed that as the gap zones fill with hydroxyapatite, the mineral crystals extend beyond the gap zones and continue as thin crystals that run between the collagen molecules within the adjacent overlap zones of the fibrils. High-resolution images indicate that the crystals that form in the initial stages of mineralization exist as platelets with a thickness of up to about 4 nm and with lengths that can be in excess of 200 nm. That is, the length of platelets in the direction parallel to the collagen fibrils' long axis can be much longer than the 40 nm length of the gap zone where mineral crystal growth is initiated. The width of the platelets varies, and platelets may aggregate laterally, but the platelet's width is typically less than its length, and the platelet's thickness is much smaller than either the length or the width. In addition, crystal platelets form on the outer surfaces of the fibrils, presumably filling spaces that were previously filled with proteoglycan gel (decorin) that established the matrix of the collagen fiber's composite structure, and in this process the collagen fiber is converted into a mineral-reinforced composite that has a dramatically increased stiffness.

This process of mineralization is illustrated in fig. 11.1. Fig. 11.1A shows the banded appearance of a collagen fibril, as described previously in fig. 9.4B, and in fig. 11.1 we see how the process of bone mineralization proceeds, starting from the basic structure of the collagen fibril. Fig. 11.1B shows the collagen-molecule overlap pattern that exists within the collagen fibril, and in this diagram the arrows that represent the tropocollagen molecules are shown in light gray so that it is possible to distinguish the protein from the black mineral crystals that form within the various levels of the collagen-fibril structure. Note that this diagram shows black bands in a number of gap zones that occur between the C- and N-terminal ends of the tropocollagen molecules. These are the small mineral platelets that form in the initial stages of mineralization, and they do not fill the gap zones fully. Electron-microscopic studies show that the initial nucleation starts at focal sites and spreads throughout the fibrils with time. Fig. 11.1C shows a later stage in this process when all gap zones contain mineral crystals, and where the crystals are expanded in length and thickness, overfilling this space and squeezing

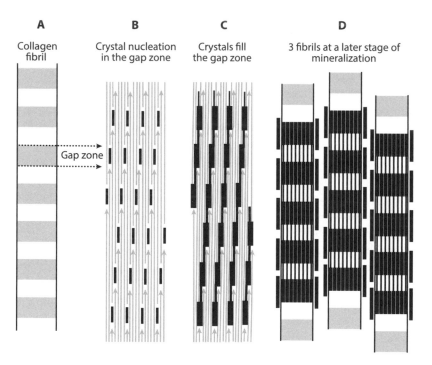

Figure 11.1. The nanoscale structural organization of bone. The figure shows the pattern of mineral deposition in the collagen fibrils of developing bone. **Panel A** shows the collagen-molecule overlap pattern in collagen fibrils, which form the scaffold for the formation of the rigid composite of bone. **Panel B** shows the location of initial hydroxyapatite-crystal formation within the gap zones of the collagen-molecule overlap structure. **Panel C** shows a later stage in the mineralization of a collagen fibril, in which the gap zones are overfilled by mineral crystals that are thicker than the gap zone. In addition, mineral crystals extend between collagen molecules in the overlap zones. **Panel D** illustrates the pattern of mineralization within and between three collagen fibrils. Note that mineral platelets are found both within and on the surface of collagen fibrils.

adjacent collagen molecules. In addition, thin crystals can be seen between the gap zones. Fig. 11.1D shows a late stage in the mineralization process, where extensive mineral deposition has taken place on the surface of several mineralized fibrils. Again, the crystals are typically platelets, and they first appear at the location of the gap zones in the collagen fibril, although in some forms of bone they may fuse both laterally and along the length of the fibril.

The nanoscale composite structure of bone discussed above is organized at larger scales into several types of composite structures. Indeed, there are several general patterns of large-scale composite structure that are seen in different parts of a bony skeleton. Of course, the mineralized fibrils in fig. 11.1D are parts of collagen fibers, with diameters of similar magnitude to the collagen fibers seen in tendon. That is, the collagen fibers have diameters of about 1–2 μm, but the packing of these fibers comes in several different structural forms at larger scales. Before we look at the structural organization of collagen fibers in bone, however, it will be useful to look at the general organization of a whole bone in fig. 11.2A, where we can identify the central shaft of this typical femoral bone that exists as a thick-walled, hollow cylinder made from the material we call bone. The plane that transects the bone near its center reveals a circular cross-section that is made from compact bone, but in most cases the cross-sectional shape is more complex so that the bone can deal with the complex bending and compression loads that act on it in its normal function, as well as provide insertion surfaces for muscles that attach directly to the bone's surface.

The term "compact bone" means that the walls of the shaft are a single, solid structure. At the ends of long bones, however, the external layer of compact bone thins considerably, and the bone structure takes on more complex shapes. Here we see a system of internal struts that act to support a thin layer of compact bone. The struts, called trabeculae, are thin columns of bone that form a meshwork that is known as cancellous bone. The cancellous bone reinforces the thin compact bone at the surface, and this combination produces a low-weight system that stabilizes the shape of the bone ends, where large compressive forces are transmitted between bones in weight bearing and during locomotion.

Returning to the structural organization of the mineralized collagen fibrils and fibers in the bone material, the simplest structural form is called "woven bone," and it is characterized by essentially disorganized arrays of wavy mineralized collagen fibers, where the orientation of fiber bundles varies in three dimensions, as shown in fig. 11.2B. This form of bone is present extensively at the early stages of skeletal development, and its most important characteristic is that it can be laid down very rapidly. Currey tells us that woven bone can be produced at rates of 4 μm per day, and in the case of fetal development and in bone fracture repair it is produced at much higher rates. However, its random collagen organization leads to reduced material properties, and generally woven bone is replaced with more organized forms of bone as development proceeds.

Fig. 11.2C shows a diagram of a more structured type of bone architecture, called "lamellar bone." Here we see a block of bone that contains five discrete layers, or lamellae, and the individual lamellae are typically 1–5 μm thick, as illustrated in this diagram. The collagen fibers and their associated mineral crystals are arranged in essentially parallel arrays, but the local direction of the collagen fibers can vary somewhat at different positions within a layer. Typically, the average orientation varies from layer

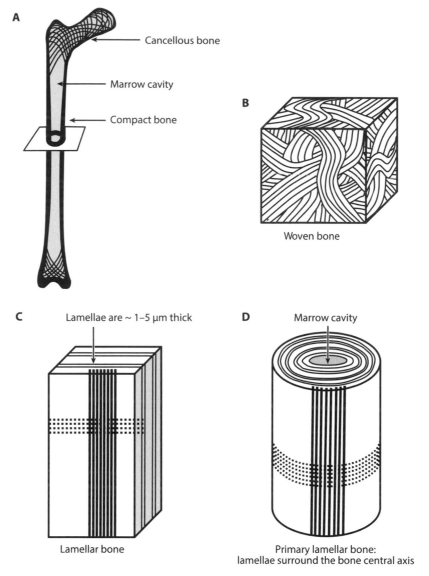

A

Cancellous bone

Marrow cavity

Compact bone

B

Woven bone

C Lamellae are ~ 1–5 μm thick

Lamellar bone

D Marrow cavity

Primary lamellar bone:
lamellae surround the bone central axis

Figure 11.2. Bone structure. **Panel A** shows the basic organization of a typical whole bone, in this case a femur. It shows the thick-walled, hollow shaft at the center of the bone, and the thin-walled bone ends that are supported by thin bone columns of cancellous bone. **Panels B and C** show patterns of mineralized-collagen-fiber organization found in bone. Panel B shows the random arrangement of mineralized collagen in woven bone that is dominant in early bone development. Panel C shows the organization of lamellar bone that typically replaces woven bone as bones mature. Here we see parallel arrays of collagen fibers that are organized into discrete layers, and are most often organized as shown here with alternating layers running at right angles to each other. Most frequently in long bones these alternating layers are organized as shown in **panel D**, where we see that the layers are arranged parallel and perpendicularly to the long axis of a bone, with the parallel layers being thicker than the perpendicular layers. This organization is characteristic of primary lamellar bone in humans and many animals.

to layer, but in most instances it alternates as shown here, with orientations from essentially parallel to the bone's long axis to essentially parallel to its transverse axis. In addition, it is important to note that the layers that run parallel to the bone's long axis are typically thicker than the transverse layers. This gives lamellar bone a plywood-like organization, but one that is biased toward longitudinal stiffness. In other bones the collagen-fiber directions may alternate from left- to right-handed helical trajectories, although this is less common.

The general form of lamellar bone is organized into larger structures in several different forms. Frequently the first stage formed when woven bone is replaced is a structural arrangement called primary lamellar bone, and in many animals this means that the bone lamellae are arranged concentrically around the central axis of a long bone; thus, the lamellae surround the marrow cavity of the bone. Fig. 11.2D illustrates this type of structural organization. This organization is called primary lamellar bone because it is created in the initial formation of a bone's structure. This aspect of bone growth is illustrated in fig. 11.3A, which shows two cross-sections of a cylindrical bone at its midshaft that represent the transition of a small-diameter bone in a young animal to a larger-diameter bone with growth. These bone sections are drawn with a white center because bones are hollow structures, where the internal compartment, the marrow chamber, is filled with a variety of cells, including blood-cell-forming tissues and, in some animals, fat deposits.

The smaller-diameter bone cross-section on the left has two sets of arrows, on its inner and outer surfaces, and these arrows are meant to indicate the direction of growth changes that occur as the young bone is enlarged during the growth of the animal. The outer black arrows indicate the tendency for growing bones to increase in outer diameter and thickness as a person or an animal gets larger, a change that is required to deal with increasing body mass. This growth occurs by the synthesis of new bone on the outer surface of the bone. The inner set of gray arrows indicate the other process that takes place as an animal increases in size; namely, that the bone material at the interior of the bone is removed to maintain a marrow cavity that also increases in size as the animal grows. This process of bone growth through the addition of new bone tissue and/or removal of old bone tissue is called "bone modeling," as indicated in fig. 11.3A.

The reason for the increase in the marrow cavity is twofold. First, it provides space for the cellular and physiological processes that occur in the marrow, and second it eliminates the mass of bone that is located near the bone's center, where it would play a minimal role in providing structural support when a bone is loaded in bending. The important issue for the current discussion, however, is that new bone is added on the outside and old bone is removed from the inside. Much of this growth will coincide with the transition from woven bone in very young animals to primary lamellar bone with growth, because the new bone on the outside is laid down with a lamellar structure. The larger bone section in fig. 11.3A is therefore shown with a series of concentric white lines that indicate the presence of primary lamellar bone in early bone growth, with lamellae that are concentric with the center of the bone's shaft.

The growth of primary lamellar bone, however, is much slower than the growth of woven bone, typically at a rate of about 1 μm per day, and in large animals it is common to see a more complex architecture that combines woven bone and lamellar bone in alternating layers, which is called "fibrolamellar" or "laminar" bone. However, in this structure the lamellar bone does not form circular lamellae that are concentric with the bone axis; rather it forms smaller cylindrical structures called "osteons" that have diameters of the order of 100–200 μm, as shown in fig. 11.3B. As with primary lamellar bone, the lamellae are 1–5 μm thick, and collagen-fiber orientation varies between layers, but the lamellae are not concentric with the bone axis; rather, they form around a local axis, and there is a central space called the vascular canal at this axis that contains

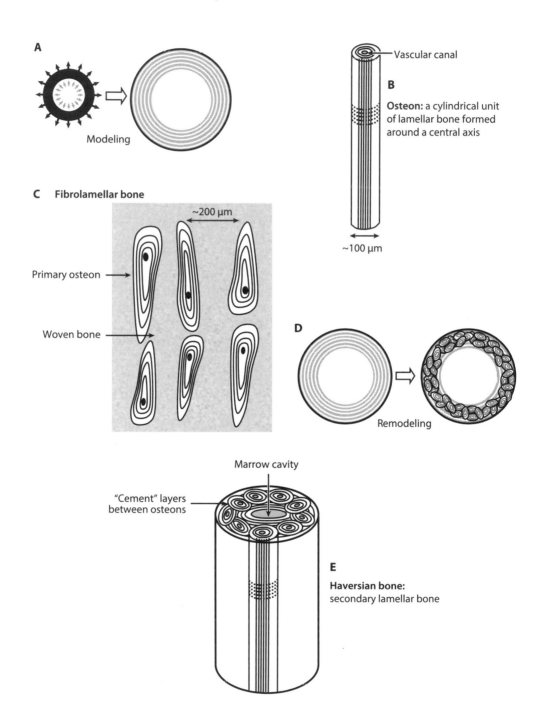

A

Modeling

B

Vascular canal

Osteon: a cylindrical unit of lamellar bone formed around a central axis

~200 μm

~100 μm

C   Fibrolamellar bone

Primary osteon

Woven bone

D

Remodeling

Marrow cavity

"Cement" layers between osteons

E

Haversian bone: secondary lamellar bone

the blood vessels that nourish this cylindrical structural unit. The osteon drawn here is shown as a simple cylinder, but they may also have a flattened cross-sectional shape.

Osteons are found in two types of bone, fibrolamellar bone and Haversian bone. Fig. 11.3C shows the basic organization of fibrolamellar bone. Here we see repeating layers about 200 μm wide that contain a layer of woven bone that alternates with a layer of flattened osteons. The osteons in fibrolamellar bone are called primary osteons because they are formed during the initial synthesis of the growing bone. Bone with this structure is found in cows, and much of the data available on bone material properties are from tests on bovine fibrolamellar bone.

Human bones and the bones of many smaller animals follow a different pattern of development. The initial growth involves the transition of woven bone into primary lamellar bone, as shown in fig. 11.3A, and during the rapid growth stages of early development this primary lamellar bone structure is largely maintained. As growth slows, however, bones respond to changes in an individual's lifestyle and activity levels, and

Figure 11.3. (*left*) Bone modeling and remodeling. **Panel A** illustrates the process of bone modeling that occurs in the growth and development of bone. This diagram shows bone cross-sections in early growth where bone size increases dramatically. The arrows around the outside of the small cross-section on the left indicate the addition of new primary bone on the outside of the existing bone. The arrows on the inside indicate the removal of bone material from the inside of the bone. Together, these processes lead to the creation of the larger bone seen on the right. The process illustrated here would likely involve the replacement of woven bone with primary lamellar bone, which is indicated by the concentric white lines in the larger cross-section on the right. **Panel B** shows the basic organization of an osteon, which is typically a cylindrical unit of bone that is made from lamellar bone arranged around a central vascular space that contains blood vessels that nourish the bone tissue. Osteons are usually about 100–200 μm in diameter, and they occur in two basic types of mature bone structure, fibrolamellar bone and Haversian bone. **Panel C** shows the structure of the primary bone formed during the growth (modeling) of bones in a large animal such as a cow or horse. It is called fibrolamellar bone, and it combines flattened primary osteons in parallel arrays that are separated by layers of woven bone. **Panel D** illustrates the process of bone remodeling, which occurs continuously through the life of an animal. In this example the bone as a whole does not change in size, but structures within the thickness of the bone are removed and replaced by new bone. The bone cross-section on the left contains primary lamellar bone, and the cross-section on the right shows the organization of Haversian bone that is formed when the primary lamellar bone is replaced by cylindrical structures, called osteons, in the process of bone remodeling. **Panel E** shows the organization of Haversian bone, which is formed in the remodeling of primary bone in both human and animal bones. Here we see more clearly the organization of the secondary osteons that are formed in bone remodeling. Secondary osteons are typically more cylindrical than the primary osteons of fibrolamellar bone, and in addition they are separated from each other by cement lines, which are not present around primary osteons.

internal changes in the structure of a bone occur. In these situations, "bone remodeling" takes place, where volumes of older bone are excavated from primary lamellar bone, and the spaces created are filled with cylindrical units that are called secondary osteons, because these osteons replace previously existing lamellar bone, as shown in fig. 11.3D. This process leads to the formation of "Haversian bone," where much of the primary lamellar bone is replaced by secondary osteons, and remodeling can continue over long periods of time. Interestingly, individual osteons can overlap with their neighbors because the process of remodeling in Haversian bone is an ongoing process (fig. 11.3E), and older osteons can be removed and replaced by newer ones. Indeed, studies suggest that intense loading of Haversian bone in live animals increases remodeling activities, which likely is an adaptive process for repairing stress-induced defects in a bone's structure. Essentially all long bones in mature humans will have been extensively remodeled into Haversian bone. Remodeling also occurs with time in large animals with fibrolamellar bone, and portions of their bones are converted into Haversian bone as described above.

It is important to note that the osteons in fig. 11.3D are shown as much larger structures than they are in reality. A typical long-bone cross-section in an adult human would have a wall thickness of the order of 5 mm, and the diameter of an osteon is only about 0.2 mm. So the diameter of the osteons drawn here is about an order of magnitude larger than it would be in a real bone. Note also in fig. 11.3E that the secondary osteons in Haversian bone are separated by a cement layer at the interface between osteons that is about 5 μm thick. This feature is not present in the primary osteons of fibrolamellar bone.

2. BONE CELLS. In our discussion of bone growth and remodeling we mentioned in passing that there are cells, blood vessels, and tissues present in and around bones, and it is important to note some of the central roles that the living components of bone play in bone biology. An individual bone is lined on its outside with a fibrous layer called the periosteum and on the inside by an endosteum, and these layers contains cells that play key roles in bone synthesis, modeling, and remodeling. Specifically, they contain cells, derived from osteoprogenitor cells, called osteoblasts, the cells that produce the collagenous matrix that forms the scaffold for bone mineralization and control the mineralization process by concentrating calcium and phosphate within in the forming bone matrix.

These cells become trapped inside the mineralized bone that they have created, and they are then called osteocytes. Osteocytes reside in small cavities (lacunae) in the mineralized bone. In woven bone the lacunae are essentially spheres, and in primary and secondary lamellar bone the cavities are oblate spheroids (flattened spheres), with their long axis being about three times the length of their short axis. Cavities of this sort have the potential to act as stress concentrators that might play a role in fracture processes of bone. We saw in chapter 5 (eqs. [5.8]) that the stress-concentration factor for a sphere is 3, and the stress-concentration factor for the 3:1 oblate spheroid cavity with its short axis parallel to the direction of the applied stress would be about 7. So these "defects" in the bone structure could contribute to fracture and fatigue processes in the bone material. However, the lacunae in primary lamellar bone are oriented with their short axis directed radially, and this means that the tensile or compressive stresses arising from the bending of a bone, which run longitudinally in the bone, will therefore run parallel to the flattened surface of the lacunae. In this case the stress concentration calculated with eqs. (5.8) would be about 1.6. Thus, the oblate lacunae in lamellar bone create smaller stress concentrations than those made by the spherical lacunae in woven bone. This orientation of the osteocyte lacunae is thus very likely an adaptation to reduce stress concentrations in bone for loading directions that receive the highest stress levels in an animal's normal activities. It may also help to explain why bone samples loaded in the radial direction are much weaker than those loaded in the longitudinal direction.

The cell lacunae are placed quite far apart in lamellar bone, but the cells are able to communicate with their neighbors through thin channels in the bone matrix called canaliculi, which are about 0.2μm in diameter. Cell processes extend into the canaliculi, and they connect to processes from neighboring cells through gap junctions at their ends. This allows molecular flow between the cytoplasm of adjacent cells. In addition, the canaliculi diameters are larger than those of the cell processes, and this provides a route for extracellular fluid movement within bone.

Another type of cell that plays a major role in bone biology is the osteoclast. These are cells that function to destroy bone. Osteoclasts are not derived from osteoprogenitor cells; rather, they are produced by the fusion of macrophages in the circulatory system into multinuclear giant cells that attach to bone and dissolve it, largely by the active secretion of acid. This accounts for the removal of bone at the interior surface of a bone during the initial growth stages (fig. 11.3A) and the excavation old bone during bone remodeling and the formation of new Haversian systems in the middle of preexisting primary bone (fig. 11.3D).

An interesting feature of secondary lamellar bone is the presence of a cement layer, at the outer surface of osteons in Haversian bone, that is about 5μm thick. This layer may play a role in the composite organization of Haversian bone. Cement layers are not, however, found between the primary osteons of fibrolamellar bone. In theory, it is possible that this layer plays a strengthening role in Haversian bone mechanics, acting as a softer, perhaps energy-absorbing "matrix," or, conversely, it may provide a structural defect leading to crack propagation. Quantitative electron micrographic studies by John Skedros and his colleagues (2005) indicate that the cement line is not deficient in its mineral content, although nanoindentation studies by Timothy Montalbano (Montalbano and Feng, 2011) indicate that the cement layer is about 30% less stiff than the lamellae of the osteon. It is possible, therefore, that the cement layer provides a somewhat weaker interface, and this may explain the observation that fatigue cracks seen in Haversian bone frequently are seen to divert around osteons, a process that may indeed contribute to bone's strength and fatigue resistance. With this knowledge of structural organization in a bone, we can now analyze the mechanics of the material that we call bone.

3. THE COMPOSITE STRUCTURE OF BONE MATERIAL. Fig. 11.4A provides stress-strain data for typical compact-bone material, along with data for the two most prominent components of bone's composite structure, collagen and hydroxyapatite. The line labeled "tendon" shows the initial portion of a tendon's stress-strain curve, and the linear slope following the toe zone has a value of 1.5 GPa in this graph. For future reference, however, we will take the stiffness of individual collagen fibrils, at about

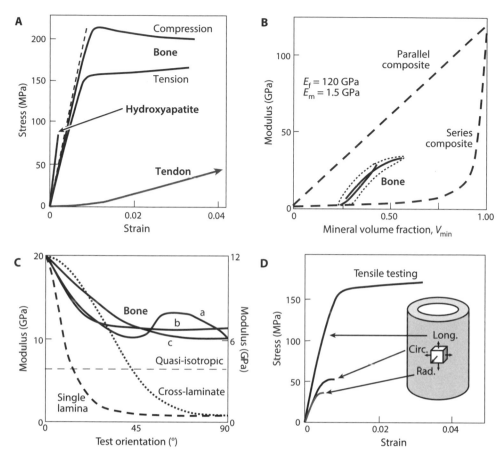

Figure 11.4. Stiffness of bone. **Panel A** shows typical stress-strain curves for bone material in tension and compression, along with stress-strain curves for hydroxyapatite and collagen, the materials that form the composite structure of bone. **Panel B** shows the effect of mineral volume fraction, V_{min}, on the stiffness of bone. The two *solid lines surrounded by a dotted line* indicate the full range of stiffness found in bone materials seen in nature. The *dashed lines* indicate the calculated stiffness of continuous-fiber composites made from a fiber and a matrix having stiffness equal to those of hydroxyapatite and collagen respectively, in both series and parallel arrangements. E_f, modulus of the fibers; E_m, modulus of the matrix. **Panel C** shows the orientation dependence of stiffness in bone, compared with calculations for the orientation dependence of stiffness in some typical fiber-reinforced-composite designs. The composite designs show greatly reduced stiffness at test angles above 45° relative to the fiber axis, but bone does not because it is reinforced by mineral platelets that reinforce in both the longitudinal and the circumferential directions in a bone. See the text for an explanation of lines *a, b,* and *c.* **Panel D** shows typical stress-strain curves for bone samples with different orientations relative to the whole bone. Longitudinal samples (long.), which run parallel to the long axis of a bone, show the highest levels of stiffness, strength, and extensibility. These properties are partially reduced in the circumferential (circ.) direction and dramatically reduced in the radial (rad.) direction, because the bone microstructure described above is designed to resist the large longitudinal stresses that are created in the normal activities of animals.

3 GPa, as a basis for understanding the nanoscale composite of bone. If any adjustment is made for the stiffness of the collagen in bone it would necessarily be upward, because the crowding of the matrix space between fibrils in a collagen fiber by the close packing of mineral platelets and by the intrusion of mineral crystals within the fibril internal crystalline structure could only stiffen the collagen. However, at this point in time, no one has been able to establish unequivocally the stiffness of collagen in bone.

The stiffness of hydroxyapatite is difficult to measure experimentally because its crystals are quite small. A theoretical analysis of the hydroxyapatite crystal structure, using bond-energy theory, estimates a stiffness of about 140 GPa, which likely places an upper bound on its modulus. Various bone researchers have estimated hydroxyapatite stiffness as 100–120 GPa, and we will assume a value of 120 GPa as a reasonable compromise. The line labeled "hydroxyapatite" in fig. 11.4A has a slope of 120 GPa. The failure point of this mineral is estimated at 80 MPa, which is the stress it would develop at a strain of about 0.0007. This is probably a reasonable estimate for the strength of largish crystals of this material. However, as we saw with glass fibers in fig. 5.3, strength will increase with smaller crystal size, and so the hydroxyapatite nanocrystals in bone may be stronger than this.

Now look at the stress-strain curves for the bone composite. These data are typical of primary lamellar bone and fibrolamellar bone from cows and humans, although there can be considerable variation in these values. Note first that the initial linear region of the stress-strain curve indicates a stiffness of about 20 GPa, with the modulus in compression and tension being essentially the same. In fact, the actual values vary with the type of bone microstructure, but we will not deal with this issue at this stage in the discussion. The other characteristic feature of the bone stress-strain curve is the dramatic yield seen at strains of about 0.007–0.01, although the decline in modulus starts at strains of the order of 0.005. Following the yield, the stress may either increase gradually, as is the case in tensile tests, or decline gradually, as is the case in compression tests. Failure typically occurs at strains of 0.02–0.04, although in some types of bone and different test orientations the failure strain can be somewhat higher or significantly lower.

The primary feature of bone structure and composition that dominates the variation in material properties is the mineral content of the bone tissue, and fig. 11.4B shows a plot of bone modulus versus the volume fraction of mineral in the bone composite. This graph is similar to the one shown in fig. 8.1B, which shows the predicted stiffness for a Kevlar/epoxy composite, where the fiber stiffness is 130 GPa and the matrix stiffness is 3 GPa, and values are given for various levels of the fiber volume fraction of parallel and series composites made from these two materials. We see a similar comparison in fig. 11.4B, where the dashed lines indicate the properties of parallel and series composites made with continuous fibers, based on the stiffness of hydroxyapatite for the fibers and of collagen for the matrix. At the bottom of this graph, in the mineral volume-fraction range of 0.25–0.5, there are two solid lines surrounded by a dotted line that indicate the full range of stiffness levels observed in materials that can be called bone. The properties of bovine and human bone would fall roughly in the center of this range, and the extreme values indicate how changes in mineral volume fraction, V_{min}, provide a mechanism to adapt the material properties of bone for various mechanical functions. We will deal with this kind of adaptive design of bone shortly, but at this point it is important to note that the stiffness levels of bone in general seem to diverge dramatically from those predicted by the equations we developed previously for continuous-fiber composites, and only at the lowest level of V_{min} do we see the bone properties come close to those of the predicted series composite. There are two reasons why bone properties do not fit the theoretical predictions. First, as we have just seen, the reinforcing stiff component of bone is not present in the form of continuous fibers; rather, the hydroxyapatite crystals are typically found in the form of thin platelets, with aspect ratios of the order of 40. Bone is a discontinuous-fiber composite! In addition, we saw in fig. 8.2 that composite properties can be strongly influenced by the orientation of reinforcing fibers in a variety of ways.

Fig.11.4C shows how calculation of reinforcement efficiency, as modeled in eqs. (8.3)–(8.5), can provide an indication of how fiber orientation can control mechanical properties. In this figure, however, the theoretical curves have been adjusted to match the measured mechanical properties of bone tested in a direction parallel to the bone's long axis. Thus, in this graph the stiffness shown at a test orientation of 0° would be the stiffness of a bone sample oriented parallel to the long axis of the bone that it was isolated from. Note that the curves in this graph, both the theoretical curves shown as dotted and dashed lines and the experimental curves shown as solid lines, start at a stiffness level of 20 GPa. The theoretical curves indicate the effect of fiber orientation in several formats, including a single lamina, balanced cross-laminates, and multilamellar quasi-isotropic systems. The solid lines for bone are derived from several studies. Line "a" is from tests using ultrasound velocity data to provide stiffness values, and line "b" is from standard tensile testing at low strain rates. Line "a" testing was carried out at essentially 10° increments of test orientation, and as a result it provides reasonable detail about variation with angle. Line "b" testing was carried out at four orientations, 0°, 33°, 66°, and 90°, and hence it does not provide as much detail. However, it is clear that the pattern of modulus versus test orientation for bone does not come close to matching the predictions of the models presented here. In particular, we see that bone stiffness does not fall to the low values predicted by the theoretical curves at test orientations above about 45°. At these orientations, bone stiffness remains at a level that is 50%–60% of the 0° level, while the theoretical curves for single lamina and balanced cross-laminates fall rapidly to the level of the series composite. The quasi-isotropic line shows constant stiffness, as advertised, but at a lower level than that seen at any orientation with bone.

Why do bone properties fail to match the theory? Or, more appropriately, why do the theoretical curves fail to match the properties of bone? The short answer is that the theory is based on the presence of reinforcing fibers, and bone is not reinforced by long, thin, rodlike fibers. In fact, the predominant form of the mineral crystals is platelets, and this difference can easily explain the discrepancies between the theory and the properties of real bone. Currey and colleagues (1994) investigated this issue by testing the orientation dependence of modulus in "bone" samples obtained from the dentine in narwhal tusk. Dentine is not exactly bone, but it is essentially the equivalent of bone, and the advantage of dentine as a source of test material is that it was possible to cut thin slices of it in which the collagen fibers are all parallel with each other and also all run in the plane of the test sample. Thus, these authors could test samples with uniform collagen-fiber orientation at fiber angles that varied continuously from 0° to 90°. That is, each test sample is equivalent to a single lamina, and if reinforced by fibers the samples should have properties like those for the single-lamina data in fig.

11.4C. The results, however, were very different. Currey et al. measured the stiffness and tensile strength, and they observed that the stiffness varied with fiber orientation as indicated by line "c" in fig. 11.4C. Note that the 0° stiffness of narwhal dentine is about 12 GPa, not 20 GPa as seen typically in bone, so the curve for dentine was scaled to match the relative scale of the curves for bone, and this allows us to make direct comparisons with the other curves in this graph. For clarity, however, the modulus scale on the right side of the graph indicates the actual stiffness values for dentine. Here again we see that the stiffness of narwhal dentine does not fall dramatically at high fiber angles like the single-lamina curve does; rather, it remains at about half the 0° level at angles above 45°.

The authors then attempted to model these properties, and they discovered that they could do this only if they used a model that allowed them to define the reinforcing elements in the dentine composite as low-aspect-ratio platelets. Their best fit for the shape of the mineral crystals was with platelets having a longitudinal aspect ratio (length ÷ thickness) of 25 and a transverse aspect ratio (width ÷ thickness) of 10. In this analysis, the length direction was the platelet length parallel to the collagen fibers, and the width was the platelet width perpendicular to the collagen. If the platelet thickness is 3 nm, the length and width of the platelet would be of the order of 75 and 30 nm respectively. These are reasonable descriptions of the hydroxyapatite crystals observed in bone, considering that the gap in the collagen fibril is approximately 40 nm long.

Why, then, do platelets lead to such different properties than those found with long rodlike fibers? The answer is quite straightforward. A fiber has a single long axis, and it only reinforces to any significant extent along this axis. A platelet, however, has two long axes, and thus it can reinforce along two of its three axes simultaneously. With this idea in mind, we can think about how the stiffness and strength should vary along the three principal axes of a bone. In this case, we will define the primary axis of the nanoscale bone material as being parallel to the predominant orientation of the collagen fibers in the substructures of the macroscopic bone. To distinguish the bone material from the bone as a macroscopic object, we will call the object a femur, as a typical kind of bone. Now we need to consider the orientation of the collagen fibers relative to the structure of a femur, as illustrated in fig. 11.2A. In both primary lamellar bone (fig. 11.2D) and Haversian bone (fig. 11.3E), the predominant orientation of collagen fibers is parallel to the long axis of the femur. So in these bone structures the long platelet length would be predominantly oriented parallel to the femur's long axis, and the shorter platelet length would be oriented predominantly in the circumferential direction around the shaft of the femur. This leaves the thickness axis of the platelets to be oriented predominately in the radial direction. The consequence of these patterns of orientation for the collagen fibers and their associated mineral platelets is that there should be quite different degrees of composite reinforcement in different directions in the femur. The organization of these axes relative to the structure of the femur are confusing, so they are illustrated in fig. 11.4D where we see a cylindrical section of a femur at its midshaft position, with a small cube of material showing these three principal axes.

In fibrolamellar bone things are a bit more complex, because the primary osteons form around blood channels that run in two dimensions. That is, they run both vertically and circumferentially in the femur's structure, and thus it seems plausible that the collagen fibers, and the long axis of the crystal platelets, will be oriented vertically in some areas and circumferentially in other areas. The flattened nature of the osteons will place the two longer axes of the platelets predominantly in the longitudinal and circumferential axes of the femur, again leaving the platelet's thickness oriented primarily in the radial axis. In addition, the osteon layers of fibrolamellar bone are separated by layers of woven bone, where there is no dominant collagen-fiber orientation, and hence platelet orientation should be quite random in these layers. Thus, the organized composite structures in the osteon layers are continuous in the vertical and circumferential directions, but they are discontinuous in the radial direction.

The consequence of these patterns of crystal-platelet organization in a bone is that, in general, the axis of greatest reinforcement should be in the direction of the long axis of a bone such as the femur described above, and it should be somewhat less in the circumferential direction. The stiffness should be lowest in the radial direction, which is the principal orientation of the platelets' thickness axis in most cases. Fig. 11.4D shows the general pattern of tensile mechanical properties for compact bone in the longitudinal, circumferential, and radial directions, and the differences in properties are very clear. The longitudinal stiffness, strength, and extension to failure are all significantly higher than those in the circumferential and radial directions. Circumferential strength is somewhat less than half that in the longitudinal direction, the failure strain is reduced as well, but the initial stiffness is only slightly reduced. The decline in properties is greater in the radial direction; the initial stiffness remains the same, but the strength and the extensibility are both further reduced relative to those in the circumferential direction.

4. NANOSCALE COMPOSITE MODELS FOR BONE. Before we look in more detail at the various mechanical properties of bones it will be useful to try to connect the general features of bone properties with the nanoscale structure described in the previous section. Studies using synchrotron X-ray diffraction on bone material that is being tested mechanically have provided some interesting insights into the structural origins of material properties. One study, by Gupta and colleagues in Peter Fratzl's lab (Gupta et al., 2006), quantified the levels of strain that develop in the bone material, in the collagen fibers and in the hydroxyapatite crystals that make up the smallest level of the bone's structural hierarchy. This was achieved by the simultaneous use of small-angle X-ray scattering to follow collagen-fibril strain and large-angle X-ray diffraction to measure hydroxyapatite-crystal strain in a bone sample that was strained in a mechanical test system. The experiments were carried out with thin, 50–100 μm thickness, slices of bovine fibrolamellar bone obtained from young cows, and the samples were maintained in phosphate-buffered saline (PBS) solution during the experiment to maintain tissue hydration. A more recent study by Hoo and colleagues (2011) carried out similar tests on larger-scale samples of bovine fibrolamellar bone, which at 4 mm thickness contained all the large-scale structural features of the whole bone, and the moist tissue samples were washed with water prior to testing, and tested in air rather than in PBS solution. The reason for pointing out these differences in tissue handling comes from the observations discussed previously in the section on collagen, where it has been observed that the cross-sectional area of tendons doubles when they are held in PBS, likely because of interactions between the buffer and the interfibrillar- and interfiber-matrix materials.

The results of these two studies are quite different. The Gupta study observed that over the initial linear portion of the stress-strain curve the ratios between the tissue strain, the collagen-fibril strain, and the mineral strain were 12:5:2. That is, the collagen-fibril strain was about 2.5 times greater than the mineral strain, and the bone-tissue strain was about 2.5 times greater than the collagen-fibril strain. Thus the tissue strain was about 6 times greater than the mineral strain. This inequality of strains within a bone's structure means that when the bone tissue as a whole is strained there will be large shear strains within and between its fibrillar, fiber, and lamellar components. The Hoo study saw something quite different. The strain ratios were 10:8:7, which means that the nanoscale strains of collagen fibrils and mineral platelets were not all that different from the macroscopic material strain. That is, the collagen-fibril strain was only 14% greater than the mineral strain, and the whole-bone strain was only 25% greater than the collagen-fibril strain. Thus, the tissue strain was only about 40% greater than the mineral strain. It is also important to note that these strain levels are averaged over all the various microstructures in the bone samples that were used, so the Hoo data should be more relevant to the properties of whole bone.

Shear deformation can exist at several levels of bone's structural hierarchy. Fig. 11.5 provides diagrams to illustrate the sites of shear within and between the mineralized collagen fibrils. Both panels show an interface between two adjacent collagen fibrils that are separated by an interfibrillar space. These diagrams are essentially blowups of fig. 11.1D, which shows the disposition of mineral crystals within and between collagen fibrils in bone. The black bars indicate the mineral platelets, and the gray arrows show the individual tropocollagen molecules, in their characteristic overlap pattern. Both panels indicate locations where molecular shear likely occurs when bone is strained. First, note the chevron-shaped dotted lines in fig. 11.5A spanning the four collagen molecules that separate the short (ca. 55 nm long) mineral platelets that are located in and extend a short distance beyond the ~40 nm gap zones. These chevrons indicate the shear of the collagen molecules sandwiched between platelets as the mineralized fibril as a whole is stretched. Because the gap zones in collagen fibrils are all arranged in perfect register, producing the 67 nm banding pattern of the fibrils, the mineral platelets are all arranged in register. As a consequence, there are no overlaps between these short platelets. Thus, the continuation of the collagen molecules above and below the layer of platelets will be loaded primarily in tension because they extend beyond the shear zones adjacent to the mineral crystals. The short platelets and their arrangement in the collagen lattice will result in the formation of a series-composite structure that will dramatically limit the stiffness of the collagen fibril. In addition, the low aspect ratio for these short platelets will limit reinforcement efficiency of the parallel-composite structure that exists at the level of the gap zone and platelets. These features, if they exist in bone, will have a large impact on bone stiffness.

In fig. 11.5B, large—approximately 110 nm long—mineral crystals are shown filling two consecutive collagen gap zones plus the intervening overlap zone. In addition, they are shown as being offset longitudinally by one gap zone relative to their neighbors on each side, a structure analyzed in the Gupta study. This strongly overlapping structure means that much, if not all, of the collagen is deformed in shear, as indicated by the fact that the chevrons cover much of the space between the mineral platelets. Thus, the tensile loading of the collagen will play a minor role in the composite behavior of this

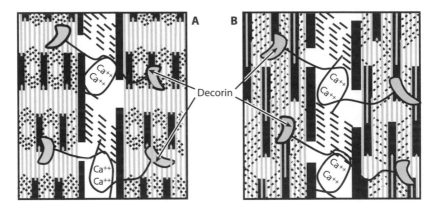

Figure 11.5. Nanomechanical models of bone. **Panel A** shows a model for the nanoscale structure within and between mineralized collagen fibrils in bone. This model assumes that mineral platelets are relatively short, extending only slightly beyond the length of the gap zones of the collagen-fibril lattice structure. Because the gap zones are fixed in perfect register in the fibril, these short platelets do not overlap, and the discontinuous-fiber reinforcement of this structure will be interrupted by zones where all stress passes through intervening segments of unreinforced collagen. Molecular shear is indicated by the *chevron-shaped dotted lines;* shear at the interfibrillar interface is indicated by the *slanted dashed lines* between the two fibrils. **Panel B** shows a model for the nanoscale structure of bone in which the mineral platelets span two gap zones and an intervening overlap zone of the collagen lattice. As a consequence, the mineral platelets overlap and will form a competent discontinuous-fiber composite, which can explain the mechanical properties of bone. Here, the *dotted chevrons* indicating shear cover much of the space between the mineral platelets. Both diagrams show mineral platelets in the interfibrillar space, which contribute to bone stiffness, as well as an organic matrix made from decorin, which may mediate the interactions between the mineralized fibrils in bone.

model, and the fibril stiffness will be determined by the shear transfer between collagen molecules and the mineral platelets. The reason that the shear loading is important is that the collagen molecules within the fibrils, which are assumed to play the role of a matrix in the mineral-reinforced-composite structure of bone, do not form an isotropic matrix phase like the epoxy in a typical fiber-reinforced composite. The shear modulus of epoxy is about 40% of its tensile modulus, making it about 1 GPa. Collagen provides a fibrillar matrix in which the overlapping parallel arrangement of the triple-helical collagen molecules creates very stiff structures in tension, while the relatively much weaker lateral bonding between adjacent tropocollagen molecules offers much less resistance to the shearing of tropocollagen molecules with their overlapping neighbors in a fibril.

A study by Yang and colleagues (2008) provides us with key information on the relative magnitude of the tensile and shear stiffness of collagen fibrils that can help us to understand the bone nanocomposite. These authors used atomic-force microscopy to measure the bending stiffness of isolated collagen fibrils from tendon, and their results indicate that dry collagen fibrils have a tensile modulus in the range of 1–4 GPa, as we discussed in chapter 9, on tendon collagen. We will assume that the collagen fibrils in bone have a tensile modulus of about 3 GPa, as we did for tendon. The shear modulus of the dry fibrils, however, was found to be 33 MPa. That is, the shear modulus of dry collagen is only about 1% of its tensile modulus. Yang et al. also measured the properties of collagen fibrils when immersed in PBS, and the change in properties was dramatic, with a drop in tensile modulus of around 20-fold to about 100 MPa, and a tenfold drop in shear modulus to about 3 MPa. We know from studies on hydrated collagen in tendon, and for hydrated isolated collagen fibrils (but not hydrated in PBS), that the whole-tendon stiffness is about 1.5 GPa and isolated fibrils are even stiffer at about 3–4 GPa. So we are likely seeing in this experiment another manifestation of the PBS effect that dramatically alters the properties of collagen-containing structures, and we should be wary of any results obtained when PBS is used. However, if we take the shear modulus for dry collagen as a starting point, we can begin to think about modeling the mechanics of the mineralized collagen fibrils in bone.

The discontinuous-fiber-composite model we used in our analysis of tendon collagen may allow us to better understand this. The input parameters we need for a model of bone are the tensile modulus of the collagen fibrils (3 GPa) and hydroxyapatite crystals (120 GPa), the shear modulus of collagen (33 MPa), the volume fraction of mineral (0.36, a value typical of human and bovine bone), and the aspect ratio of the mineral platelets. The aspect ratio of the platelets is likely going to be a key variable in this analysis, but other studies suggest that it falls in the range of 20–50. Looking to our two models for the nanoscale structure of bone in fig. 11.5, we see two possibilities, in which the platelets range in length from about 55 to about 110 nm. If we assume a platelet thickness of 2.5 nm, which is a plausible first guess, and we use a 55 nm platelet length as indicated in fig. 11.5A, the aspect ratio is 22, and the predicted modulus for the mineralized fibril is about 8 GPa. This is well below the actual stiffness of bone. In addition, we must consider the consequence of the nonoverlapping pattern of platelets created by the structural organization of gap zones in the collagen fibril that imposes a series structure on this nanocomposite design. The discontinuous-fiber-composite model assumes a random disposition of overlapping fibers, and the nonoverlapping structure of this model will further reduce the stiffness from the level calculated by the model. A reasonable guess for a corrected stiffness is a value in the range of 3–4 GPa. Thus, the structure in fig. 11.5A is not able to explain the observed properties of bone fibrils, or for that matter whole bone. Significant overlap of mineral crystals in the fibrils is crucial for the development of bone properties.

However, if the platelets are 110 nm long, as suggested in the Hoo study, and 2.5 nm thick the aspect ratio is 44, and if the adjacent platelets overlap as shown in fig. 11.5B, we get a stiffness of 18 GPa, which is quite close to the actual stiffness of whole bone. Fig. 11.6 shows a plot of the fiber-matrix shear stress (dotted line) and the fiber tensile stress (solid line) predicted for the model with the 110 nm long platelets by use of eqs. (8.9) and (8.10), and it is clear that this structure has low reinforcement efficiency (38%) because the transfer of stress from the matrix to the fiber continues along essentially the full length of the fiber. It may be useful to compare this figure with fig. 8.3B, which shows a similar plot for a Kevlar-epoxy composite in which the Kevlar fibers have a somewhat lower aspect ratio of 30 and a fiber-matrix modulus ratio of

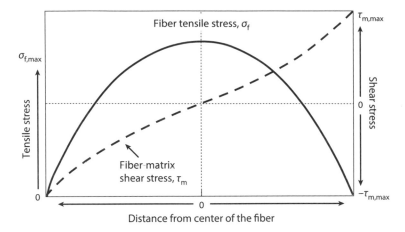

Figure 11.6. A discontinuous-fiber model. The calculated distribution of reinforcing-fiber tensile stress and fiber-matrix shear stress in the discontinuous-fiber model shown in fig. 11.5B. Comparison with fig. 8.3B shows that the tensile and shear stresses do not reach a plateau along the fiber, owing to lower reinforcement efficiency in the bone composite that arises from the very low shear stiffness of the collagen matrix.

130. Together, these parameters bring the reinforcement efficiency to 90% because of the higher matrix modulus and a higher fiber volume fraction (70%). It is clear in fig. 8.3B that the shear-stress transfer is complete at about a third of the length of the Kevlar fiber. The low shear modulus of collagen in the bone composite brings its modulus ratio up to 3600, and with a lower fiber volume fraction, this brings the reinforcement efficiency down to a much lower level. However, this level of reinforcement very nearly explains the actual stiffness of whole bone.

Since the X-ray diffraction data of Hoo et al. (2011) indicate that there is about a 40% difference between the reinforcing-mineral stiffness and the whole-bone stiffness, the current version of this model would predict a whole-bone modulus of about 11 GPa, which is significantly lower than the actual stiffness of whole bone but is reasonably close. However, small adjustments to the model in fig. 11.5B could explain the tensile stiffness of whole bone in the linear portion of its stress-strain curve. For example, reducing platelet thickness and/or increasing platelet length will increase aspect ratio, and modest changes in these parameters could bring the predicted modulus values into line with the measured values for whole-bone stiffness. In fact, the 1993 study by Landis and colleagues of initial mineral deposition in turkey tendon, discussed previously, indicated that the mineral platelets could be longer than the 110 nm indicated in fig. 11.5B. So, aspect ratios could be higher, and calculated stiffness could be increased to the range seen in bone.

However, in making these calculations we have assumed that the shear modulus of the overlapping tropocollagen molecules in the fibril is 33 MPa, which is the measured value for dry collagen fibrils. With the actual value for the shear modulus of the collagen in hydrated bone fibrils not known, we can only speculate that the intrusion of large-scale mineral platelets may squeeze the tropocollagen lattice in a way that increases the effective shear stiffness of the collagen "matrix" of this nanocomposite. Remember that the adjacent collagen molecules in the lattice structure of a collagen fibril are offset relative to their neighbors, and this means that a mineral crystal that extends beyond the length of the gap zone must proceed between tightly packed collagen molecules in the overlap zones that extend above and below the gap zone. The mineral crystals, which may be 2.5 times thicker than the center-to-center spacing of the tightly packed collagen molecules in the fibril, will thus strongly squeeze the collagen molecules together, particularly in the overlap zones. This squeeze could increase the effective shear modulus of the collagen layers in the composite and may also increase the tensile stiffness of the collagen to levels that bring the calculated stiffness close to the observed properties of bone. Only time will tell if these assumptions are correct.

Returning to the two panels of fig. 11.5, at the center of each diagram we see a close view of the interface between the two collagen fibrils, and we see that there are mineral platelets on the outer surface of the fibrils. In addition, other structures span the space between the fibrils, and these structures have been drawn to represent decorin, the proteoglycan that is known to form a soft gel that functions as a "matrix" and binds the collagen fibrils together into the collagen fibers of tendons and ligaments. The arrows identify the decorin core protein of this complex, which is shown bound to the surface of the collagen fibrils. A single proteoglycan chain is attached to each core protein and spans the interfibrillar space, where it interacts and binds with a proteoglycan chain from another decorin molecule on the adjacent fibril, through Ca^{2+} ions that create ionic bonds with the negatively charged sulfate groups on the proteoglycans. Note also the slanted dashed lines in the interfibrillar space. These are meant to indicate the shearing at this interface between the collagen fibrils. Shear at this level is thought to explain, in part, the discrepancy between the tissue strain and the collagen-fibril strain, which in the case of wet bone without PBS, is about 25%. This calculation, however, applies only to the initial, totally reversible, portion of the bone stress-strain curve where the observed stiffness is at the levels we have calculated above. Bone is known to start to yield, likely by the failure of this interfibrillar interface, at stress levels that are well below the yield point that we have identified at stresses of about 150 MPa and strains of 0.007–0.01, where we see a dramatic drop in whole-bone stiffness. In reality, the initial linear slope of the stress-strain curve begins to decline subtly at stresses below 100 MPa and strains of 0.005–0.007. Interestingly, these are the strains the Gupta synchrotron X-ray study identified as the level where mineral and collagen strains began to fall, as tissue strain continued to rise, and this likely indicates the failure of the interfibrillar structures described above and the large-scale sliding of adjacent fibrils relative to their neighbors. The Hoo study observed similar behavior in their PBS-free tissue sample, but they saw an even more dramatic shift, with declines in mineral and collagen strain occurring at stresses of 60–80 MPa and strains of about 0.002. Thus, the stress and strain levels where the bone nanostructure begins to pull apart are well below the macroscopic yield point seen in whole-bone samples. However, the test protocols used in both of these X-ray diffraction studies involved strain rates of about 3×10^{-4} s^{-1}, and at this low strain rate the effective stiffness of the bone samples may be somewhat lower than the stiffness levels typically attributed to bone, as given in fig. 11.4A. The discussion of whole-bone properties that follows will provide information about strain-rate effects. Regardless of these details, it is clear that bone's nanostructure begins to "fail" at very modest stress and strain, and this process is likely to have an impact on the way that bone is used in animal skeletons.

Another study by Gupta and others from the Fratzl lab (Gupta et al., 2007) used a mechanical testing protocol to characterize the energetics of deformation and failure at the interface between fibrils at large strain. That is, the study involved careful analysis of the mechanics of bone specimens taken to strains beyond the yield point. Their X-ray studies, described above, had shown that the pattern of strain in the components of bone shifted dramatically at the yield point, and that for postyield deformation the strain of both the mineral and the fibril phases decreased and then remained at a constant level until fracture. Thus, in postyield extension the rise in strain is due entirely to shearing at the interfibrillar interface, and similar interfaces between collagen fibers and interfaces at larger scales. Because this shear occurs with no increase in stress it must be associated with some sort of shear flow or viscoelastic process.

Gupta et al. probed the energetics of the postyield strain by measuring the change in the stress required to maintain the shear flows at different temperatures and strain rates. Typically, lower strain rate and higher temperature decreased the postyield stress, and vice versa, for this viscoelastic process. The analysis allowed these authors to infer two parameters that characterized the molecular processes that occurred during postyield strain: the activation enthalpy, ΔH_{act}, for the molecular movements responsible for the postyield strain, and the activation volume required for these molecular movements. Their analysis indicated an activation energy of about 1 electron volt, which is approximately 100 kJ mole^{-1}, and an activation volume of approximately 1 nm^3. Thus, the process of shear flow in the interfibrillar matrix during postyield deformation involves a nanometer volume that certainly fits within the matrix space between fibrils, and an activation energy that is considerably larger than the energy to break hydrogen bonds but about a third of the energy required to break a single covalent bond. Their conclusion was that the best explanation for the energetics of shear flow at this and perhaps other interfaces involves the breaking of a small number of ionic bonds, likely those between Ca^{2+} ions and the sulfated proteoglycan chains that are attached to decorin, or similar molecules, and the creation a soft solid in the interfibrillar space between the mineralized collagen fibrils. Thus, the yield process is likely controlled by the shearing of the interfibrillar matrix. Since this is basically a viscous process, shearing at this interface provides a mechanism for dissipating the strain energy that is released when a flaw in the bone's microstructure grows. Thus, the interfibrillar interface provides a potentially important crack-stopping mechanism. Similar mechanisms may also exist at interfaces on larger size scales, such as those between collagen fibers, between lamellar structures, and between Haversian structures such as the cement line. Indeed, bone's complex microstructure offers many opportunities for controlling the propagation of cracks, and hence for controlling fracture.

One of the most interesting aspects of the bone stress-strain curve is the large-scale strain that occurs after the yield. Following this yield, bone strain continues, with a modest rise in stress in tensile tests and with a modest fall in stress in compression tests, to failure strains that typically range from 0.02 to 0.04, as shown in fig. 11.4A. This rather large postyield strain is a key feature of bone design, as it provides a variety of mechanisms that toughen bone and limits the probability of a catastrophic fracture. Thus, we expect that bone microstructure provides a number of mechanisms for limiting the growth of cracks. This is indeed the case, and observation of fractured surfaces and of defects created in the postyield deformation of unfractured specimens gives us important clues to how bone hierarchical micro-structure contributes to its toughness.

In our general discussion of composite mechanics in chapter 8 and in fig. 8.6 we identified processes that allow composite materials to limit crack growth and increase fracture toughness. Several of these mechanisms play important roles in the structural design of bone. Once a tensile-test specimen of bone has been strained beyond its yield point, there are obvious changes that occur in the surface structure of the specimen that are easily visible to the unaided eye. In certain zones, most obviously near a precut notch of a specimen designed to measure the fracture toughness of bone, the bone surface color changes because of the accumulation of micrometer-scale defects that are generated in advance of and around the notch tip. This change in surface appear-ance arises from the scattering of light from the many stable microcracks that develop during the postyield elongation of a test sample. Microcracking caused by the presence of weak interfaces, called the Cook-Gordon effect, is though to provide a mechanism that can prevent catastrophic crack propagation in composite materials. As described in chapter 8, when a composite contains a relatively weak interface between a rein-forcing fiber and its surrounding matrix, this weak interface can open up in advance of a crack that is propagating across the reinforcing fiber, creating an opening with a large radius of curvature that lowers local stress concentration and dissipates the strain energy released at the crack tip, and hence halts crack growth. The numerous inter-faces that exist in bone microstructure provide many opportunities for this kind of mechanism to prevent catastrophic bone failure from the growth of initial cracks dur-ing the yield process. The creation of additional microcracks continues through the postyield extension, and the toughness of bone actually increases through this process. The catastrophic failure of test samples, or of whole bones in an accident or fall, occurs when microcracks accumulate to the point where they can interconnect, creating stress concentrations that overcome the structure, resulting in the catastrophic failure of the whole sample or bone.

Various ideas about the fracture toughening of bone have been reviewed in the ar-ticle by Wang and Gupta (2011). Many of the weak interfaces exist at the larger scales of bone structural hierarchy. For example, as mentioned above (2. Bone Cells), the cement lines that surround secondary osteons in Haversian bone seem to form weak interfaces that redirect growing cracks around osteons, and through this preserve the continuity of the cylindrical osteon as a structural unit. By weakening this interface, the ultimate fracture process may involve a fiber pullout that consumes a significant fraction of the strain energy release that is required for catastrophic crack propagation. In addition, the interfaces between individual lamellae in primary lamellar and Haversian bone are under-reinforced with mineral platelets, and hence they may provide additional weak interfaces for sacrificial crack growth. Perhaps the most important location for weak interfaces that occur at the nanostructural level is illustrated in fig. 11.5. This is the interface between mineralized collagen fibrils that we described above. A crack propa-gating at right angles to the long axis of mineralized collagen fibrils will arrive at this interface, which will have had its proteoglycan matrix sheared in the postyield straining of the material in the process described above, and the weakened interface will likely open as the growing crack approaches, another example of the Cook-Gordon effect that will likely stop the crack from proceeding beyond this interface.

To place these kinds of fracture-toughening processes into the context of a bone structure, we need only establish the orientation of the composite structures in a bone with the direction of greatest stress and hence the most likely direction for crack propagation in that bone. The largest stresses that develop in bones in their normal function in an animal's skeletal system are typically oriented parallel, or nearly parallel, to the long axis of the bone. These stresses can arise from end-to-end compression, but most frequently from bending. Direct end-to-end tensile loading is relatively rare, and such tensile stresses are typically low relative to those generated in bending and compression. However, in direct compression, and most importantly in bending, the compressive and tensile stresses are high, and these are the stresses that break bones. That is, the most important stresses that develop in normal movements are predominantly oriented in the longitudinal direction, as defined in fig. 11.4D, and tensile stress in this direction will act to propagate cracks primarily in the radial direction. It is not surprising that most of the weak interfaces that exist in the various levels of the bone micro- and nanostructure are oriented at right angles to this direction of crack propagation. Thus, the most dangerous direction for crack propagation is the direction having the greatest number of weak interfaces that provide crack-stopping and energy-absorbing processes. Indeed, the hierarchy of lamellar structures at various levels is integrated into a system focused on preventing the kind of fractures that are most likely to occur in the normal activities of animals. The most interesting aspect of this system is that as the tensile surface of a bone in bending approaches its yield stress, and nano- and microcracking begin, the actual fracture toughness of the bone material increases, and this increased toughness is due to the creation of crack-stopping defects in the bone's structure that restrict further crack growth. Only after an additional 2%–3% of bone strain do the individual microcracks coalesce, and catastrophic failure of the bone occurs.

The other consequence of focusing the crack-stopping mechanisms on preventing propagation across the long axis of a bone is that the mechanical properties of bone in other directions are compromised by this design. These compromises have already been demonstrated in fig. 11.4D, where we see that the circumferential and especially the radial mechanical properties of bone material are dramatically reduced from those in the longitudinal direction. This is because loading in these directions is normally much lower than that in the longitudinal direction. This matching of bone's nano- and microstructure and resulting material properties with the direction of normal loading in living animals is a superb example of the high quality of mechanical design in this key vertebrate structural material. With this background on the nano- and microscale mechanics of bone material, we can proceed to document the material properties of bone at the macroscopic level.

5. THE MECHANICAL PROPERTIES OF BONE. Table 11.1 was constructed with data from Currey's book (2002). The table compares properties for human bone, which will have been extensively remodeled into Haversian bone, and bovine bone, which has been separated into fibrolamellar and Haversian structural organization. There are a number of comparisons that can be made. First let's compare the Haversian bone from humans and cows. The data suggest that the longitudinal stiffness of bovine Haversian bone is greater by about 25% than that of human Haversian bone, and this difference is seen in both tension and compression. It is also clear that there are large differences in

Table 11.1. Mechanical Properties of Mammalian Bone

| | | Human Haversian Bone | | Bovine Bone | | |
| | | | | Haversian | | Fibrolamellar |
	Test Orientation	Tension	Comp.	Tension	Comp.	Tension
Modulus (GPa)	Long.	13	18	23	22	26.5
	Circ.	13	12	10	10	11
	Radial	13	12	10	10	11
Strength (MPa)	Long.	133	205	150	272	167
	Circ.	53	131	54	170	55
	Radial			39	190	30
Yield stress, σ_{yield} (MPa)	Long.	114		141		156
	Circ.			0.007		
	Radial			0.007		
Breaking strain, ε_{max}	Long.	0.031	0.019	0.02		0.033
	Circ.	0.007	0.05	0.007		0.007
	Radial			0.007		0.002

Data from Currey (2002).
Circ., circumferential; comp., compression; long., longitudinal.

the tensile and compressive strengths in this comparison. That is, the tensile strength of the bovine bone is 13% greater than that of the human bone, and the compressive strength of the bovine bone is 33% greater. This kind of difference between animal bones and human bones is seen in other data sets as well. Further, the tensile yield stress is higher in the bovine tissue, but the failure strain is somewhat higher in human bone. On balance, the bovine Haversian bone seems to have somewhat better properties than human Haversian bone.

Next, let's compare the differences in tensile and compressive properties for these two bone types individually. The longitudinal stiffness of human bone in both tension and compression is about 40% higher than that in the circumferential and radial directions. Differences in strength, however, vary significantly between tension and compression and with orientation. That is, for human bone the longitudinal compressive strength is about 55% greater than the longitudinal tensile strength. In tension the longitudinal strength is 2.5 times greater than the circumferential strength, and in compression the longitudinal strength is 56% greater than the circumferential strength.

The longitudinal stiffness of bovine Haversian bone is similar in tension and compression, and the stiffness levels in circumferential and radial directions are about 40% of the longitudinal stiffness in both tension and compression. Differences in strength, however, are more variable. Longitudinal strength in tension is only 55% of the longitudinal strength in compression. The strength ratios for tests in different directions are as follows. The circumferential strength is about 35% of the longitudinal strength in tension and about 63% of the longitudinal strength in compression. The radial strength is about 25% of the longitudinal strength in tension and about 70% in compression.

Next, look at the properties of the bovine fibrolamellar bone, where we show data only for the tensile direction. Here we see that tensile stiffness is 50% higher than that of human Haversian bone and 15% higher than that of bovine Haversian bone. Tensile strength is also considerably higher than that of the human and bovine Haversian bone, by 26% and 11% respectively. Thus, it seems that the mechanical properties of fibrolamellar bone are somewhat better than those of Haversian bone, particularly when the comparison is made with human Haversian bone. We will see shortly that this difference between fibrolamellar bone and Haversian bone is revealed in other modes of mechanical testing.

The properties of bovine fibrolamellar bone in the circumferential and radial directions are similar to those seen for the two types of Haversian bone. For fibrolamellar bone the longitudinal stiffness is 2.4 times greater than the stiffness in the circumferential and radial directions. Its longitudinal strength is 3 times greater than that in the circumferential direction and 5.6 times greater than that in the radial direction. It is also important to note that the failure strains in the circumferential and radial directions are dramatically lower in tension. Finally, the stress-strain curves shown in fig. 11.4A were drawn to reflect the tensile properties of bovine fibrolamellar bone, with properties that are similar to those indicated in table 11.1.

Another important property for bone is its toughness in fracture, and fracture toughness has been measured in a number of studies. A study by Koester and colleagues (2008) summarizes the results of earlier studies of bone toughness and also provides new data on experiments that track fracture toughness in the initial growth of cracks from small-scale defects. The previous studies had concentrated on large-scale samples, and the results obtained suggested that the toughness, as indicated by the stress intensity factor, K (see eq. (5.10), table 5.1), falls in the range of 5–10 MPa m$^{\frac{1}{2}}$ for transverse cracks—that is, cracks propagated across the long axis of a bone and hence across the dominant orientation of the collagen fibrils, fibers, and lamellae in the material, as discussed above. The toughness of bone in the longitudinal direction, with cracks running parallel to the long axis of a bone and its collagenous structure, is lower, with K typically in the range of 2–5 MPa m$^{\frac{1}{2}}$. The Koester group, however, carried out tests on very small samples with correspondingly small initiating cracks, and they followed the progress of crack growth by conducting their test inside an environmental scanning electron microscope (ESEM). The "environmental" here means that the device can take micrographs of specimens that are wet, by allowing for a gaseous environment in the specimen chamber. The ESEM allowed these authors to view small structural detail in a bone sample soaked in Hank's balanced salt solution, a solution with ionic composition that closely matches that of tissue fluids. Their results provide key insights into bone-fracture processes at the microscale. At crack extensions of 100–200 μm the K values were in the range of 1–5 MPa m$^{\frac{1}{2}}$, but as the crack extended further, to the limit allowed by the ESEM at about 500 μm, the K values reached levels of about 25 MPa m$^{\frac{1}{2}}$. That is, the toughness of bone at this microscale is actually quite a bit higher than observed in previous measurements on macrosamples. If you look back at table 5.1, you will note that the strain-energy release rate, G or J, for bone is listed as 1.5 kJ m^{-2}, but in the Koester study the J-integral value for the energy required to create fracture surfaces in a growing flaw is about 32 kJ m^{-2}. Thus, these new data have bone toughness actually exceeding that of horse-hoof keratin, which was previously thought to be much tougher than bone.

What are the structural origins of this exceptional toughness of bone? They are the cumulative effects of the various interfaces that we have identified above, where crack-stopping and energy-absorbing processes can occur. The structural changes that were visible in the ESEM were much larger than the nanoscale interactions we considered above at the level of interfibrillar sliding, or the interfiber effects. On the scale of tens and hundreds of micrometers Koester et al. observed dramatic deviations in the direction of crack propagation from its initial direction running transversely to the collagen direction in the bone composite. These deviations will have occurred at interfaces between micro- and nanoscale structures such as fibrils, fibers, lamellae, and osteons. Indeed, the group saw dramatic shifts in crack directions at the cement layers around secondary osteons. So the complex hierarchical structure of bone truly does result in a composite material with a very elegant design.

Having tables of values for material properties can be very satisfying, but they do not tell us very much about how bones actually perform under the varying conditions that might occur in a living, moving animal. Like other biological materials, bone has time-dependent material properties, and these are manifest in a number of ways. First, long-term loading can lead to the phenomenon of fatigue failure, and loading at different strain rates can lead to viscoelastic effects in which stiffness and strength may increase and/or decrease. Therefore, we need to consider the time dependence of material properties. There have been a number of studies that have measured these kinds of effects, and they provide us with some interesting insights into the way that bone functions dynamically in an animal's skeleton. As you will remember from the section on tendon collagen in chapter 9 (7. The Structural Design of Tendons and Their Fatigue Lifetime), the fatigue resistance of a material can be measured both under long-term static loading at a fixed stress level and by repeated load cycling to a constant stress level. In both cases, the loading is continued until the material sample fails, and both protocols have been used with bone.

Interestingly, structural fatigue in bone, unlike that in tendon, is determined primarily by load time, and cyclic loading does not appear to increase fatigue rate, as it did with tendon. This fact was nicely documented in a study by Peter Zioupos and colleagues (2001), in which they compared the fatigue of bone samples under tensile cyclic-loading regimes at two different cycle frequencies, 0.5 and 5 Hz. Samples were loaded to various stress levels, which were kept constant over the course of the test, and cycle number at failure was the result of each test. The results were clear: samples cycled at the lower frequency failed with fewer cycles than samples cycled at the higher frequency, but at the same stress amplitude. However, if the number of cycles to failure were converted into total time to failure, the fatigue data for the two frequencies merged into a single relationship. Thus, in the case of bone there appears not to be a dynamic (i.e., cyclic) component to the fatigue process; fatigue is determined primarily, if not exclusively, by the total time under stress. This is different from what we saw with dynamic fatigue in tendon (fig. 9.14), which occurred faster than static fatigue at the same stress level.

Fatigue properties have been measured in a number of studies, but the data in fig. 11.7 are presented because they also show the effect of bone microstructure on the fatigue process. The data presented in this figure come from a study by Kim and colleagues (2007), and they carried out their tests in tension because fatigue in tension proceeds more rapidly than in compression. The figure shows the effect of load-cycle peak stress on the fatigue lifetime, and the samples were all cycled at 10 Hz. Thus, with the information from the previous paragraph that there is no cyclic-loading effect with bone, we can convert cycle number into static fatigue time in seconds by dividing by 10. You will see that the graph shows a logarithmic scale for cycle number to failure on the left axis, and log time (s) to failure on the right axis. There are four solid lines, and they are labeled with the initials, Fp, Fi, Hp, and Hi, to indicate two bone types, fibrolamellar (F) and Haversian (H), which each came in two structural variations. These variations were in the orientation of the fibrolamellar and Haversian structures relative to the long axis of the bone from which the samples were obtained. The samples came from bovine humerus and radius bones, and they were machined parallel to the long axis of the bone. Samples were selected for having either all Haversian or all fibrolamellar structure, and in some instances the microstructures were oriented parallel to the long axis of the bone (p), and in others they were inclined at average angles of about 26° to the bone's long axis (i).

With this nomenclature in mind, look at the figure and you will see, as expected, that the number of cycles to failure decreases as the peak stress is increased, but there is an interesting pattern in the data from the four sample groups. The two lines at the top, which are the most fatigue resistant, are labeled Fp and Fi. The two lines below are labeled Hp and Hi. The lines appear to run reasonably in parallel, and they make a quite good fit to the individual data points, which are not plotted here. They tell us that fibrolamellar bone with its microstructure running parallel to the long axis of the bone is more fatigue resistant than the Haversian bone that is created when the bone is

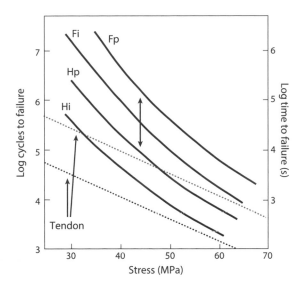

Figure 11.7. The fatigue behavior of bovine fibrolamellar and Haversian bone. The *solid lines* indicate the fatigue behavior of the bone samples, and the *dotted lines* indicate the fatigue behavior of tendon samples. Fi, fibrolamellar sample with microstructures inclined at 26° to the long axis of the bone; Fp, fibrolamellar sample with microstructures parallel to the long axis of the bone; Hi, Haversian sample with microstructures inclined at 26°; Hp, Haversian sample with microstructures parallel. See text for details of this figure.

remodeled. Thus, the fatigue testing shows us the same thing that we saw in table 11.1, that fibrolamellar bone has better properties (in this case greater fatigue resistance) than Haversian bone. The difference is quite substantial, as the vertical separation between the Fp line and the Hp line, indicated by the double-ended arrow, is more than an order of magnitude of cycle number, likely about 20-fold. That is, fatigue proceeds at about 20 times the rate in remodeled bone than primary bone at a given stress amplitude. It is also interesting that bone samples with their microstructures inclined to the direction of the applied stress are compromised relative to samples where the microstructure is parallel to the stress axis; in this case by a factor of about four- to fivefold.

Fig. 11.7 also has two dotted lines, and these represent the fatigue behavior of tendons obtained from wallabies, as discussed previously in the section on tendon in chapter 9 (7. The Structural Design of Tendons and Their Fatigue Lifetime). These lines were obtained from static fatigue tests at the stresses indicated on the bottom axis of this figure, with fatigue times indicated by the log-time axis on the right side. Assuming that the time scale for the dynamic bone fatigue can be determined as cycle time multiplied by cycle number, we can compare bone fatigue with tendon fatigue, and it appears that bone may be more fatigue resistant than tendon. The two tendon lines represent a high-safety-factor and a low-safety-factor tendon (in this case the tail tendon and the plantaris tendon from wallabies), and the low-safety-factor plantaris tendon's properties are shown by the upper of the two dotted lines. Thus, high-quality, low-safety-factor tendons like the plantaris tendon fatigue at rates that fall in the middle of the range for the bone samples, and the low-quality, high-safety-factor tendons like the tail tendon fatigue much more rapidly. At this point we cannot explain the more rapid fatigue of the high-safety-factor tendons, but perhaps it has something to do with their collagen-fibril length and the sliding of fibrils or fibers relative to their neighbors. It is tempting to speculate that the presence of mineral crystals in the spaces between fibrils and fibers in bone may provide greater resistance to sliding at these interfaces than the decorin gel alone in these spaces in tendon.

Currey (1975) tested bovine bone samples at a range of strain rates to quantify the time dependence of bone properties in a different way. The strain rates ranged from 10^{-4} to 10^{-1} s^{-1}, which for samples that fail at a strain of 0.04 would take from 400 to 0.4 s to break, and he separated his samples' results by estimating the percentage of remodeling from fibrolamellar to Haversian bone. The two lines drawn on the graph in fig. 11.8A present the data for all samples with 100% fibrolamellar (black circles) and 100% Haversian structure (open boxes), and the effect of Haversian bone structure is clearly apparent again. That is, there is a large difference in the tensile strength of Haversian bone that is superimposed on the effect of strain rate on bone strength, and it appears that the strength of Haversian bone is lower by about 30–40 MPa at all strain rates.

Keeping with the viscoelastic behavior of bone, we might ask how bone behaves at high strain rates, as might occur in falls or car crashes. However, before we look at the data it will be useful to consider what range of strain rate is actually relevant to bone function in the normal activities of humans and animals. Data from strain gauges bonded onto the limb bones of experimental animals by various authors suggest that strain rates in normal locomotor activities, from slow walking to maximal running or galloping, fall within the range of about 0.004 to 0.08 s^{-1}, and strain rates in maximal events, such as in jumping, can reach levels of 0.15 s^{-1}. Thus, data from Currey's experiments shown in fig. 11.8A should provide information for the normal range of loading, in which we do not expect to see bone failure in a single catastrophic loading event. However, in exceptional circumstances, such as a fall or perhaps in a motor vehicle accident, strain rates can be much greater, and it is possible that in these cases strain rates can be as high as 25 s^{-1}. For failure strains of the order of 0.04, failure at this strain rate would occur in about 1 ms, and we need additional high-strain-rate tests to quantify the behavior under these conditions.

Fig. 11.8B shows stress-strain curves from compression testing over a broad range of strain rates up to values of 1000 and 1500 s^{-1}, and it is clear that bone properties show very significant changes at high strain rates. In this figure the solid lines came from a study by Adharapurapu and colleagues (2006) on bovine fibrolamellar bone, and the dashed lines came from a study by McElhaney (1966) on an embalmed human femoral bone. Both were tested in compression. The stress-strain curves from the two samples are quite similar, and it is clear that tests at strain rates in the 300 to 1500 s^{-1} range indicate that bone is impressively both stiffer and stronger in compression than bone in the range we have tentatively indicated as the relevant range of strain rate for fractures in living animals and people (i.e., 0.001 to 25 s^{-1}). For perspective, the time to failure for bone at a strain rate of 1500 s^{-1} and a failure strain of 0.03 is 20 μs, and this time scale is likely only applicable for failure from a ballistic projectile.

More recently, Hansen and his colleagues (2008) measured the strain-rate dependence of human (Haversian) bone, and they presented their data in the context of all the other studies that have been published over the years on the strain-rate dependence of bone material properties. Figs. 11.8C and 11.8D show data for the stiffness and strength of bone over the full range of strain rates, along with Hansen's new data for human Haversian bone. In these figures, the solid lines show the general trend of the data for the previous studies, which include data for tests in both tension and compression. Thus, in fig. 11.8C the upper solid line shows the general trend for the modulus of bone in compression, and the lower solid line shows the trend for the modulus of bone

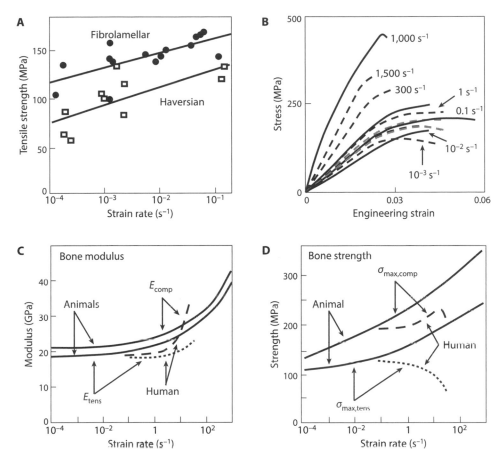

Figure 11.8. Material properties of bone I. **Panel A** shows the strain-rate dependence of tensile strength of bovine fibrolamellar and Haversian bone. Note that strength increases by about 50% over a 2000-fold increase in strain rate that roughly spans the range of strain rate seen in the normal behavior of people and animals. Note also that fibrolamellar bone is about 40% stronger than Haversian bone over the full range of strain rate. **Panel B** shows the strain-rate dependence of human bone (*dashed lines*) and bovine bone (*solid lines*) in compression as a function of strain rate, at strain rates that reach 10^4 times higher than the highest strain rate in panel A. Stiffness and strength continue to increase, although failure strain declines at the highest strain rates. **Panels C and D** illustrate the general trend of strain-rate effects on bone material, both in tension and compression. Panel C shows data for stiffness and panel D shows data for strength. The *solid lines* represent the trends of data combined from a number of different studies dominated by animal tissues, and the *dotted* and *dashed lines* indicate data for human Haversian bone. See text for details. E_{comp}, compression modulus; E_{tens}, tensile modulus; $\sigma_{max,comp}$, breaking stress in compression; $\sigma_{max,tens}$, breaking stress in tension.

in tension, both as a function of strain rate. Similarly, in fig. 11.8D the solid lines show the trend for bone strength in compression and tension. It is important to note that the data represented by these solid lines are dominated by experiments with animal bones, including experiments on cow, horse, pig, and dog bones in the various studies, although the test on human embalmed bone is included in the full data set. The new data for human bone are shown as the dotted and dashed lines in the two panels, where the dashed line shows the properties in compression and the dotted line shows the properties in tension. Here we see some interesting differences that may reflect a more general pattern of difference in the properties of human Haversian bone from the bone in a variety of animals. All four of the lines for the human bone fall somewhat below the solid lines, but three of the four show the same general trend. That is, both stiffness lines and the compressive-strength line show an upward trend with increasing strain rate, but the tensile-strength line shows a distinct downturn with increasing strain rate. This trend is surprising; particularly the magnitude of the decline in strength. However, it is also important to note that one of the seven studies included in the general trend line actually saw a decline in tensile strength of bovine fibrolamellar bone over a similar range of strain rate as that used by Hansen. However, that study had used dry bone in the mechanical tests, and this might account for the fact that the study showed a trend that was opposite to those of the other studies.

At present, therefore, it is not possible to determine whether the behavior seen by Hansen is characteristic of all human bone. The Hansen study used the two femurs from a single 51-year-old individual. The data set was robust (24 test samples), but perhaps this individual had bone with unusual properties. The impact study by McElhaney on embalmed human bone in fig. 11.8B clearly shows a different trend, with increasing stiffness and strength with strain rate. However, this could be because of the effect of the chemical treatment in the embalming process. Thus, it remains an open issue whether human Haversian bone shows significantly different properties in tensile impact than bones from animals. However, it seems clear from the studies discussed above that bovine Haversian bone has significantly reduced stiffness and strength compared with bovine fibrolamellar bone. It also appears that human Haversian bone may have reduced properties relative to those of Haversian bone from animals, but we need more data for human bone to confirm this trend.

The apparent differences in mechanical properties that we have seen over the past several tables and figures make a strong case for there being significant differences in the mechanical properties of fibrolamellar and Haversian bone, with Haversian bone consistently having lower strength. If this difference is real, it raises the question of why animals would have evolved a process that over time transforms a strong bone into a weaker bone. To think about this question we need to revisit the concepts of modeling and remodeling, in light of the information discussed above on the structural basis for bone's properties. Remember that bone modeling is the process of adding and/or removing new bone on inner or outer surfaces to accommodate changes in the mechanical requirements for the bones within an animal's skeleton. Thus, as body size and mass increase in animal growth, bone diameter and thickness must increase to keep pace with this growth. New bone is added on the outside of a growing bone, and old bone is removed on the inside, as we saw in fig. 11.3A.

Bone modeling is thought to occur through cellular mechanisms that sense the stress and strain levels that develop during normal activities in an animal, with increasing stress/strain inducing cellular mechanisms that produce new primary bone, and decreasing stress/strain inducing the removal of existing bone. Changes in activity levels or changes in gravitational forces can increase or decrease bone deposition or removal by modeling, as we see with increased exercise in sport training or decreased exercise in bed rest. Astronauts spending many weeks or months at microgravity in the international space station show considerable bone loss, and they must exercise frequently to reduce this loss. So, bone modeling represents processes that add or remove primary bone tissue, and this is distinct from the process of remodeling where existing bone is excavated internally, within the center of the compact bone, and replaced with a different bone microstructure. That is, the new bone contains secondary osteons that collectively form Haversian bone.

What, then, is the function of the Haversian bone? This remains an open question, but one possibility is that bone remodeling is a response to the accumulation of structural damage that occurs within a bone through high-intensity, but otherwise normal, activities. That is, activities that do not actually break a bone but generate microcracks, as described above, can over time lead to the accumulation of microdamage to the point where normal activities could cause a catastrophic failure in a bone. This occurs in what are called overuse injuries, where an animal or person may continue to exercise at a high intensity for too long, and then a bone breaks under stress levels that are well below the expected failure stress. However, in most cases, periods of intense exercise that may induce microcracking within the interior of a bone are separated by periods of rest and repair, which keep the levels of microcrack damage to a safe level. Presumably, bone cells sense the accumulating structural damage, and the process of bone remodeling is initiated. It seems plausible that replacing a volume in a bone that has a growing accumulation of microdamage, even if the replacement ends up having reduced mechanical properties, is a better option than allowing a better bone microstructure to continue accumulating microdamage that will ultimately result in a bone fracture. With this idea in mind, we should consider the stress levels that occur in varying levels of exercise intensity. That is, it seems appropriate that we should consider the relationship of these properties to their function in a living, locomoting organism.

What are the levels of stress and strain that occur in the various activities seen in animals and people? Data for in vivo stress levels have been recorded from animals with strain gauges attached to their limb bones, and peak strains can be converted into peak stress with knowledge of bone stiffness or modulus of elasticity, E. Data are available for a number of animals, and fig. 11.9A shows reasonable estimates that are appropriate for a range of animals and people. In low-intensity locomotion, like slow walking, the measured strain levels are of the order of -0.0005, but in intense activity such as a horse at a maximal gallop the strains can reach levels of the order of -0.003. In this case, the minus sign indicates the strains are compressive, because the highest strains are typically seen on the compression side of a long bone in a limb. Taking 23 GPa as a reasonable value for the modulus of animal bone in compression, these strains indicate stress levels of about 10 MPa in low-intensity locomotion, and in a full gallop the stress can reach levels of 60–80 MPa. In maximal movements, such as a horse or dog jumping over a barrier, stress levels can increase to 80–100 MPa. In fig. 11.9A these values are indicated next to a stress-strain curve for a typical bone sample in compression.

Now look at the shaded triangle next to the arrows indicating the stress levels associated with locomotor activities. The height of this triangle can be taken as the full range of stress that is seen in normal activities of the animals, and therefore we can take the area under the stress-strain curve to this level as an indication of the elastic-energy

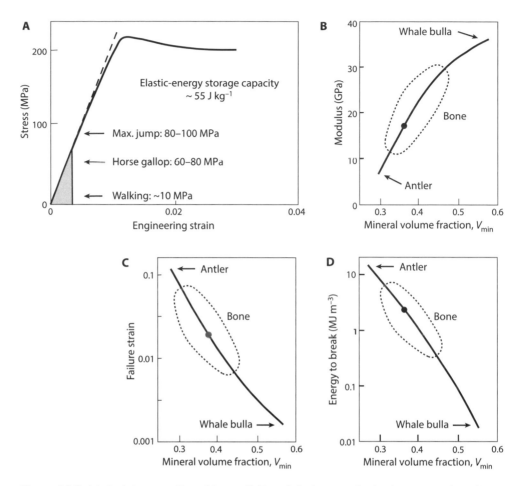

Figure 11.9. Material properties of bone II. **Panel A** shows a typical compression stress-strain curve for bone, along with data for the average compressive stress measured in the bones of animals in different forms of locomotion. The shaded triangle near the origin of the graph provides a visual indication of the elastic-energy storage capacity of the bone material in an intense activity such as galloping. As indicated, the elastic-energy storage capacity is estimated at about 55 J kg^{-1}. **Panels B–D** show the range of stiffness, failure strain, and energy to break for a range of bone samples with different values of mineral volume fraction. The actual materials span from deer antler at low V_{min} to whale bulla at high V_{min}. The *dotted loop* indicates the range of human and animal bone, and the *dot* near the center indicates the value for human bone. See text for details.

storage capacity of bone that is used in normal animal movement. The result of this calculation indicates that the elastic-energy storage capacity of bone, based on maximal sustained (i.e., galloping) activities, is about 60–80 kJ m^{-3}, and with a typical bone density of 2000 kg m^{-3}, the energy storage per unit mass is about 55 J kg^{-1}. This puts the elastic-energy storage capacity of bone at a level that is about 7% that of the collagen in spring tendons. The conclusion we can draw from this comparison is that it is quite unlikely that elastic-energy storage capacity has played a large role in the evolution of bone material or in the formation of bone shape and size. On the other hand, elastic-energy storage has clearly played a key role in the evolution of tendon collagen and in tendon structure.

It is also interesting to consider the relationship of the in vivo bone stress and strain values with the fatigue behaviors seen in fig. 11.7. The fatigue lifetime, in either seconds or cycles, is very long at the stress levels seen in low-intensity activities such as walking, where stress levels are in the range of 10–20 MPa. In fact, the lifetime is so long at this stress level that researchers don't bother to carry out the exceptionally long test needed to determine lifetime. However, in intense activities such as galloping, with stress levels of the order of 60–70 MPa, fatigue failure is quite rapid, and in maximal jumps the fatigue lifetime is very short indeed. Unfortunately, fatigue studies are not normally run at 100 MPa, but the lifetimes should be impressively short, and it is well known that overtraining can cause "stress fractures," which are essentially failures from accumulated fatigue damage.

6. THE ADAPTATIONS OF BONE. The final issue to deal with is the adaptation of this material we call bone to a variety of differing functions. To this point we have focused on what we might call "normal" bone, which is the bone found in the skeletons of humans and other mammals, and the properties we have described so far likely fit for most mammals, with relatively minor adjustments. Other animals, however, have different requirements for their bones that may need significant adjustment in structure and composition to meet different mechanical functions. Currey has been the leader in identifying and analyzing the structure and properties of bone materials that have evolved for a range of different functions. To a good first approximation, the diversity of bone and bone-like materials has been achieved by changes in the mineral volume fraction, although there are also changes in the basic microstructure.

Figs. 11.9B, C, and D show the trend of data for bone properties taken from a wide range of animals, which are primarily mammals, but also include data for several bird species. The three figures show the general trend for the tensile modulus, E, the failure strain, ε_{max}, and the energy to break, W_{break}, all as functions of the mineral volume fraction (V_{min}). These are three material properties that scale strongly with mineral volume fraction. Interestingly, the tensile strength of these bones does not scale well with mineral volume fraction, and as a consequence these data are not plotted. The central dotted loop in each of the three figures represents the distribution of a variety of bone materials from a range of mammals. At the two extremes are red-deer antler, with the lowest V_{min} of about 0.28, and whale tympanic bulla bone, with the highest V_{min} of about 0.56. The black dot on the trend line is the approximate position of human bone. Generally, there are many data points within the central dotted loop, with the whale tympanic bulla being an outlier that is well separated from the other bone materials. Fig. 11.9B shows the tensile modulus data on a linear scale because the range of stiffness values is relatively narrow, from about 7 GPa for immature antler to 34 GPa for whale bulla. The failure strain data (fig. 11.9C) and the W_{break} data (fig. 11.9D) are plotted on log scales because the range of data for these parameters in much broader; the ε_{max} data span a range of 60-fold, and the W_{break} data span a range of 800-fold. The materials at the two extremes provide some interesting insights into the role of mineral platelets in the functional design of bone and bone-like materials.

First, let's consider the behavior of antler, which is an unusual form of bone because it does not function inside the soft tissues of an animal. Rather it is an exposed structure, and it is routinely used in mating-combat behaviors that involve direct impact between the antlers of sparring males. That is, this material has evolved to survive impacts with other rigid objects, and the key property for this function is an exceptionally tough material that is resistant to fracture. Antler has a microstructure that consists of tightly packed primary osteons that run parallel to the long axis of the antler shaft.

Early studies of antler material properties were based on tests with wet tissue, since this was thought to be the appropriate condition for comparison with bone, which does indeed function in a fully hydrated state. Mature red-deer antler has a tensile modulus of about 7 GPa and a tensile strength of 158 MPa when wet. This modulus is about one-third that of human and bovine bone, but the tensile strength is quite similar to that of typical mammalian bones. The failure strain, however, is much greater at 0.114, which is three to four times greater than the failure strain of typical mammalian bone. This combination of reasonable strength and high failure strain gives the wet antler material an energy to break, W_{break}, of 9 MJ m^{-3}, which is about three times higher than that of human bone.

A major determinant of these properties is the mineral volume fraction, V_{min}, which at about 0.29 is 25% lower than V_{min} for human bone and about 40% lower than V_{min} for bovine bone. Fracture surfaces from broken test samples reveal a somewhat different surface topology. The fracture surfaces of normal bone are usually fairly rough, although this can vary between samples, and at small scales there is clear evidence of the internal microstructure involved in the toughening mechanisms of bone. That is, osteon fragments of Haversian bone often appear to have been pulled out of their surrounding microstructures. The antler fracture surface, however, is exceptionally rough, with what look like relatively long segments of the primary osteons projecting from the fracture plane. Likely this arises from a reduced extrafibrillar mineral deposition, and the pullout of these structures likely contributes to the toughness of this kind of "bone."

A more recent study by Currey and colleagues (2009) suggests that a more functional measure of antler material properties requires that bone samples be tested under environmental conditions that reflect the condition of the antler material when deer actually use their antlers in combat. Antlers emerge from the skull as bone-like structures that are covered with a thick tissue layer called "velvet," which is actually the periosteum. This tissue supplies nutrients to support the rapid growth of the antlers during the period from early spring to late summer, and the antlers become mature and the velvet is lost at the end of summer, when the mating season begins. These authors studied the mechanics of antler from the red deer, *Cervus elaphus*, from southern Spain, and they carried out environmental tests on isolated antler samples to determine the hydration state of the antler material when it is used in mating combat. Their results indicate that the antler material has significantly reduced water content relative to that of fully hydrated antler at the time of its use, and this reduced hydration dramatically alters the mechanical properties of the antler. Table 11.2 shows the results for elastic modulus, bending strength, and fracture toughness in quasi-static tests and impact tests, where fracture toughness is the energy required for crack elongation and is

Table 11.2. Mechanical Properties of Red-Deer Antler and Bone

Material (Condition)	Bending Modulus (GPa)	Bending Strength (MPa)	Quasi-static Work of Fracture (KJ m^{-2})	Work of Fracture in Impact (KJ m^{-2})
Antler (wet)	7.3	116	31	
Antler (dry)	17.5	352	23	48
Femur (wet)	22.4	263	10	7

Data from Currey et al. (2009).

similar to the J-integral term described in chapter 5, on fracture mechanics. The bending modulus of the dry antler material (17.5 GPa) is more than twice that of the wet material (7.3 GPa), but it is about 22% less than the stiffness of the deer's femoral-bone material (22.4 GPa). There are significant differences in strength and fracture toughness as well. The bending strength of the wet antler material (116 MPa) is one-third that of the dry material (352 MPa), and is about 44% that of wet bone.

The differences between the wet antler and wet bone likely arise from significant differences in mineral volume fraction, as described above, although we do not have data for mineral volume fraction in the actual samples tested. The differences between wet and dry antler, however, are even more dramatic than the differences that arise from mineral-volume-fraction comparisons between antler and bone. Drying of the antler increases its stiffness almost 2.5-fold, and it increases bending strength by three-fold. Interestingly, the quasi-static work of fracture is somewhat reduced when antler is dried, but its magnitude is still 2.3 times greater than that of deer femoral bone. More importantly, the work of fracture in impact, at 48 KJ m^{-2} for dry antler, is 7 times greater than that of wet deer femoral bone. The exceptional levels of bending strength and impact resistance of the dry antler are the crucial properties for this material. They are achieved by a combination of reduced mineral volume fraction and by the reduced hydration of the collagen-mineral composite.

At the other end of the mineral-volume-fraction range we find the whale tympanic bulla, a bony structure that surrounds the ossicles of the middle ear, leading to the cochlea, which is surrounded by a similar bone, the periotic bone. These bones do not play a normal structural role like other bones; rather, they create an acoustic environment in the whale ear that allows the animal to have directional hearing in its aquatic environment. This requires that the sound vibrations that arrive at the whale's cochlea come through the acoustic canal leading to the ear, and not from the conduction of sound through the soft tissues of the head or the bone of the skull. The problem of sound conduction through the tissues is important in aquatic animals because their environmental medium, water, has a density and stiffness similar to that of the soft tissues of the head, and therefore sound can be conducted through the head to the ear with small loss of intensity. The bones of the middle and inner ear have evolved to provide rigid structures that are more effective than normal bone in reflecting sound waves propagating through the head. Most of the bones of the ear are small and complex in shape, but the bulla is large enough that it is possible to make samples for mechanical testing. As shown in table 11.3, the whale bulla bone has an exceptionally high mineral volume fraction of 0.56, and as a result both its stiffness and its density are higher than those of normal bone. This makes the bulla bone, and presumably the periotic bone, a more efficient reflector of sound transmission through the skull into the inner ear. The physics of this situation arises from the acoustic-impedance mismatch between the soft tissues of the head that surround the bulla and the bulla itself. Acoustic impedance, Z, is determined by the density and stiffness of a material, as follows. For a solid material, like bone, $Z_b = \sqrt{\rho E}$, where ρ is the density and E is the elastic modulus of the bone material. For a fluidlike connective tissue $Z_t = \sqrt{\rho K}$, where K is the bulk modulus of water, which is similar to its elastic modulus but quantifies the ability of a fluid to resist compression in all directions to reduce the fluid's volume. These equations give a value for the acoustic impedance of the tissue of $Z_t = 1.7 \times 10^6$, and for the bulla bone, $Z_b = 6.7 \times 10^6$. The relative reflection of sound waves at an interface between connective tissue and bulla bone, R_{rel}, is determined by:

$$R_{rel} = \left[\frac{Z_1 - Z_2}{Z_1 + Z_2} \right]^2, \tag{11.1}$$

and the relative reflection at a tissue-bulla interface will be 0.48. That is, 48% of the incoming sound intensity will be reflected at the bulla surface. For a bulla made from cow bone, the relative reflection would be 37%, which means that the high-density and high-stiffness whale bulla will reflect about 30% more sound than would one made from cow bone.

The other consequence of the exceptionally high mineral content and stiffness of the whale bulla bone is that its strength and toughness are dramatically reduced. The strength of the bulla bone is only about one-fifth that of normal bone, and because the failure strain of the bulla bone is about 0.002, the energy to break, W_{break}, is reduced by about 100-fold in comparison to bovine and human bone. Of course, these changes in mechanical properties do not really matter because the bulla is not designed to support large loads; it only functions to shield the middle ear structures from external vibrations.

Table 11.3 also shows some data for the mechanical properties of bird bones, and it is interesting and likely significant that bird bones appear to be quite a bit stronger than bones from mammals. Note that the mineral volume fraction for the bird bones and mineralized tendon are similar to the values for human and bovine bone, but on average the tensile strength of the bird bones, at about 240 MPa, is about 60%–80% greater than that of mammalian bones; hence the specific strength of bird bones should be about 60% higher than that of mammalian bone. Here we see a possible adaptation to the requirements of flight, although it is not known what aspect of bone structure in birds provides this improvement in strength. However, it is relatively easy to understand the functional driving force for the evolution of greater specific strength in bird bones. The demands of flight place greater value on improved properties with lower weight, and this is exactly what we see for these bird bones. Interestingly, Currey (2002) also provides values for the properties of several bones from king penguins, which are flightless birds, and their bone materials have strengths that are intermediate between those of mammals and those of flying birds. Clearly, these comparisons are not sufficient to make sweeping conclusions about the designs of bird and mammal bones, but

Table 11.3. Variation in Bone Properties with Mineral Content

Material	Modulus (GPa)	Tensile Strength (MPa)	Energy to Break, W_{break} (MJ m^{-3})	Mineral Volume Fraction	Density (kg m^{-3})
Whale bulla	34	27	0.03	0.56	2.47
Cow femur	22	150	2.5	0.41	2.06
Human femur	13	133	2.8	0.36	
Sarus crane (ossified tendon)	18	271	12.1	0.32	
Sarus crane (tarsometatarsus)	23	218	2.0	0.34	
Sarus crane (tibiotarsus)	24	254	4.1	0.38	
Flamingo (tibiotarsus)	28	212	1.4	0.38	

Data from Currey (2002).

they do indicate a direction for future research into biomaterial adaptations of bone in the form and function of vertebrate animals. Indeed, Erickson and colleagues (2002) have taken data from a variety of sources to evaluate the evolutionary origins of bone's material properties in a broad range of vertebrate animals, and their general conclusion was that "material properties of the first long bones 475 Ma were conserved throughout evolution" (p. 268). They have pulled together a useful collection of data for a broad range of animals that may suggest useful systems for more detailed analysis of mechanical adaptation in vertebrate bone. Finally, Currey (2010) provides a review article on mechanical adaptations in "some less familiar bony tissues," which presents an even broader range of intriguing avenues for future study. It will be interesting to discover how changes in bone mineral content and microstructure design have been used to modulate material properties in the evolution of vertebrate animals.

This brings us to the end of our discussion of bone, which has illustrated the design of a rigid composite that is stiffened by inorganic-crystal platelets that fill the role of the reinforcing fibers found in artificial composites. Similar mineral-reinforced composites are also found in a number of invertebrate phyla, but they are typically reinforced by calcium carbonate. These kinds of material are most prominent in the mineralized skeletons of coelenterates, such as the reef-building corals in the subclass Scleractinia, the shells of mollusks, and in the echinoderms, but they are also found in other invertebrate phyla. Most of these materials have very high mineral content, which leaves relatively little room for toughening organic layers, but perhaps the most famous of these calcium carbonate–reinforced composites is found in the shells of bivalve mollusks and in the shell of the chambered nautilus in the material called "nacre" or "mother-of-pearl." Nacre turns out to be the invertebrate analog of bone because it presents a mineral-crystal-reinforced composite material with a high mineral volume fraction and a toughness that approaches that of bone. Wang and Gupta (2011) also consider the toughening mechanisms in nacre, and readers can access the literature on nacre through this review. Our next case study will be directed at rigid polymer-based composites, and the material discussed will be insect cuticle, a classical fiber-reinforced composite in which the reinforcing fiber is formed from a linear polymer of N-acetylglucosamine, which is called chitin, and the matrix is formed by proteins of varying composition and structure.

Chapter 12

The Structural Design
of Insect Cuticle

1. THE EVOLUTION OF INSECT CUTICLE. In the introduction to the discussion of collagen in chapter 9 we stated that this exceptional protein fiber played a key role in the evolution of the animals that appeared in the earliest stages of metazoan evolution in the early oceans. Indeed, pre-Cambrian and early Cambrian fossils indicate a diversity of soft-bodied marine animals, from sponges to coelenterates and a variety of wormlike animal groups, in which skeletal structure was dominated by soft connective tissues based on collagen fibers and associated soft-matrix materials. Thus, collagen-based materials likely existed more than 580 million years ago (Ma). However, in the Cambrian explosion, which started about 550 Ma, we see evidence in the fossil record for animals with mineralized shells and skeletons, such as mollusks, echinoderms, and mineralized coelenterates in the form of corals. However, the appearance of mineralized collagen—i.e., bone—did not occur for another 65 million years with the evolution of jawless fish at around 485 Ma.

Other animals appeared in the early Cambrian oceans (ca. 550 Ma) with bodies that were supported by other rigid structural materials, including the first arthropods. The arthropods evolved an exoskeletal body plan based on a fiber-reinforced-composite design that combined a high-stiffness polysaccharide fiber called chitin and a protein matrix. This composite exists as a continuous structure that is external to all of the living tissues of these animals, and it is present in a wide variety of forms. These include rigid structures such as limbs, wings, the head, and the thorax, as well as softer and more flexible structures that enable the movement of the rigid segments relative to each other. The rigid cuticle provided the early arthropods with full-body armor that gave them an important advantage in their competition with the soft-bodied invertebrates. In addition, internal structures such as the lining of the gut and the tracheal system that delivers oxygen to tissues and removes carbon dioxide are also lined with cuticle. We call this material arthropod cuticle, and it has provided the basis for the skeletal system of one of the most successful animal groups, the insects. Indeed, insects have achieved exceptional levels of diversity in the terrestrial environment, and they play many key roles in a wide variety of terrestrial ecosystems.

Before we get to the mechanical-design details of this remarkable material, it will be useful to consider the process by which insects and other arthropods are able to accommodate a rigid exoskeletal material with their growth and development, as there are a number of possibilities for combining rigid structural materials with collagen-based soft tissues. Soft-bodied animals, in general, can continually add to and remodel their collagen-based soft tissues as they grow, and soft-bodied animals that incorporate external mineralized structures, such as coral skeletons, mollusk shells, and echinoderm tests, are able to increase their body size by the continuous addition of soft tissue and by the accretion of new mineral onto preexisting mincralized structures. They can do this because their cellular tissues are not completely enclosed within continuous, rigid skeletal structures.

There is, however, an important group of animals in the superphylum Ecdysozoa that encase their entire soft-tissue bodies in a continuous external cuticle, and in many cases this cuticle is very rigid. This group of animals includes the arthropods, which, as mentioned above, have a continuous cuticle that encases their entire body, as well as the nematodes, and a number of similar minor phyla. A key feature of these animal groups is that they all require the process of ecdysis, or molting, to allow them to increase in body size. That is, these animals must degrade and shed their exoskeleton before they can increase the dimensions of their internal living tissues. And in preparation for increasing the size of their internal tissues, they must first create a new exoskeleton that can protect these soft tissues once the old exoskeleton has been shed. They do this by creating the new exoskeleton within the old exoskeleton, which is then in part digested and incorporated into the living animal, and the remaining remnant is discarded. The most important feature of this process is the creation of a new exoskeleton that can provide space for the growth of new soft tissues in the next growth stage. That is, the new cuticle must be able to expand, and this expansion is caused by the movement of internal body fluids or by inflation with air or liquid that is taken in from the external environment. Since internal fluid movements and inflations involve relatively small pressures and forces, it is crucial that the new cuticle be quite soft and hence easily deformed, but it must then be transformed into rigid materials that will form the stiff, strong exoskeleton required for these exceptionally active animals.

The molting process begins with the separation of the old cuticle from the layer of epithelial cells that were responsible for its synthesis, a process called apolysis. Enzymes are then released by these cells into the space formed, and the inner layer of the cuticle, called the endocuticle, is digested. The materials released in this digestion are absorbed by the cells, leaving the old exocuticle and epicuticle layers intact. These remaining layers will be discarded as the molting process proceeds, but the undigested cuticular layers remain in place at this stage to protect the animal as the cells release new structural materials into the space under the old cuticle where the new cuticle will be assembled. This includes the creation of the thin, waxy, and chitin-free external epicuticle, which forms a relatively impermeable barrier to prevent water loss from the living tissues of the animal, followed by a layer called the exocuticle that contains the cuticular composite of chitin fibers embedded in a protein matrix. Once these layers are assembled, the process of ecdysis begins, and the remaining old cuticle is split open and discarded. The insect then emerges with its new cuticle, which it inflates, and the new cuticle is hardened over a period of hours. After the exocuticle is hardened, additional chitin-protein composite material is added to the cuticle's internal surface to form the endocuticle that in most cases makes up the majority of the cuticle's thickness.

The key features of the newly synthesized cuticle are that the chitin-protein composite is in a very much softened state, owing primarily to the high water content of the protein matrix that binds short chitin microfibrils together in its composite structure, and, importantly, that the new cuticle is actually larger than the old one. This larger size is accommodated by the fact that the new cuticle is folded and wrinkled so that it fits inside the old cuticle. It is the combination of low stiffness and wrinkled shape that allows the animal to increase its size following the molting process. However, this new exocuticle is much too soft to provide the animal with the skeletal support that it needs to carry out its normal activities. The exocuticle must be stiffened by the drying and the chemical modification of its matrix proteins and by the addition of new cuticular material, the endocuticle, to its internal surface. Together, these processes produce a robust exoskeleton that supports the flight and locomotor systems. With this general background in hand, we can begin our analysis of the structural design of insect cuticle. The book *Insect Physiology and Biochemistry* (2008), by James Nation, provides much more detail on cuticle synthesis and molting (ch. 4, 91–122), and Julian Vincent's book *Structural Biomaterials* (3rd edition, 2012) provides an excellent source of detailed information on insect cuticle mechanics. We will begin our analysis of insect cuticle by looking in detail at the structure and the properties of chitin, the polymer fiber that forms the basis for the composite design of insect cuticle.

2. THE CRYSTAL STRUCTURE OF CHITIN AND CELLULOSE. To this point, the tensile materials we have considered, collagen and spider silk, are made from rigid protein fibers, and the rigid composite of bone contains collagen fibers reinforced by mineral crystals. Arthropod cuticle is reinforced by the polysaccharide polymer chitin, a linear polymer of the sugar N-acetylglucosamine, and chitin is a remarkable polymer indeed. Chitin is structurally very similar to cellulose, the polysaccharide polymer that dominates the plant world and is likely the most abundant polymeric fiber on the planet. We start our structural analysis of insect cuticle with a comparison of these two polysaccharide superfibers, because this comparison will allow us to more fully appreciate the role that ecdysis plays in the structural design of insect cuticle and ultimately in the limits that cuticle design imposes on the size of insects. Briefly, the chemical and crystalline microstructure of chitin allows insects to produce short and very thin (i.e., low aspect ratio) chitin fibrils with stiffness and strength that are equivalent to those of the high-aspect-ratio cellulose fibrils seen in plant cell walls. It is the presence of these low-aspect-ratio chitin fibrils that allows insects to accommodate the requirements that a continuous exoskeleton imposes on the development of its cuticle.

Figs. 12.1A and 12.1B show the repeating chemical units that make up the polymers of cellulose and chitin. Cellulose is made entirely from glucose, and the polymer repeat unit contains two glucose rings linked by a β-1→4 glycoside linkage. Note that the two glucose rings are rotated by 180° about the polymer long axis, which means that the polymer forms a flat ribbon inside the crystals that assemble to form the cellulose fibrils in plant cell walls. Similarly, chitin is made from a glucose derivative in which an N-acetyl group replaces the hydroxyl group at the C-2 carbon of each glucose ring, to form N-acetylglucosamine. The glucose rings are also rotated by 180° in the crystal

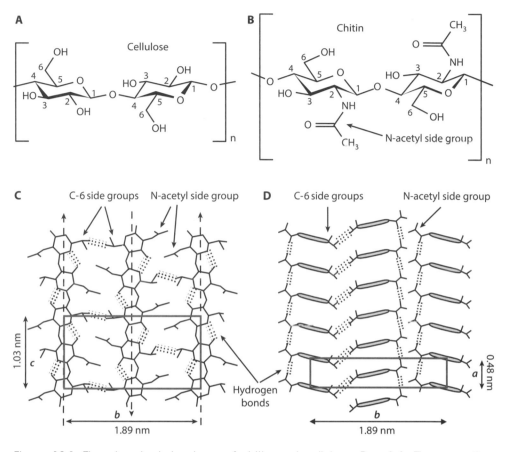

Figure 12.1. The chemical structures of chitin and cellulose. **Panel A.** The repeating chemical structure of cellulose. **Panel B.** The repeating chemical structure of chitin. **Panel C.** The crystal structure of α chitin, viewed in the crystal's *b–c* plane, showing individual chitin chains running in the direction of the *c* axis in directions that are antiparallel to their adjacent chains. The *solid box* in this panel indicates the dimensions of the crystal unit cell in this plane. The *dotted bars* in this diagram indicate the location of hydrogen bonds in the crystal structure. **Panel D.** The crystal structure of α chitin, viewed perpendicularly to the *b–c* plane, showing the chitin chains end on. That is, the chains emerge from the plane of the page, toward the viewer.

structure of chitin fibrils. Both cellulose and chitin exist in the form of nanoscale fibrils, and the chain conformations can take on a number of different patterns. However, the patterns seen in the cellulose in plant cell walls and the chitin found in insect cuticle are different. Cellulose fibrils contain fully extended cellulose molecules that are all arranged in parallel with their neighbors, in a form that is called cellulose I, whereas the chitin molecules in insect cuticle are believed to be arranged in an antiparallel arrangement, in a form called α chitin. That is, adjacent chitin chains run in opposite directions to their immediate neighbors.

Figs. 12.1C and 12.1D show the crystalline organization of the α-chitin polymer chains in the chitin microfibrils that exist in insect cuticle, and it is distinguished from other forms of chitin by the antiparallel arrangement of the chitin chains, as indicated in fig. 12.1C where we see a single sheet of chitin chains. This diagram shows the arrangement of three N-acetylglucosamine chains within the crystalline structure of a chitin microfibril, and the bold dashed arrows running vertically in this diagram indicate the alternating direction of the three chain axes. The box drawn toward the bottom of this diagram indicates the dimension of the crystal unit cell of α chitin in this plane, where the unit cell is the basic repeating three-dimensional structure within a crystalline object. The double-headed arrows at the side and below show the dimensions of the unit cell in this plane. Note that the unit cell in the direction of the polymer chains (the c axis of the unit cell) contains two N-acetylglucosamine rings as the basic repeating unit in this polymer. The c-axis dimension of the unit cell is 1.03 nm, which means that there are about 2000 sugar rings per micrometer of chitin-molecule length. This dimension is essentially identical in the crystal structure of cellulose. In this view we also see the hexagonal shape of the sugar ring, with the N-acetyl group and the C-6 side group projecting to the right and the left of the polymer chain's axis. The unit cell in the direction between adjacent polymer chains (the b axis) contains two chains because the individual chains run in opposite directions. The organization of the hydrogen-bonding network (dotted stripes) that stabilizes this crystal structure is based on the recent study by Michal Petrov and colleagues (2013).

Now look more closely at the left chain in fig. 12.1C, which runs upward, and you will we see that the top sugar unit in this chain has its large N-acetyl group pointing to the left, and its shorter C-6 side group pointing to the right. This pattern is reversed at each level as we move down the chain because of the 180° rotation of sugar units in the polymer chain. The hydrogen bonds form both between sugar rings within an individual chitin chain and between adjacent chains. The intrachain bonds play a key role in maintaining the 180° rotation of the sugar units, and hence they establish the ribbonlike conformation of the polymer chains that enables the close packing of these extended chains in sheets that lie in front of and behind the sheet shown in this diagram. The hydrogen bond between the hydrogen of the C-3 hydroxyl group and the oxygen of the C-5 carbon in the same chain is one of the strongest and most abundant hydrogen bonds in the crystal structure.

Note also that the chain to the right of this chain points downward, and its side groups face similar side groups on the adjacent chain. This provides opportunities for interchain hydrogen bonds that tie the chains together into a rigid sheet. The gaps between the chains show the location of one of the two dominant interchain hydrogen bonds that interconnect the chitin chains in a chitin microfibril. This bond connects the C-6 carbons of adjacent sugars. However, the bond actually connects on a diagonal through the crystal structure, to another C-6 carbon in the chain next to it and one sheet back, which will become clear when we look at the bonding in a cross-section of the crystal structure in fig. 12.1D.

In fig. 12.1D we see a view looking down the long axis of 19 chitin chains that emerge from the page, and we see that the ribbonlike structure of the individual chains allows very close packing between the sheets of chitin chains shown in fig. 12.1C. As indicated, the a-axis spacing of these sheets is only about 0.5 nm, and this provides numerous opportunities for the formation of interchain hydrogen bonds. In this diagram the ring portion of the sugar unit is indicated by a shaded ellipse, with the groups at the ends of the ellipse indicating the N-acetyl group and the C-6 carbon with its attached hydroxyl group. Look first at the interface between the left and center columns of chitin chains. Here we see that the C-6 side-chain groups face each other and form hydrogen bonds that are oriented on a diagonal because they run at about 45° to the sheets of chitin molecules. These are the interchain hydrogen bonds described above and shown in fig. 12.1C. In addition, we see a large number of hydrogen bonds oriented vertically in this diagram, which interconnect the N-acetyl side chains in every sheet in the crystal, and these bonds are particularly abundant and strong. They play a dominant role in stabilizing this crystal structure. It should be clear, therefore, that the extensive hydrogen-bond network in the chitin microfibrils will make them very stiff and very strong.

The structural stabilization of cellulose microfibrils is quite similar to that of chitin microfibrils, but with one important difference. That is, there are intrachain hydrogen bonds, as well as interchain hydrogen bonds, formed by the C-6 side groups, similar to those shown for chitin in fig. 12.1C, but cellulose crystals have essentially no equivalent of the strong intersheet hydrogen bonds between the N-acetyl groups in the chitin crystal structure. However, it is believed that the sheets of cellulose molecules are held together by much weaker van der Waals forces and a small number of weak hydrogen bonds. The recent study by Chen and colleagues (2014) provides an excellent overview of the hydrogen bonding in cellulose I crystals. The importance of this difference between chitin and cellulose is that the extensive three-dimensional hydrogen-bond network in chitin allows low-aspect-ratio chitin microfibrils to have mechanical properties that are very similar to those of the high-aspect-ratio cellulose microfibrils that form the reinforcing structures of plant cell walls, as we will soon see.

3. THE STRUCTURE OF CHITIN MICROFIBRILS. With this background on the crystal structures of chitin and cellulose microfibrils, we can begin to understand the design of a reinforcing-fiber system that will allow insects to create composites that can change their properties easily from low stiffness to high stiffness during the molting process. The way this is achieved is by use of a high-stiffness fiber, but to form a discontinuous-fiber composite with low-aspect-ratio fibers so that properties can be changed dramatically by altering the stiffness of the matrix material. This type of design is just the opposite of what we saw in collagen, where the fiber aspect ratio was extremely high, of the order of 10^4 to 10^5, making the reinforcing efficiency of the composite high, even with extremely soft matrix materials. Thus, the stiffness of a collagen fiber is largely unaffected by the modulus of the matrix material (see fig. 9.15). Again, a comparison of chitin and cellulose may be instructive because they both have similar microfilament structures, but they appear to have quite different aspect ratios. First, we need to establish the dimensions of chitin fibrils in insect cuticle, and then we can compare these with the dimensions of cellulose fibrils in plant cell walls.

Electron micrographic (EM) studies of insect cuticles reveal that they contain fibrillar structures that are about 2.8 nm in diameter, and these are surrounded by varying amounts of an amorphous-looking matrix. Fig 12.1D shows the crystal structure for α chitin looking down the long axis of the chitin chains, as discussed above, but it has also been drawn to represent what is believed to be the cross-sectional structure of the actual chitin microfibril that provides the reinforcing elements in the insect-cuticle

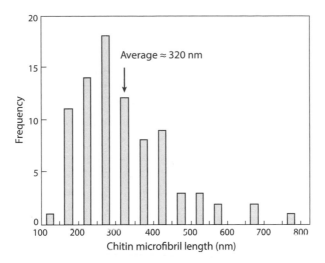

Figure 12.2. Chitin microfibril length. The length distribution of chitin microfibrils isolated from the abdominal cuticle of a beetle. The average microfibril length is 320 nm.

composite, as suggested by Atkins (1985). That is, to the best of our current knowledge, insect-cuticle chitin microfibrils contain 19 individual antiparallel chitin chains.

Julian Vincent suggests in his book *Structural Biomaterials* (2012) that the chitin microfibrils in insect cuticle are about 300 nm long, and he attributes this number to the Atkins paper as well. Unfortunately, there is no mention of chitin-microfibril length in the Atkins paper, but Eric Hillerton's 1980 paper on chitin microfibrils isolated from the soft abdominal cuticle of the assassin bug *Rhodnius prolixus* provides an EM image of isolated chitin microfibrils whose lengths can be measured. Fig. 12.2A shows the distribution of fibril lengths taken from this image, and it indicates an average fibril length of about 320 nm, with a range that spans 200–800 nm. That is, a typical chitin microfibril has a diameter of 2.8 nm and a length of 320 nm, and thus it has an aspect ratio of about 110, which is a relatively low value. If the chitin polymers that make up these microfibrils run the full length of the fibrils, the average chitin molecule would have an average chain length of about 640 N-acetylglucosamine units, with a range of 400–1600. However, it is possible that the chitin polymers form overlapping arrays within the microfibrils, and thus the chitin molecules may actually be shorter than the microfibril length. Alternatively, the antiparallel chitin chains may fold back on themselves, and the chitin molecules may actually be longer than the chitin microfibrils that they form.

Similar information can be provided for the cellulose microfibrils in plant cell walls. The chemical analysis of cellulose molecules from plant cells indicates that individual molecules are of the order of 10,000 glucose units in length. This makes the molecules about 5 μm long, or about 15 times longer than the chitin microfibrils in insect cuticle. It is well established that the cellulose molecules in a plant-cell cellulose microfibril are present as fully extended parallel molecules, and thus they clearly determine a minimum length for cellulose microfibrils that is much greater than that of chitin microfibrils. However, if the cellulose molecules are arranged in overlapping arrays, the microfibrils may be many times longer than the length of the cellulose molecules. Typical cellulose microfibrils observed by electron microscopy have diameters in the range of 2.5–5 nm. Thus, the minimum aspect ratios for cellulose microfibrils will fall into the range of 1000–2000, assuming that the length of the cellulose microfibril is equal to the extended length of the cellulose molecules. However, the aspect ratios could be considerably higher if the microfibrils contain overlapping cellulose molecules, which is very likely the case. Indeed, it is likely that cellulose microfibrils have aspect ratios that are considerably greater than the minimum levels given above. These exceptionally high values for cellulose-microfibril aspect ratios make it likely that the discontinuous-fiber-composite structure in plant cell walls will achieve very high levels of reinforcement efficiency with relatively modest levels of matrix shear stiffness.

4. THE STIFFNESS AND STRENGTH OF CHITIN MICROFIBRILS. At present, we do not have clear experimental data for the tensile stiffness or strength of chitin microfibrils. However, given the similarity of the chitin crystal structure with that of cellulose, it seems reasonable to assume that their properties are quite similar. Julian Vincent's book (2012) provides an estimate for the stiffness of cellulose microfibrils at about 140 GPa, and he suggests that the presence of the intersheet hydrogen bonds in the α-chitin crystal structure likely make chitin's stiffness somewhat greater than that of cellulose. However, Nikolov and colleagues (2010) estimate that the axial stiffness of a chitin microfibril is actually closer to 120 GPa. Given the lack of direct experimental data, we will take 120 GPa as a conservative estimate of the axial modulus of chitin microfibrils. Plant fibers that contain high volume fractions of well-oriented cellulose microfibrils (e.g., hemp and flax fibers) are known to have stiffness levels of the order of 30 and 90 GPa in the wet and dry states, respectively (table 12.1). Since the dry stiffness of these fibers is about three-quarters of the stiffness we have estimated for the crystalline microfibril, the cellulose microfibrils likely occupy a similar fraction of a flax fiber's cross-sectional area. So these assumed stiffness values seem reasonable, if not exact. To put the stiffness of the chitin and cellulose microfibrils into perspective with other reinforcing fibers, high-modulus Kevlar 49 has a tensile stiffness of about 130 GPa, steel has a modulus of 200 GPa, and high-modulus carbon fibers can reach stiffness levels of the order of 600 GPa. In a comparison with other tensile fibers from animals, chitin is truly impressive, with chitin microfibrils being 30–40 times stiffer than individual collagen fibrils and about 120 times stiffer than collagen tendon. Chitin microfibrils are about 12 times stiffer than spider silk.

The situation for the tensile strength of cellulose and chitin microfibrils is quite similar. Plant fibers with well-oriented cellulose microfibrils have strengths in the hydrated and dehydrated states that approach 0.9 GPa, (table 12.1), and adjusting for nonfibrillar cross-sectional area gives strengths for the cellulose microfibrils in the range of about 2 GPa. Cellulose microfibrils achieve this very remarkable strength in spite of the lack of intersheet hydrogen bonding in the cellulose crystal structure, because the individual cellulose molecules are extremely long. That is, the long cellulose molecules have very long overlaps where large numbers of weak intersheet bonds can tie adjacent sheets of cellulose molecules to each other very strongly. As we will soon see, the highest tensile strength reported for an insect cuticle that contains well-oriented chitin microfibrils (an insect "tendon") has a tensile strength of about 0.6 GPa. This implies

Table 12.1. Structural and Mechanical Properties of Chitin and Cellulose

Fiber	Modulus (GPa)	$\dfrac{E_{dry}}{E_{wet}}$	Strength (GPa)	Microfibril Angle	Extensibility (%)	Degree of Polymerization
Hemp (wet)	35	2		2.3°		9300
Hemp (dry)	70		0.9		1.7	
Flax (wet)	27	3.5	0.9	6°	2.2	7000–8000
Flax (dry)	95		0.9		1.8	
Insect tendon	19		0.6	0°	≈3	
Cellulose microfibril	≈140		≈2		≈1	
Chitin microfibril	≈120		≈2		≈1	≈600

Data from Wainwright et al. (1976).

that the tensile strength of chitin microfibrils is comparable to that of cellulose micro-fibrils despite the relatively short length of the chitin molecules, likely because of the strong intersheet hydrogen bonds formed by the N-acetyl groups in the α-chitin crystal structure, as discussed above. It also suggests that chitin microfibrils are constructed from chitin molecules that run the full length of the microfibril. The tensile strengths of most normal cuticle samples, however, are considerably lower than the value for the tendon, with typical values in the range of 0.08 to 0.2 GPa. Thus, the low-aspect-ratio chitin microfibrils might have similar strengths to the cellulose microfibrils, but given the shortness of the chitin microfibrils, they do not impart the same strength levels to cuticle composites. To put the strengths of chitin and cellulose microfibrils into per-spective with other fibers, their strengths are about half that of Kevlar, Spectra, and high-modulus carbon fibers, which have strengths of about 4 GPa, and they are about 20 times stronger than collagen fibers, 10 times stronger than bone, and 2 times stron-ger than spider silk.

5. THE ORGANIZATION OF CHITIN MICROFIBRILS IN THE CUTICLE COMPOSITE. Clearly, a material reinforced by chitin microfibrils has excellent po-tential for the creation of high-stiffness composites, but it remains for us to consider the larger-scale structural organization of the cuticle composite before we can translate nanoscale morphology and mechanics into macroscopic mechanical properties of the insect exoskeleton. To understand the design of the cuticle composite, we need to know a number of key features, including the volume fraction of fibers in the composite, the organization of the fibers, and the mechanical properties of the matrix materials that bind the fibers together to form the composite. Fiber volume fraction has been estimated from a number of EM studies of cuticle ultrastructure. Perhaps the most comprehensive analysis was carried out by Charles Neville and colleagues in 1976. The study involved high-resolution electron microscopy of various cuticle types, and the results suggest that volume fractions of fibers likely vary from about 0.05 to perhaps as high as 0.3. These authors provided a quantitative analysis of the chitin volume frac-tion of a cuticular structure from the locust called the prealar arm ligament, which is a composite of chitin microfibrils and the rubberlike protein resilin. Unfortunately, this

form of insect cuticle is much softer than the rigid cuticles that form the majority of the skeletal structures of insects, but it provides us with a good starting point for estimating the volume fraction of chitin in rigid cuticles.

They chose to calculate volume fraction by estimating the diameter and spacing of chitin microfibrils seen in electron micrographs. Their estimate of fibril diameter was 2.8 ± 0.05 nm, and they assumed that the fibrils were packed in a hexagonal array of six fibrils surrounding a central fibril. Interestingly, they decided to use the smallest center-to-center spacing that they observed in a "typical" hexagonal array, with a value of 5.4 nm, when they carried out their analysis. They calculated the chitin volume fraction for this cuticle in the dry state as 0.2, and if we apply corrections for the loss of matrix volume during sample preparation based on the swelling of the resilin matrix, we can estimate the volume fraction of chitin in the hydrated, in vivo state. Because resilin functions as a highly hydrated, rubberlike protein, with a protein volume fraction of about 0.45 (water makes up about 55% of its volume) the volume correction will be about 2. Thus, their calculated fiber volume fraction is likely about twice that of the hydrated material, making the actual volume fraction about 0.1. However, had they chosen to measure an average spacing of chitin fibrils, they would have come up an even lower volume fraction, because their EM images showed fibril spacing that was up to twice the minimum value that they used in their calculation. Thus, a more reasonable estimate of the actual chitin volume fraction in this particular material in its hydrated state is likely in the range of 0.05.

High-resolution images of rigid cuticles from various insect structures showed similar chitin-microfibril diameters and center-to-center spacing, but the matrix materials in these rigid cuticles are relatively dry in their functional state, and therefore correction for shrinkage during sample preparation would be much smaller. This suggests an upper limit for the fiber volume fraction in mature, stiff cuticle will fall in the range of 0.2 to 0.25. However, the chitin volume fraction during the initial deposition of cuticular proteins in the formation of new cuticle may be considerably lower because the cuticular proteins will be highly hydrated. We will see the mechanical consequences of these changes in matrix hydration in 7. The Mechanical Properties of Rigid Cuticle, where we analyze the changes in cuticular stiffness during the molting process

Before we proceed to the analysis of material properties, we need to establish the patterns of chitin-microfibril organization in the composite structures found in insect cuticle. It turns out that the patterns of fiber reinforcement in insect cuticle are generally quite conventional, although there is one pattern that is unusual. The unidirectional reinforcement seen in many artificial composites is a common pattern seen in cuticles, particularly in a structure that requires enhanced stiffness in a preferred direction. For example, cuticular "tendons" (these structures are called apodemes in the arthropod literature, to distinguish them from collagen tendons) that attach muscles to a limb segment are organized with chitin microfibrils running parallel to the long axis of the tendon. This organization is consistent with the tensile loading applied to the tendons by their attached muscles. The tubular limb segments that form insect legs and many other body parts, however, experience a variety of loading regimes that place tensile, compressive, and shear stresses on these structures. Hence, more complex reinforcement patterns are required to reinforce them appropriately. In many limb structures, the chitin-microfiber-reinforcement patterns come in two alternating forms. These are

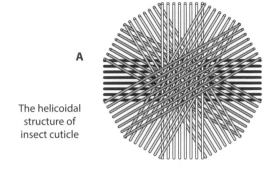

A

The helicoidal
structure of
insect cuticle

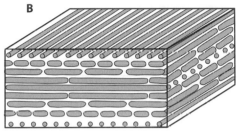

B

Figure 12.3. The helicoidal structure of chitin microfibrils in insect cuticle. **Panel A.** Top view of six layers of microfibrils, each rotated by 30°. **Panel B.** Perspective view of nine layers, each rotated by 22.5°.

layers of parallel-fiber structure, with the fiber axis running parallel to the long axis of the limb segment, and a "quasi-isotropic" fiber organization.

Recall from chapter 8 and fig. 8.2C that quasi-isotropic composites can be formed by having layers with fibers running in as few as three different directions. In the case of insect cuticle, a common form of quasi-isotropic-composite structure involves the creation of a multilevel composite where each level in the composite contains a single layer of chitin microfibrils, and in each successive level the chitin microfibrils are rotated by a small angle relative to the layers above and below it, as illustrated in fig. 12.3. Fig. 12.3A shows a top view of a series of fiber layers, where each layer is rotated by 30° relative to the layers above and below it, giving six layers for 180° of rotation. Fig. 12.3B shows a three-dimensional view of a block of cuticle where the end and side views of the structure indicate this rotation of the layers. In this diagram each layer is rotated by 22.5° relative to the adjacent layers, giving eight layers for 180° of rotation. These diagrams are simplified versions of the actual structure, as shown by Neville and Luke (1969), who observed that the rotation between layers was of about 8°. Thus, there are about 22 layers per 180° of chitin-fibril rotation, and with a layer thickness of about 5.4 nm, 120 nm of composite thickness is required for 180° of microfibril rotation. This pattern of chitin-fibril organization is called helicoidal, as the orientation of fibrils through the composite's thickness follows a "helical" pattern. However, it is important to note that the chitin microfibrils themselves are not helical; they are straight. Interestingly, it is believed that this structural organization self-assembles in a liquid-crystalline process, which is possible because the short chitin microfibrils released by the cells that produce them are able to reorient themselves through their interactions with the "fluid" phase in the extracellular space.

In addition to the parallel and the helicoidal patterns, insects are able to create a structural analog of plywood-type structures, where chitin microfibrils are present in alternating layers that are rotated by about 90° to each other. However, the structure is not truly orthogonal, and the pseudo-orthogonal structure involves a combination of parallel and helicoidal organization. That is, a number of layers will have similar orientation directions and hence appear to be "parallel," and then a small number of layers will form in a short series of helicoidal layers that rotates the direction of chitin-fibril orientation by about 90°, followed by multiple parallel-fibril layers. Typically, the parallel layers are much thicker than the helicoidal layers, giving an overall structure that resembles a 90° cross-ply composite, but in reality it is a pseudo-orthogonal structure.

These basic structural motifs for chitin-microfibril organization in cuticle are assembled into interesting patterns in different locations in the insect exoskeleton. Perhaps the best-studied structure is found in the cuticle of a locust's hind leg, which plays a key role the animal's powerful jump. Fig. 12.4A shows a diagram of a locust and fig. 12.4B shows a close-up of its hind leg. Fig. 12.4C shows the pattern of cuticle microstructure observed in the tibial segment of the locust hind leg, which is based on studies by Charles Neville (1970). The tibia is a long, slender tubular structure, and this structure requires appropriate reinforcement to deal with the considerable bending forces that develop during the locust's jump. The small circular diagram to the upper left in fig. 12.4C shows the cross-sectional shape of the tubular tibial cuticle, and the small rectangle at the top of this cross-section indicates the location of the cuticular structure shown in the full diagram to the right. This diagram indicates the four principal layers through the thickness of the cuticle, starting with the epicuticle on the outside (top), followed by the exocuticle, the endocuticle, and the layer of hypodermal cells on the inside (bottom) that synthesize all the cuticular components, including the chitin microfibrils and the matrix proteins of the cuticle composite. Not included in this diagram are membrane-bound cellular extensions that extend through the endo- and exocuticle layers, called pore canals, which are believed to function during cuticle formation to deliver waterproofing lipid materials to the exocuticle.

The epicuticle, which is shown as a thick black line, is the waxy, water-impermeable layer that coats the structural cuticle below. The exocuticle and the endocuticle below the exocuticle are indicated as alternating white and gray bands, and the hypodermal cells are shown at the bottom. The gray bands indicate the location of parallel cuticle, and the white bands indicate the location of helicoidal cuticle. Thus, the exocuticle is shown as white because it is entirely helicoidal, and the endocuticle appears with banding because it contains alternating layers of parallel and helicoidal cuticle. A crucial feature of this organization is that the gray layers of parallel cuticle contain chitin microfibrils that run along the long axis of the leg. Thus, these are layers of chitin-microfibril orientation that preferentially stiffen the leg along its long axis.

An interesting aspect of this cuticle structure is that the alternating dark and light bands in the endocuticle are controlled by diurnal light changes, where the animals produce the parallel-fiber cuticle during the day and the helicoidal cuticle at night. Thus, the pattern of bands provides a record of the timing of cuticle deposition during the life of an animal. The cuticle sample illustrated in fig.12.4C would have been laid down over about 11 days following ecdysis. As we will soon see, the exocuticle is laid down prior to the emergence of the animal from its old cuticle, and this layer is then chemically modified and cross-linked over the day following the animal's emergence

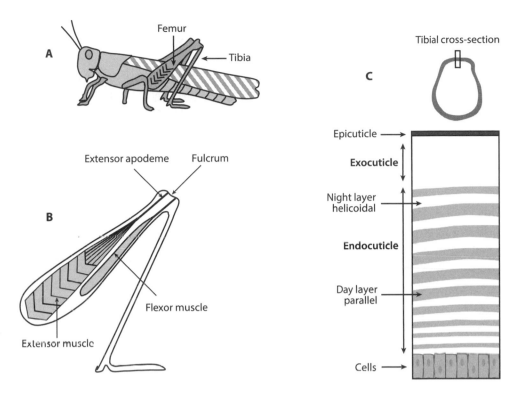

Figure 12.4. Structure of locust leg cuticle. **Panel A.** Diagram of a locust, showing the femur and tibia of the animal's jumping leg. **Panel B.** The internal structure of the muscle and apodeme (tendon) within the femoral segment of the locust jumping leg. **Panel C.** The composite structure of the insect cuticle of the locust tibia. The circular diagram on the lower left shows the cross-section of the tibia, and the small box at the top indicates the location of the structure drawn to the right, which indicates the structural layers through the thickness of the tibial cuticle.

from its shed cuticle. The 10 band pairs below the exocuticle correspond to 10 days of endocuticle production.

6. THE PROTEIN MATRIX AND ITS SCLEROTIZATION. To complete our description of the structural design of insect cuticle we must consider the matrix that binds the chitin fibers together to form a mechanically competent composite. Thus, we need to consider the chemical and biophysical properties of the proteins that form the matrix, and the chemistry of any protein modifications that stabilize these proteins. Unlike the materials we have considered to this point (collagen, silk, and bone, where the chemical composition of the constituent materials is well documented), much less is known about the matrix proteins of insect cuticle. Not because there is a lack of data on the chemical structure and diversity of cuticle proteins, but rather because the diversity and the complexity of the matrix proteins are so great. There are, however, some interesting trends seen for the variety of proteins that have been analyzed.

We will start by considering the amino-acid composition of cuticle matrix proteins. The huge variation in insects means that there are an enormous number of different kinds of cuticle. Even in a single species, the variety of cuticle types is large, and each type typically contains a number of different proteins. For economy of this discussion, we will consider three general classes of cuticle. The first type is the rigid cuticle that forms the majority of the cuticular limb segments and the rigid plates that make up the majority of the skeletal structures of the head, thorax, and abdomen of adult insects. Secondly, in addition to this basic form of rigid cuticle, we see specifically modified forms of rigid cuticle, where increased stiffness and hardness are achieved by additional chemical modifications and by the incorporation of metal ions within the cuticle matrix. A third form of insect cuticle, which is actually a broad spectrum of materials, is the soft cuticles, such as the arthrodial membranes that span gaps between rigid cuticular segments and allow rotational and translational movements of these segments. In all cases, the reinforcing fiber is chitin, and we will assume that the general features outlined above for chitin-microfibril structure, properties, and organization within the cuticle composite apply to all forms of insect cuticle. In addition, larval insects, like moth and butterfly caterpillars, have soft bodies, and their cuticles must be much less stiff than the rigid cuticles seen in mature adult insects.

The first, and perhaps most important, feature of cuticle matrix proteins is that their amino-acid composition is very hydrophobic. That is, the majority of their amino acids have nonpolar side chains that typically aggregate with other nonpolar side chains to exclude water and create hydrophobic domains. The hard-cuticle matrix proteins are particularly hydrophobic, and the more extensible parts of the exoskeleton, such as the intersegmental membranes, are somewhat less hydrophobic, as are the soft body-wall cuticles of larval insect forms, such as caterpillars. Table 12.2 lists the amino-acid composition of a number of rigid and soft cuticles from the locust, along with the amino-acid composition of the vertebrate rubberlike protein elastin. The data for elastin are given because elastin is the most hydrophobic structural protein known, and its properties may give us some insights into the water-binding capacity of cuticle proteins. That is, the comparison between elastin and cuticle proteins may allow us to understand the role of water, and more specifically of water loss through hydrophobic mechanisms, on the mechanical properties of cuticle.

Looking first at the column for elastin, we see that the hydrophobic amino acids, which are in bold type, make up 93% of its amino-acid content. The smallest amino acid, glycine with its single hydrogen as a side chain, is included in this grouping of hydrophobic amino acids because it has two hydrogens attached to the α carbon of the peptide backbone, making this a CH_2 group. Thus glycine does not add any potential hydrogen-bonding or ionic-bonding sites to the protein. Proline is also a very abundant amino acid in elastin, and it is known that glycine- and proline-rich sequences dominate the network chains of this rubberlike protein. Elastin is also quite rich in alanine, a key amino acid in the crystalline domains of spider silks, but in elastin the alanines form short α-helical segments that are associated with the cross-linking domains of this protein.

The effect of hydration on the elastic properties of elastin is known to be quite dramatic. We have already seen in fig. 3.8 that removing 15% of the bound water causes a large rise in the dynamic stiffness of elastin. In a different study, Margo Lillie (Lillie and

Table 12.2. Amino-Acid Composition of Insect Cuticle

Amino Acid	Bovine Elastin	Locust Tibia	Locust Abdominal Sclerites	Locust Intersegmental Membrane	Locust Neck Membrane
Aspartic acid/ asparagine	6	39	39	40	48
Glutamic acid/ glutamine	15	39	50	65	63
Threonine	9	27	30	32	34
Serine	10	49	52	64	52
Proline	**120**	**102**	**113**	**106**	**104**
Glycine	**324**	**93**	**122**	**195**	**157**
Alanine	**223**	**327**	**263**	**159**	**207**
Valine	**135**	**77**	**76**	**85**	**69**
Isoleucine	**26**	**34**	**32**	**43**	**40**
Leucine	**61**	**56**	**56**	**60**	**56**
Tyrosine	**7**	**97**	**81**	**69**	**80**
Phenylalanine	**30**	**1**	**12**	**10**	**16**
Lysine	7	23	23	25	28
Histidine	1	10	16	17	16
Arginine	5	33	33	31	31
Hydroxyproline	11				
Desmosine	8				
Total number ϕ	926	787	755	727	729
% ϕ	93	78	76	73	73
Total number of amino acids	998	1007	998	1001	1001

Data from Andersen et al. (1986).

Amino-acid composition of insect cuticle samples from various locations on the body of the locust *Locusta migratoria*. Hydrophobic amino acids are in bold; ϕ, hydrophobic residues. Data for methionine and tryptophan are not given because they were below detection limits.

Gosline, 1990) determined that purified aortic elastin at 37°C and 99.5% relative humidity (RH) had a water content of 0.45 kg per 1 kg of dry protein, and when hydrated at 95% RH, the water content falls to 0.28. Extrapolation of this trend to 100% humidity indicates that at full hydration the water content is 0.55 kg per 1 kg dry protein. Thus, about half of the full water content is removed from the elastin protein network at very high humidity levels because this water is very weakly bound. In fact, there is repulsion, not attraction, between water and the many hydrophobic amino-acid side chains in this protein, and this loosely bound water enters the elastin network because of the positive entropy of mixing, not because of attraction. The consequence of this loss of very loosely bound water was a large rise in the dynamic stiffness of elastin, increasing the storage modulus by about 15-fold, from 1 to about 15 MPa. Further loss of water requires more dramatic decreases in relative humidity because the remaining water is more tightly bound to polar sites in the elastin protein, including the very few charged amino-acid side chains and the numerous polar peptide groups in the protein backbone. Lorenzo Gotte and his colleagues (1968) showed that when the water content falls to about 0.15 kg per 1 kg of dry protein (at 30% RH) the stiffness rises to about 600 MPa, and it reaches 1 GPa in fully dry elastin.

Now look at the amino-acid content of the locust-cuticle proteins shown in table 12.2, where you will see that the hydrophobic-amino-acid contents of these proteins are nearly as high as that of elastin. That is, the proteins from hard cuticles, such as locust tibia and abdominal sclerites, have hydrophobic-amino-acid contents of the order of 75% to 80%, and the proteins from softer forms of insect cuticle, such as intersegmental membrane and neck membrane, have slightly lower hydrophobic-amino-acid contents, at levels on the order of 73%. So in all forms of insect cuticle we should expect that a large fraction of loosely bound water in newly synthesized cuticle will be easily removed at quite high levels of relative humidity. This loosely bound water should plasticize the matrix and give newly synthesized cuticle very low stiffness. However, when this water is lost, the matrix stiffness will rise and the cuticle will become more rigid.

Interestingly, there appears to be considerable difference in the ratio of alanine to glycine in the proteins of hard and soft cuticle. Hard cuticle (tibial cuticle) has an alanine/glycine ratio of 3.5, and the soft intersegmental membrane cuticle has a ratio of 0.8, which is similar to that of elastin, at 0.7. Glycine favors more flexible protein chains, as in elastin, and alanine favors more rigid proteins, as in spider silk. So these differences in composition might, in part, explain the differences in the stiffness of hard and soft cuticle. The question is: how soft should the matrix in insect cuticle be in newly synthesized cuticle and in the mature cuticular material? We will address this below in analyses of the discontinuous-fiber reinforcement in the composite structures of insect cuticles.

Evidence from studies of elastin provides additional interesting insights. Recall from chapter 6 that rubberlike materials are stabilized by the addition of chemical cross-links to flexible polymer chains, and this process creates soft, extensible, elastic materials. Lightly cross-linked rubbers can be quite soft, and highly cross-linked rubbers can be quite stiff. Un-cross-linked polymers made from flexible linear polymers (i.e., rubber polymers), however, can be very viscoelastic in conditions that bring them near to the center of their glass transition. That is, their properties can change dramatically over time, with loading frequency, at different temperatures, and at different solvent levels because the polymer chains can slip apart over extended loading times (see fig. 3.6C). The data discussed above for elastin indicate that the stiffness of highly hydrophobic elastomeric proteins can increase by about 1000-fold from the fully hydrated state to the fully dry state. However, if insect cuticles are intended to have a constant rigidity once they are partially dehydrated to their functional range, it may be crucial that they are appropriately stabilized by covalent modifications and by the formation of cross-links. Before we move on to chemical modifications, we need to consider the 20%–25% of polar amino acids in the cuticle proteins. There is clearly a more significant component of polar amino acids in cuticle proteins than in elastin, particularly in soft-cuticle proteins, and it turns out that the majority of these polar amino acids are found in unique sequences that are associated with the binding of the matrix proteins to chitin microfibrils.

Work on the sequence structure of cuticle proteins has revealed the presence of sequence blocks called "Rebers and Riddiford consensus blocks," or "R-R consensus elements." These are known to encode chitin-binding domains into the cuticle proteins (see Rebers and Willis, 2001), and they are quite rich in polar amino acids. Thus, they likely explain the somewhat lower hydrophobicity seen in table 12.2 for the amino-acid compositions of various types of soft locust cuticle. Svend Andersen (1986, 2001) has shown in two studies that the protein between the R-R consensus elements is best described as having extensive regions of unordered, possibly random-coil, secondary structures, much like that of elastin. His initial study used an environmentally sensitive fluorescent probe that is more fluorescent in a nonpolar environment created by a cluster of hydrophobic-amino-acid side chains than in polar environments, as seen for a fully denatured globular protein where amino-acid side chains are well exposed to water. His results clearly indicate that the cuticle proteins do not resemble the compact structure of typical globular proteins; rather, they behave as partially disordered but relatively compact structures.

More recently, Andersen (2011) has refined this conclusion with computer analysis of a range of cuticle-protein sequences, and this analysis has revealed very useful information about the general organization of these proteins. He used computer algorithms developed to predict the secondary structures of various domains within proteins that are dissolved in water or in neutral buffered dilute saline. His analyses were able to distinguish the R-R consensus blocks from the "network chains" that made up the remainder of the cuticle proteins. Interestingly, the network chains between the R-R consensus blocks are shown to have high probability of forming disordered structure, particularly in the glycine-rich proteins found in soft cuticles. The more alanine-rich proteins in hard cuticles showed features of semiordered structures. The R-R consensus blocks are typically organized into two to five short segments that fold back and forth to form a small β sheet that is thought to bond tightly to the surface of chitin microfibrils.

The presence of the hydrophilic R-R consensus blocks that bind to chitin raises the issue of how this binding actually happens. Fig. 12.5A shows a structure that was originally proposed by Fraenkel and Rudall in 1947 and that was based on their early X-ray diffraction studies and the discovery of an extended protein structure that formed when hair keratin was stretched. They called this structure β protein, which we now recognize as a β-pleated-sheet structure, the protein structure we have seen in spider silks. Their insight was that the fiber repeat structure in purified chitin from insect cuticle arose from two sugar rings, giving a repeat length of 1.03 nm, and the repeat structure for their β protein had a repeat distance of 0.69 nm. Thus, if a β-pleated-sheet structure is aligned with a chitin microfibril, hydrogen bonds will form between the peptide groups of every sixth amino acid to the C-6 hydroxyl group of every fourth chitin sugar ring. This interface provides the crucial link between the reinforcing chitin fiber and the protein matrix, which is essential for establishing the composite organization of insect cuticle. The general conclusion we can take away from these data is that the matrix proteins bind tightly to chitin; but how do they link to each other? This brings us to some additional chemistry; namely, the chemistry of cuticle-protein modification and cross-linking, and to a discussion of the relative importance of chemical cross-linking and water content in determining the mechanical behavior of insect cuticles.

Insect-cuticle protein modification has been known since the 1940s, when it was proposed by Mark Pryor that the hardening of insect cuticle was caused by a process called quinone tanning. The chemistry of this process involves the conversion of the

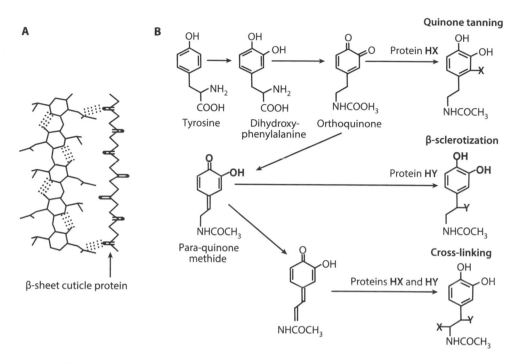

Figure 12.5. Insect cuticle chemistry. **Panel A.** Diagram of the bonding of the β sheet in a cuticle protein to the surface of a chitin microfibril. **Panel B.** Simplified diagram of the chemical reactions that lead to the formation of cross-links in insect-cuticle proteins. See text for details.

amino acid tyrosine into dihydroxyphenylalanine, which is then converted in several steps to an orthoquinone, a compound that can react with amino acids in proteins to create covalent modifications of the matrix proteins, and in some cases can produce cross-links that interconnect proteins. The chemistry of these reactions, which are more generally referred to as sclerotization, has been reviewed by Svend Andersen (2010), and it is referenced here to provide interested readers with insights into the subtleties and complexities of the chemistry. For our purposes, the simplified diagram in fig. 12.5B provides the essentials. The orthoquinone reaction, also called quinone tanning, is one of the two major kinds of sclerotization reaction found in insect cuticle, and the second was discovered and documented by Svend Anderson in the 1970s. This second form of sclerotization is called β sclerotization, because the linkage of protein to the para-quinone methide is through the β carbon of the original tyrosine, as shown in the central section of fig. 12.5B. Quinone tanning produces a product that darkens the cuticle, which in some cases becomes quite black. β sclerotization produces a colorless product. Various levels of these two reactions can leave the cuticle clear (all β sclerotization), to various levels of gray to black depending on the relative abundance of the quinone reaction. A third possibility exists for the formation of cross-links that will create covalent linkages between and within proteins. This reaction involves the conversion of the para-quinone methide into another compound that is able to bond to two different protein sites. Thus, the protein tanning/sclerotization reactions can add dihydroxybenzene groups onto cuticle matrix proteins, and they can also create covalent cross-links between distant sites on proteins.

These sclerotization reactions have the potential to make the cuticle matrix quite rigid if they are present in great abundance. Early bowling balls were made of rubber, but to get the rigidity needed for a bowling ball they needed to have an extremely high cross-link density. Alternatively, the total removal of water from elastin can increase its stiffness by 1000-fold, and thus dehydration has the potential to provide a more effective method of stiffening cuticle. Indeed, it is now recognized that the stiffening of insect cuticle following a molt is a two-stage process that starts with drying, and is in some cases completed by the introduction of the covalent cross-links described above. Julian Vincent and his colleagues have demonstrated in a number of studies (see Vincent, 2012) that reacting unsclerotized cuticle samples with the sclerotizing compound catechol (this is the dihydroxybenzene ring group of dihydroxyphenylalanine) creates reactions like those shown in fig. 12.5B. Interestingly, reaction at high concentrations of catechol can increase the dry weight of a cuticle sample by a third, indicating that a great deal of catechol can be bound to cuticle proteins, and Andersen's article (2010) indicates that the sclerotization reactions are observed with several different amino acids, including glycine, alanine, tyrosine, lysine, and histidine, so there is considerable potential for the sclerotization reactions described above to dramatically alter the chemistry and physics of the cuticular proteins. It is also interesting that the addition of catechol does not significantly change the stiffness of the cuticle when it is immersed in water.

This suggests an important feature of the chemical modification scheme of fig. 12.5B, in which the orthoquinone and para-quinone methide react extensively with various amino acids in the cuticle proteins, but primarily at single sites on the protein and not therefore forming many cross-links between two separate protein sites. In this single-site process, a large hydrophobic group becomes attached to the protein, and for lysine and histidine this addition of a hydrophobic group causes them to lose their positive charge as well. Thus, sclerotization on this scale can dramatically increase the hydrophobicity of the proteins of the cuticle matrix, and this increase in hydrophobicity will cause the proteins to lose water. Thus, sclerotization will reduce the water-binding capacity of cuticle proteins, regardless of whether they form cross-links. Indeed, it is now believed that both of these aspects of protein sclerotization are required for the maturation of the cuticle's mechanical properties. At present the consensus is that most of the stiffening is caused by dehydration, but, in rigid cuticles at least, covalent cross-links may be required to eliminate undesirable viscoelastic creep or stress relaxation under long-term loading. An analysis of the changes in the properties of cuticle in the later stages of a molt cycle will provide useful insights into the processes involved in establishing the functional properties of mature cuticle.

7. THE MECHANICAL PROPERTIES OF RIGID CUTICLE. We start our analysis of cuticle's mechanical properties with a discussion of a paper by Hepburn and Joffe (1974) that measured the changes in material properties of locust cuticle during the later stages of the molt from the final larval instar to the adult locust. These authors chose to use the rigid cuticle from the hind-femoral segment (see fig. 12.4A), and they followed changes in the structural and material properties over the first 14 days following ecdysis, when the adult animal emerges from its shed exoskeleton. The

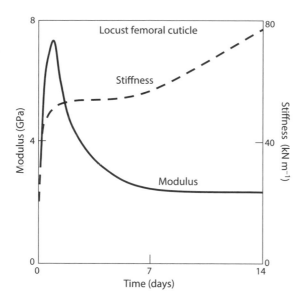

Figure 12.6. Elastic modulus and stiffness of the locust femoral cuticle. Data collected during 14 days following the animal's emergence from its cast exoskeleton.

results of their study are presented in fig. 12.6 and table 12.3. Look first at the figure, which plots data for the tensile modulus of the femoral cuticle material (solid line) and the tensile structural stiffness of the femoral test samples (dashed line), both in the longitudinal direction. The tensile-modulus data indicate that the cuticle-material stiffness increases rapidly over the first day, from about 3 to almost 7 GPa, and then it falls gradually over the next week to a value that is less than half the peak stiffness. The start of this data set indicates properties of the nearly fully formed exocuticle, which is likely to have dried considerably during the period before the start of the data set. The roughly twofold increase in stiffness during the first day is due primarily to the sclerotization of the exocuticle matrix proteins, which is essentially complete by the end of the first day. Thus, the stiffness of the fully formed exocuticle is about 7 GPa, and we will take this value as a measure of the mature helicoidal exoskeletal material in the adult locust. However, after day one the measured modulus for the whole cuticle falls to about one-third of the peak modulus, because the endocuticle that is laid down in subsequent days has a considerably lower modulus than the mature, cross-linked exocuticle.

Now look at the dashed line labeled "stiffness." This property is a measure of the tensile stiffness of the cuticle sample as a structure, measured as kN m^{-1}, and it was calculated as the modulus multiplied by the thickness of the test sample. That is, it has the units of force per unit sample width, N m^{-1}. During the first day the rise in femoral stiffness parallels its rise in tensile modulus, but in the days that follow the stiffness continues to rise because new endocuticle is added to the leg thickness and hence contributes to the cuticle's stiffness. Table 12.3 gives values for tensile modulus and tensile stiffness over time, and the fourth column gives values for the thickness of the cuticle, derived by taking the ratio of the stiffness divided by the modulus. With knowledge of the change in sample thickness over time, it is possible to estimate the stiffness of the endocuticle, and values for the endocuticle's elastic modulus are given in the fifth column. The values are somewhat variable, but on average it appears that the endocuticle's

**Table 12.3. Mechanical Properties of the Femoral Cuticle
from the Locust *Locusta migratoria migratorioides***

Age (Days)	Cuticle Stiffness (kN m⁻¹)	Cuticle Modulus (GPa)	Cuticle Thickness (μm)	Endocuticle Modulus (GPa)	Cuticle Strength (MPa)	Failure Strain
0	25	3.2	7.6		80	0.028
0.25	47	6.3	7.5		230	0.044
0.5	52	6.6	8.1		235	0.047
1	56	6.6	8.4		223	0.048
2	58	4.0	13.9	1.1	157	0.042
7	62	2.1	28.8	0.5	86	0.048
14	82	2.1	37.6	1.0	78	0.049
				Average = 0.9		

Data are from Hepburn and Joffe (1974).

Day zero is at ecdysis, when the adult locust emerges from its shed cuticle. Data for the modulus of the endocuticle are calculated as described in the text.

modulus is about one-seventh that of the mature exocuticle, at a value of about 1 GPa. How can we explain these differences?

First we need to account for the differences in cuticular microstructure in the exocuticle and endocuticle. The exocuticle has an entirely helicoidal microstructure, and on average the endocuticle has roughly half helicoidal and half parallel microstructure. Thus, if both layers were built from the same "material" (that is, the same volume fraction of chitin, protein hydration, and protein sclerotization), we should expect the endocuticle to be roughly twice as stiff as the exocuticle. This difference arises from the fact that a parallel composite will be about three times stiffer than a quasi-isotropic composite of equivalent composition (see fig. 8.2A).

However, if we could compare entirely helicoidal versions of exocuticle and endocuticle with their actual material components of volume fraction, protein hydration, and cross-linking, we should see larger differences in modulus because we know that the endocuticle's stiffness is actually about one-seventh that of the exocuticle. So adjusting the endocuticle stiffness for the conversion of its parallel cuticle into helicoidal cuticle, the all-helicoidal exocuticle should be about 10 times stiffer than an all-helicoidal endocuticle. This means that there must be quite significant differences in the stiffness of the protein matrix and/or in the volume fraction of the chitin microfibrils in the exocuticle and endocuticle composites. Current electron microscope images of insect cuticle do not indicate major differences in chitin volume fraction between the exo- and endocuticle, but this has not been investigated in detail. Therefore, it seems likely that matrix cross-linking is reduced and/or that matrix hydration is higher in the endocuticle, changes that are likely due to the differences in matrix-protein composition discussed above.

Table 12.3 also includes data for the tensile strength of locust femoral cuticle and failure strain as a function of time following ecdysis. The roughly threefold rise in tensile strength from about 80 MPa at day zero to 230 MPa on day 1 is consistent with both

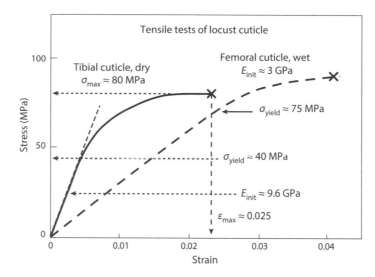

Figure 12.7. The tensile mechanical properties of locust leg cuticle. The *solid curve* shows the properties for the cuticle of the tibia, and the *dashed line* shows a prediction of the properties of the cuticle from the femur. E_{init}, initial modulus; ε_{max}, breaking strain; σ_{max}, breaking stress; σ_{yield}, yield stress.

the loss of water and the addition of cross-linking in the exocuticle. The fall in strength over the next week indicates that the tensile strength of the endocuticle is also considerably less than that of the exocuticle. It is possible to estimate the endocuticle strength, as was done for stiffness above, but given that material failure is a much more complex process than material stiffness, it is not likely that this analysis would provide valid comparisons. However, the trend of the data supports the notion that the endocuticle is not only less stiff than the exocuticle, but is considerably weaker as well.

Unfortunately the Hepburn and Joffe study does not provide stress-strain curves for their tensile tests, but the classic study by Jensen and Weis-Fogh (1962) provides tensile-stress-strain curves for cuticle from a locust's hind tibia, which shows a similar range of properties. A typical stress-strain curve from this study is provided in fig. 12.7 by the solid line labeled "tibial cuticle, dry". Here we see that the initial linear portion of the curve extends to a yield stress of about 40 MPa, with an average initial modulus of 9.6 GPa, a value that is about three times higher than that of the earliest femoral cuticle samples in table 12.3 after a several days of endocuticle deposition. Beyond the yield point at 40 MPa the stress continues to rise, but at a declining rate, and failure occurs at a stress of about 80 MPa and strain of about 0.025. Interestingly, the tibial strength does not reach the high levels seen in the femoral cuticle at the end of day 1 (see table 12.3), likely because the experiments on the tibial samples came from animals that were older and had a considerable development of the softer and weaker endocuticle. There also appears to be a large difference in extensibility, with the femoral material failing at strains of the order of 0.045, and the tibial samples failing at a strain of 0.025. The origins of these differences in stiffness and extensibility are likely due to differences in the hydration level of test samples. The dashed line in fig. 12.7 is an educated guess of the stress-strain curve that would apply to a naturally hydrated tibial cuticle sample from a relatively mature locust, based on data from the study by Hepburn and Joffe (1974). That is, the stiffness and extensibility are set to those from table 12.3 at about 4 days following ecdysis, where stiffness is about 3 GPa and extensibility is about 4.5%. Note that the failure stress is indicated as being about the same as the tibial sample, but the yield stress is considerably higher at about 75 MPa. This change in the yield stress is inferred by the lower stiffness and greater extensibility of the cuticle, as shown in table 12.3.

A recent study by Dirks and Taylor (2012) clearly documents the effect of water content on the properties of insect cuticle, and these authors show that locust tibial cuticle reaches a stiffness of the level measured by Jensen and Weis-Fogh (ca. 9 GPa) only after an isolated tibial sample has been left to dry in air at room temperature for about 2 hours. They followed the rise in stiffness with time and concluded that the best value for locust tibial cuticle at natural hydration levels is about 3 GPa for animals that had added significant amounts of endocuticle to their exocuticle after the molt. Thus, their analysis suggests that the data set produced by Hepburn and Joffe (1974) was likely produced with test samples that had hydration levels that were quite similar to in vivo levels. As a result, we can feel confident that the data set in table 12.3 and fig. 12.7 provides appropriate values for the stiffness and strength of insect hard cuticle.

There is another very useful early study of locust cuticle carried out by Henry Bennet-Clark (1975) in his analysis of the biophysics of jumping by the desert locust, which will provide us with an estimate of the tensile stiffness of parallel-fiber hard cuticle. He investigated the elastic mechanism that powered the jump, which turned out to comprise primarily the tibial extensor apodeme (tendon) of the hind leg, as illustrated in fig. 12.4B. Before discussing the properties of this tendon it will be useful to explain its function, which is to provide a spring that stores the majority of the energy used to power the locust's jump. The spring energy is stored in the extensor tendon during a period of about a quarter of a second when the tibial extensor muscle is coactivated with the tibial flexor muscle. The tendons attached to these muscles insert on the two sides of the fulcrum of the tibial joint, as shown in this figure. The prejump flexed posture of the leg is shown, and in this position the extensor tendon is attached very close to the fulcrum, whereas the flexor tendon is attached well away from the fulcrum. This gives the flexor tendon a higher mechanical advantage that allows the smaller flexor muscle to prevent the extensor muscle from shortening. As a result, elastic-strain energy is stored in the stretch of the extensor tendon. This stored energy is released when the flexor muscle relaxes, allowing the extensor tendon and muscle to contract, and in this process elastic energy released from the tendon extends the tibial segment of the leg and powers the jump.

Bennet-Clark made estimates of the stiffness, tensile strength, and extensibility of this tendon in tests with isolated apodemes, and he found the following properties. The apodeme's stiffness was about 20 GPa, and its tensile strength was about 0.6 GPa. Recall our previous discussion of cuticle fiber volume fraction, where it was reported that EM observation showed that all chitin microfibrils run parallel to the long axis of tendons. Thus, we now have an estimate for what is likely the maximal stiffness of parallel-fiber insect cuticle, which is approximately 20 GPa. However, we need to consider the hydration level of the tendon test samples, as they were apparently tested in air. Given the discussion above about the stiffness of leg cuticle, it seems likely that this is an overestimate of the true stiffness, and, as shown below where we consider composite-material models for cuticle, a more reasonable estimate suggests that its in vivo stiffness is about 15 GPa. The value for tensile strength is likely not strongly affected by hydration level, and we will accept 0.6 GPa as a reasonable estimate for the in vivo strength. With these values for stiffness and strength, we can estimate extensibility assuming a linear stress-strain curve to failure, which suggests a failure strain of about 0.04. With this information, we can estimate the strain-energy storage capacity of this tendon as about 12 MJ m^{-3}. Assuming a cuticle density of about 1300 kg m^{-3}, we can express the energy storage capacity in joules per kilogram, as we have done for vertebrate tendon, and the result is about 9 kJ kg^{-1}, which is 12 times greater than that of collagen-based spring tendons and about 80 times greater than spring steel. This is a truly remarkable material indeed. How, then, can we model these kinds of behavior?

Fig. 12.8A shows the results of calculations for the composite structure in locust exocuticle. It is based on the discontinuous-filament-composite model presented in chapter 8. The model values chosen for this analysis are based on a chitin-microfibril modulus of 120 GPa and a microfibril aspect ratio of 110. The fiber volume fraction was set at 0.20, a value that likely applies to mature, rigid cuticle. The results of this analysis are plotted as modulus versus aspect ratio to provide a clear indication of the consequences of the short-fiber construction of insect cuticle. The line labeled "parallel composite" should apply to the structural organization of the locust tendon, and the line labeled "helicoidal composite" should apply to the locust femoral exocuticle. In this graph the shear stiffness of the matrix has been adjusted to reach 6 GPa in the helicoidal composite, which is close to the value obtained in mechanical tests of the locust's femoral exocuticle at the end of the first day following ecdysis, before any endocuticle had been laid down. To get this match, we input a value for the matrix shear modulus of 30 MPa, which implies a matrix tensile stiffness of about 90 MPa. At this level of matrix stiffness, the reinforcement efficiency for the composite is about 64%, as can be seen from the parallel composite's stiffness of 15 GPa, relative to the curve's asymptote of about 24 GPa. This is an interesting value for the matrix modulus, because it may be somewhat smaller than you might have expected from our previous discussions of matrix materials in other rigid composites. Artificial composites made with an epoxy matrix will have a matrix shear modulus of about 1 GPa, more than 30 times greater than the estimated shear modulus in rigid insect exocuticle.

Keeping with our use of elastin as an initial model for cuticle matrix proteins, we can estimate the consequences of a fully hydrated and a fully dry shear modulus. In the fully hydrated state elastin's tensile modulus is about 1 MPa, making its shear modulus about 0.3 MPa. In making this assignment for shear stiffness, we are assuming that the shear modulus is about one-third of the tensile modulus, which is likely a reasonable assumption for elastin. With a matrix of fully hydrated elastin the parallel composite would have a stiffness of about 0.6 GPa, and if the cuticle matrix was made from fully dry elastin, with a stiffness 1000 times greater than that of wet elastin, the stiffness of that parallel-fiber hard-cuticle composite would be about 22 GPa. Thus, with dry elastin the stiffness of the composite would be about 50% greater than that of the real material, and if the cuticle matrix were made from fully hydrated elastin the stiffness of the composite would be only 4% of that of the real material.

Elastin is reasonably well cross-linked, with about 70 amino acids forming the random chains between cross-links, but this level of cross-linking gives a shear stiffness

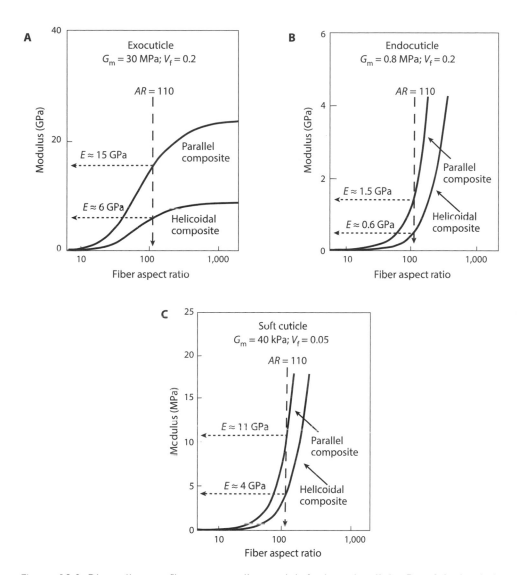

Figure 12.8. Discontinuous-fiber-composite models for locust cuticle. **Panel A.** Analysis of a discontinuous-fiber-composite model for locust exocuticle. The parameters listed in the figure were adjusted to create properties that match those of the parallel composite in locust apodeme, and the helicoidal cuticle in the locust femur. *AR*, aspect ratio; *E*, elastic modulus; G_m, shear modulus; V_f, volume fraction of the fibers. **Panel B.** Analysis of a discontinuous-fiber-composite model for locust endocuticle. The parameters listed in the figure were adjusted to create properties that match those of the parallel and the helicoidal cuticle in the endocuticle of the locust femur. **Panel C.** Analysis of a discontinuous-fiber-composite model for locust exocuticle prior to ecdysis. The parameters listed in the figure were adjusted to levels that may exist prior to cuticle dehydration and sclerotization.

of fully hydrated elastin that is well below the shear modulus required to create the measured stiffness of the locust's tendon. Recall, however, that when the loosely bound water is removed from elastin, its dynamic stiffness is increased about 15-fold, making its dynamic shear stiffness about 4.5 MPa. This level of matrix shear stiffness is reasonably close to the level required in the model to achieve the observed composite stiffness of 20 GPa. However, this increase in shear stiffness is seen in elastin at loading frequencies of about 100 Hz, which falls right in the middle of its glass transition where its material properties vary dramatically with strain rate and temperature. This is a totally inappropriate mechanical behavior for the matrix of the rigid composite in insect cuticle. Rigid cuticle needs to be a stable material with properties that do not vary with time or strain rate, and, fortunately, there is an excellent data set that confirms that this is indeed the case.

The dynamic stiffness of the locust tibia in bending was measured in a study by Steve Katz (Katz and Gosline, 1992), who found that the flexural stiffness remains constant over a broad range of dynamic-loading frequency, from 0.1 to 200 Hz, with a material stiffness of about 5 GPa in adult animals. Over the time scale of the jump impulse (ca. 25 ms) the resilience of the structure was in excess of 95%, and at the lowest frequency the resilience was only slightly reduced. Clearly, the matrix proteins in tibial cuticle are well beyond the midpoint of their glass transition, but how this is achieved remains to be determined.

The high alanine-to-glycine ratios seen in the matrix proteins of rigid cuticles, described in table 12.3, suggest that the cuticle proteins likely do not have the inherent flexibility seen in the elastin protein. Hence the loss of water from the hydrophobic cuticle proteins may have a more dramatic effect on stiffness. That is, the alanine-rich proteins in hard cuticle may contain semiordered regions that stiffen much more dramatically with initial dehydration, transforming the matrix into mechanically stable solids. In addition, it is likely that the hydration levels in mature cuticle are much lower than those that exist in elastin at 95% RH, where only half the bound water is removed. Indeed, it seems likely that allowing the water content to fall to lower levels will push the matrix properties to greater stiffness and less viscoelastic behavior. It is clear, however, that many aspects of the physical state of the matrix proteins in insect cuticle remain to be determined. For those interested in understanding the relative roles of hydration and cross-linking in cuticle, Eric Hillerton and Julian Vincent published several excellent studies in the 1970s and 1980s that describe a variety of experimental approaches to quantifying the relative roles of these two processes (e.g., Hilleron and Vincent, 1983; Hillerton et al., 1982). The short answer to this question, however, is that protein dehydration through sclerotization plays the dominant role in determining the stiffness of insect cuticles, and, as we shall soon see, protein cross-linking plays the dominant role in determining another important mechanical property, the hardness of insect cuticle.

The analysis above likely gives us a reasonable understanding of the mechanical properties of a typical hard exocuticle. What about the endocuticle? The analysis given in table 12.3 suggests that the endocuticle is less stiff than the exocuticle, with a stiffness of about 1 GPa, and there will certainly be differences between the parallel and helicoidal versions of this cuticle. Fig. 12.8B shows the results of a discontinuous-fiber-composite model for the properties of the endocuticle in the grasshopper femur. In this model the fiber volume fraction remains the same, at 0.20, but the matrix shear modulus is lowered by a factor of 37 to 0.8 MPa, which is about the stiffness of a typical rubber band. The plot shows stiffness values for parallel and helicoidal microstructure of 1.5 and 0.6 GPa, respectively. With a 50:50 mixture of helicoidal and parallel chitin microstructure, as seen in the locust's endocuticle, this would predict an average endocuticle stiffness of about 1 GPa, which is consistent with the analysis in table 12.3. Interestingly, now the graph for modulus versus aspect ratio does not include the asymptote leading to 100% reinforcement efficiency. Indeed, the reinforcement efficiency for the endocuticle composite has fallen to about 6%, which is quite low indeed. Clearly the endocuticle composite is less robust than that in the exocuticle.

Before we leave these composite models it will be useful to return to the original hypothesis that the basic design of insect cuticle revolved around the need for a structural material that would be laid down as a soft, expandable material that could be converted into a rigid composite to provide the skeletal stiffness and strength required for a vigorous flying machine. How can we model the properties of the initial soft versions of the developing cuticle? The real answer is that there is not sufficient information to make concrete proposals, but we can make some guesses and see where they take us. As a starting point, let's assume that the soft, well hydrated developing cuticle has a stiffness that is somewhat greater than that of a rubber band, because this soft cuticle needs to be inflated to its mature size before it is hardened. That is, it likely has a stiffness in the low-megapascal range. Next we need to adjust the fiber volume fraction to reflect the fact that this soft cuticle will require a dramatically lower matrix shear modulus, and this will require a higher matrix water content than we have assumed previously. As a result the fiber volume fraction must be lower than that in the mature cuticle. How big these shifts are remain unknown, but if we reduce the fiber volume fraction by a factor of four, from 0.20 to 0.05, and we set the matrix shear modulus to 40 kPa (a reduction from the 30 MPa matrix shear modulus in the mature exocuticle), we produce composites with stiffness of about 4 MPa for helicoidal cuticle and about 11 MPa for parallel cuticle, as shown in fig. 12.8C. Clearly, the large increase in water content implied by the reduced volume fraction softens the matrix dramatically, and at this early stage there is no evidence for sclerotization or cross-linking of the cuticle proteins to have occurred. In this state the reinforcement efficiency of the composite is about 0.002, but this is exactly what is required, an extremely soft composite that can be stiffened by dehydration and sclerotization.

With the data and analysis presented above, we now have a reasonable understanding of the structural basis for the stiffness and to some extent the strength of locust cuticle. Two studies by Dirks and colleagues (Dirks and Taylor, 2012; Dirks et al., 2013) document the fracture toughness and fatigue properties of locust wing and leg cuticles to provide us with a relatively complete analysis of locust cuticle's material properties. In the first study they determined the modulus and fracture toughness of the hind-leg tibial cuticle in bending. The leg sample for modulus was intact, and the leg sample for analysis of fracture toughness was notched. They also made careful comparisons between the "fresh" cuticle and air-dried cuticle. Fresh cuticle was tested immediately after removing the leg from the animal and thus should have water content similar to that in the living animal. All test samples were from adult females 14 days after their final molt. Thus, the data apply to a leg with 13 day/night layers of helicoidal-parallel

Table 12.4. Fracture-Mechanics Properties of Insect Cuticle and Other Rigid Materials

Material	Strain-Energy Release Rate, G or J (kJ m^{-2})	Elastic Modulus, E (GPa)	Breaking Stress, σ_{max} (GPa)	Critical-Flaw Length, L$_{crit}$ (mm)	Stress Intensity Factor, K (MPa m$^{\frac{1}{2}}$)
Insect cuticle (fresh)	5.6	3	0.08	1	4
Insect cuticle (dry)	0.7	9	0.21	0.3	2
Horse-hoof keratin	23	3	0.04	27	8
Bone	1.5 (32)	20	0.16	0.5	5 (25)
Wood	10	10	0.1	12	12
High-tensile steel	10	210	1.5	0.6	45
S-glass fiber	0.01	80	4.7	0.00005	0.84
Glass, bulk	0.01	80	0.07	0.015	0.84

Insect cuticle data from Dirks and Taylor (2012) and Dirks et al. (2013).

The values in parentheses for the strain-energy release rate and stress intensity factor of bone are for wet, as opposed to dry, samples (see chapter 11).

endocuticle microstructure. The results of these tests are shown in table 12.4, and are presented along with values for other rigid biomaterials and some artificial structural materials. The modulus value for "fresh" cuticle, at 3 GPa, is very similar to that obtained by Hepburn and Joffe (1974) for 14-day-old locust cuticle, and the dry cuticle has three times the stiffness, at 9 GPa, as mentioned above. The fresh cuticle stiffness falls close to the stiffness of normally hydrated hoof keratin. Fresh cuticle and hydrated horse-hoof keratin are, however, a factor of seven and two lower than the stiffness of bone, likely because the cuticle's microstructure is half helicoidal. Recall that locust apodeme, which is a parallel-fiber composite, has a stiffness of about 15 GPa, about three-quarters the stiffness of bone. Thus, with similar parallel microstructures, bone and insect cuticle have very similar stiffness levels. However, because insect cuticle is not reinforced by mineral crystals, its density will be considerably lower than that of bone, and for a flying animal minimizing weight is crucial.

The strength of fresh cuticle is only about one-third that of dry cuticle, which may seem surprising, because from the discussion above it seems reasonable that the animals could simply allow water content to fall to achieve a significant increase in strength. However, the measured values for the strain-energy release rate, G, and the stress intensity factor, K, indicate that the key property that has been selected through the evolution of insect cuticle is the resistance to fracture. Thus, the cuticle's toughness as indicated by G is eight times greater for fresh cuticle than for dry cuticle, and the toughness as indicated by K differs by a factor of two. Clearly, cuticle hydration is modulated to maintain toughness at the expense of strength. The table also provides estimates of the critical flaw length, L_{crit} for these materials. Note that L_{crit} for fresh cuticle is 1 mm, which is very similar to the diameter of an adult locust's tibia. Thus, it is extremely unlikely that structural defects leading to fracture will ever occur in the tibia, or for that matter any other structural components in the locust's body. L_{crit} for

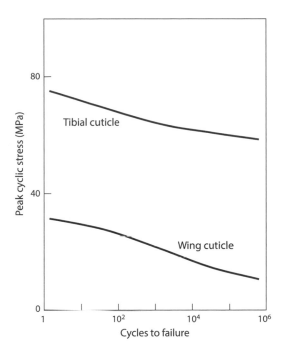

Figure 12.9. The fatigue behavior of locust tibial cuticle and wing cuticle. The chitin-reinforced tibial cuticle is more fatigue resistant than the wing-membrane cuticle, which is not reinforced by chitin microfibrils.

dry cuticle is 0.3 mm, so it is possible that a locust tibia made with dry cuticle might actually fracture. Clearly, controlled cuticle hydration is essential for the locomotor mechanics of adult locusts.

The second study by Dirks and colleagues (Dirks et al., 2013) deals with the cyclic fatigue lifetime of locust cuticle. Test specimens were loaded at cyclic frequencies of 2–3 Hz at stress levels that ranged from the ultimate tensile strength to values below this level, and fatigue lifetime was determined as cycles to failure. The tibial samples had an average ultimate tensile strength (σ_{max}) of 77 MPa, and the wing samples had an average σ_{max} of 31 MPa. The tibial samples showed a slow decline in cyclic failure stress with increasing cycle number, with failure stresses dropping by about 24% at 100,000 cycles; the wing samples showed a more rapid decline in the cyclic failure stress, with failure stress dropping by more than 50% at 100,000 cycles (fig. 12.9). It is not clear, however, why the wings show more rapid fatigue. Wings are made from wing membranes that are supported by wing veins. The wing membranes, however, do not contain chitin microfibrils and are thought to be made from a sandwich of the two epicuticle layers. That is, wing membranes do not have exocuticle or endocuticle layers, but the wing veins, which support the wing membranes, are made from the full spectrum of cuticle layers. The authors conclude that the better performance of the tibial cuticle arises from the presence of chitin-fiber reinforcement, as we have discussed extensively above.

One final aspect of cuticle mechanics is provided by an analysis of cuticle modulus as a function of size in locusts. To this point we have been dealing with the properties of adult locusts, but Steve Katz (Katz and Gosline, 1994) measured the scaling of cuticle modulus during the progression of the animal through five juvenile instars into the adult. To this point you may have assumed that the properties of the material that makes up the cuticle remain constant, but it turns out that the locust alters the stiffness of the cuticle material quite considerably during its development through these six instars. Tibial stiffness was measured in dynamic-bending tests, and the average storage modulus, E', was shown to scale as $E' \mu M^{0.313}$. Since the body mass of the locust increased by 110 times, from 18 mg for first instars to 2.5 g for adult animals, this scaling relationship indicates that E' increases more than fivefold through the animal's development. At present we do not know how these changes in modulus are created, but the function of this increase in stiffness likely has occurred through evolution as a mechanism that allows the animal to achieve the desired leg stiffness during growth with a reduced requirement for cuticle mass. That is, if the material stiffness remained constant over development from the first instar to the adult, the adult tibia would require a greater mass of cuticle than actually exists in the adult locust's leg. The precise consequence of this fivefold increase in material stiffness is not simply a fivefold saving in cuticle mass for the adult. The stiffness of the leg is determined by both the modulus of the material and the second moment of area of the tibial cross-sectional shape. Analysis of this issue will require structural engineering theory that is beyond the scope of this chapter, but this is an issue that should be assessed.

8. HARDNESS TESTING AND NANOINDENTATION STUDIES. To this point we have considered the mechanical properties of insect cuticle through studies of the cuticles of locusts. There are remarkably few other conventional mechanical tests of rigid cuticle, likely because most insects are very small, and it is difficult to carry out conventional mechanical tests on very small objects. There are, however, other techniques that can be used to quantify material properties of small cuticle samples. Hillerton, Reynolds, and Vincent (1982) used a Vickers hardness protocol, which involves the use of a weighted diamond indenter fitted to a microscope, to assess the hardness of various kinds of insect cuticle. The indentation hardness is determined by measuring the dimensions of the permanent depression that has been created when the indenter is removed from the surface of the sample. Generally speaking, we can consider the measured hardness to provide a reasonable estimate of the yield stress that would be seen in tensile tests, like those shown in fig. 12.7 for locust tibial and femoral cuticle. However, it is important to note that the yield-stress levels from hardness tests are not identical to the yield stress observed in tensile tests because the indenter imposes multiaxial strains in compression, as opposed to the uniaxial strains in tensile tests. Hillerton and colleagues used both wet and dry cuticle samples, and they observed large differences in hardness with location. They also examined the correlation of hardness with cross-link density and with hydration. Table 12.5 provides values for the hardness of various areas of the cuticle from two species of locusts.

The data in the table show the dramatic difference between the hardness of normally hydrated (wet) cuticle and fully dry cuticle. The several samples that were measured wet had average hardness values of about 0.1 GPa, or 100 MPa, which is about twice the yield stress seen in the tensile test of the locust tibia. Drying results in a two- to threefold increase in cuticle hardness. These hardness values imply a yield stress of about 0.1 GPa in tensile tests of wet cuticle, and yield stresses of about 0.2–0.3 GPa in dry cuticle. The values for hardness of wet cuticle samples, which were hydrated at in vivo levels, were limited to locations on the tibia and femur, but values for dry cuticle were taken from a broad range of sample sites, including the legs, mandibles, thorax,

Table 12.5. Values for the Hardness of Wet and Dry Cuticle Samples from Two Species of Locust

Cuticle	Hardness Wet (GPa)	Hardness Dry (GPa)
Tibia, articulation, dark (S)	–	0.26
Tibia, articulation, light (S)	0.12	0.29
Tibia, proximal (S)	0.08	0.24
Tibia, distal (S)	0.11	0.26
Femur, head, dark (S)	0.10	0.31
Femur, head, light (S)	–	0.30
Mandible, left, cutting edge (L)	–	0.36
Mandible, right, cutting edge (L)	–	0.32
Mandible, right, trailing edge (L)	–	0.20
Dorsal mesothorax (S)	–	0.33
Abdominal tergites (S)	–	0.24

Data from Hillerton et al. (1982).

The letter following the cuticle type indicates the species of locust for that specimen, with L for samples from *Locusta migratoria*, and S for *Schistocerca gregaria*.

and abdominal tergites. The dry-cuticle hardness values range from 0.20 to 0.36 GPa, and the variation is likely a reflection of the function of the particular cuticle sample. For example, the hardness of the cuticle at the articulation surfaces at the end of the tibia, at 0.26 to 0.29 GPa, are higher than those of the proximal and distal tibial shaft, at 0.24 and 0.26 GPa. The hardness of the femoral head is even higher, at 0.30 GPa, and the hardness seems to be influenced somewhat by the presence of dark cuticle, which likely indicates a high level of quinone tanning. The left mandible cutting edge has the highest hardness, at 0.36 GPa, and the mandible's trailing edge has a hardness that is less than that of the tibial shaft. The mesothorax forms part of the rigid box that functions as an elastic structure in the flight mechanism of the locust, and it exhibits high hardness, at 0.33 GPa, which likely indicates that this cuticle is able to function at high stress levels during flight. The abdominal tergites have hardness values that are similar to those of the leg shafts. The authors also looked for correlations between hardness and the two major parameters that control cuticle properties, cross-linking and hydration. They showed that hardness is strongly correlated with the extent of cuticle cross-linking, but that hardness is not correlated with hydrophobicity of the cuticle's proteins. Thus, they conclude that cross-linking plays a major role in determining the hardness of cuticle, and hence the yield strength of the material, and that, as mentioned before, water content is the primary determinant of cuticle stiffness.

More recently, we have seen experiments using nanoindentation techniques that can measure cuticle hardness and modulus of elasticity on an even smaller scale. These techniques use a nanoindenter that functions in a manner similar to the indenter of the Vickers hardness test, and then employ atomic-force microscopy to measure the dimensions of the indentation created when a cuticle sample is probed. The device also allows the indenter to measure the force-displacement profile as the indenter is inserted into the sample and as it is removed. The force-displacement profile on removal allows the measurement of the material's compression modulus. Barbakadze and colleagues (2006) carried out one of the first nanoindentation studies of insect cuticle, in which they investigated the properties of cuticle from a beetle. Their results are interesting, and in addition they provide a nice description of the nanoindentation method and analysis. They used their system to test a structure of the head, called the gula, which forms part of the head-thorax articulation system of the animal. Scanning electron microscopy showed that the microstructure of the gula cuticle is quite unusual, because the outer part of the exocuticle has chitin fibrils that are oriented perpendicularly to the cuticle's surface. Then, below this layer, the chitin fibrils are oriented parallel to the cuticle surface, as is seen in other cuticles. These authors tested the cuticle in its "fresh" wet state and in an air-dried state. Their measured elastic modulus for the wet cuticle was about 2 GPa, and for the dry cuticle about 9 GPa. The hardness value, or yield-stress, was about 0.2 GPa for the wet cuticle and about 0.6 GPa for the dry cuticle. These values for hardness are about twice those listed in table 12.5 for cuticle samples from locusts. This may seem surprising, but it is probably a result of the vertical orientation of chitin fibrils in the outer exocuticle. Indentation stiffness and hardness in this case would be a test of a parallel-composite structure (loading parallel to the long axis of the chitin microfibrils), whereas the indentation testing of normal cuticle where the chitin fibrils run parallel to the cuticle's surface would be a test of a series-composite structure (see fig. 8.3). Clearly, the structural organization of insect cuticles can be more complex than we have discussed previously, and there is likely much to be learned. That said, nanoindentation studies, in combination with additional information about cuticle structure, offer the possibility of many novel insights into cuticle design in the future.

Another nanoindentation study, by Klocke and Schmitz (2011), showed further capabilities of this method in a study of wet and dry sternal cuticle from a locust. They probed deeply through the thickness of the cuticle with their indenter to reveal properties of the different layers of the cuticle. That is, they could distinguish the series-composite properties of the exocuticle from those of the endocuticle, as well as those for a layer identified as the mesocuticle that lies between these two layers. In addition, they made cuts across the thickness of the cuticle and probed at right angles to the cuticle surface to determine the cuticle parallel-composite properties as well. The results were as we might expect from data discussed above.

In cuticle probed normal to the cuticle surface, the modulus in the wet state showed clear differences in the series-composite properties of endocuticle and exocuticle, with the exocuticle having a modulus of about 2 GPa and the endocuticle having a modulus of about 0.7 GPa. These values are similar but somewhat lower than those shown for locust cuticle in table 12.3, where the parallel-composite modulus of the exocuticle is about 7 GPa and the parallel composite modulus of the endocuticle is estimated at about 1 GPa. The hardness values for cuticle probed normal to the cuticle surface show a similar trend to those shown in table 12.5, with wet exocuticle having a hardness of 0.15 GPa and endocuticle having a hardness of about 0.03 GPa. That is, both modulus and hardness differ considerably between exo- and endocuticle in the wet state. The situation is quite different in the air dried state, where both stiffness and hardness are much higher, and there are essentially no differences between exocuticle and endocuticle. The dry-cuticle modulus observed was 5 GPa, and the dry-cuticle hardness was about 0.25 GPa.

The situation for indentation testing at right angles to the cuticle surface produced quite different results. Unfortunately it was not possible to determine the parallel-composite properties of the exocuticle because of a surface defect between the exocuticle and the embedding medium that held the samples. So data are only given here for endocuticle. In the wet state the parallel-composite modulus for endocuticle is about 2.5 GPa, and the hardness is about 0.035 GPa, whereas in the dry state the parallel-composite modulus for endocuticle is about 10.5 GPa, and the hardness is about 0.3 GPa. Overall, therefore, the data from nanoindentation studies are in general agreement with data from more conventional mechanical tests, and they show the exceptional promise of this method for investigating the mechanics of cuticle on a much finer scale.

One final issue about hard cuticle is the reinforcement of the organic chitin-protein composite by the inclusion of inorganic minerals. As we saw in table 12.5, the highest hardness values were found in dry cuticle from the cutting edge of the mandibles (i.e., jaws) of locusts, and there is considerable interest in understanding the structural origins of this enhanced hardness. Eric Hillerton and Julian Vincent (1982) showed that the cutting edge of mandibles in a wide variety of insects from five different orders shows high levels of zinc and manganese, which are thought to increase the hardness of these structures. Interestingly, the insects with zinc and manganese in their mandibles were almost all herbivores. More recently, Cribb and colleagues (2010) showed that in exceptional cases the jaws of insects can have truly remarkable levels of hardness without the addition of inorganic mineral components. They used nanoindentation testing to measure the hardness of mandibles from larval and adult jewel beetles. The larva of this beetle is known to bore through very hard wood, and these researchers discovered that the mandibles of the larvae are exceptionally hard and stiff. They also discovered that the larval mandibles achieved these exceptional levels of hardness without the addition of inorganic minerals. Specifically, dry larval mandible cuticle had a hardness of 1.6 GPa, while in the fully wet state its hardness was only slightly lower at about 1.5 GPa. Note that this level of hardness is about four times higher than the highest value observed for dry mandibles from locusts. The modulus of the wet larval mandible was 13 GPa, and this level is essentially identical to the modulus for the dry material. It is interesting that there are such small differences in hardness and stiffness between wet and dry materials. The properties of the adult mandible were quite impressive, but they were significantly lower than those of the larva, with hardness of 1 GPa and modulus of 11 GPa. Interestingly, the adult mandible was shown to contain manganese and other inorganic ions that are commonly found in the hardened mandibles of other insects. Thus, the larval jaws are hardened to exceptional levels without inorganic reinforcement. Clearly, nanoindentation methods are providing exciting insights into the subtleties of materials design in insect hard cuticles.

9. THE STRUCTURAL DESIGN OF SOFT CUTICLES. To this point we have dealt with the rigid cuticles of insects, but clearly an insect can not be made entirely from rigid cuticle. Movement of limbs, wings, and body requires that rigid body parts must be interconnected by deformable materials, and in arthropods these materials

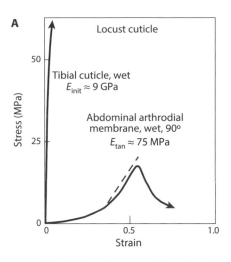
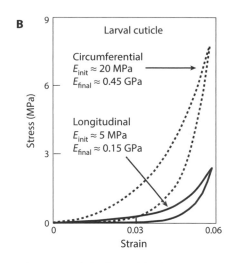

Figure 12.10. Tensile properties of adult and larval insect cuticle. **Panel A.** Tensile properties of the locust abdominal arthrodial membrane, tested in water. The stress-strain curve for locust tibial cuticle shows the initial portion of the stress-strain curve in fig. 12.7, for a comparison with a hard cuticle. **Panel B.** Tensile-load cycles of the larval cuticle from the caterpillar of the hawk moth, *Manduca sexta*. E_{final}, final modulus; E_{init}, initial modulus; E_{tan}, tangent elastic modulus.

are constructed from similar chitin-protein composites, but the proportions of the reinforcing chitin fibers and the stiffness and cross-linking of the matrix proteins must be adjusted to achieve different levels of stiffness, strength, and extensibility. Unfortunately, there are relatively few data for this kind of cuticle. Fig. 12.10A shows the stress-strain curve for the intersegmental membrane of locust abdominal cuticle strained in the transverse direction. This curve is based on information from the study by Hepburn and Chandler (1976), but it was inferred from data on the tangent modulus, tensile strength, and extensibility, as given in table 12.6. Note that the properties vary with test orientation, with tests perpendicular to the animal's long axis having greater strength and stiffness than those with test orientations parallel to the long axis. Thus, it appears that there is quite strong transverse orientation of chitin microfibrils. These numbers should be taken with caution because all samples were tested while immersed in insect saline solution, and hence tissues are likely more highly hydrated than they would be in vivo. But the precise values are not crucial because, as shown in fig. 12.10A, there are very large differences in stiffness and extensibility between these softer cuticles and the hard cuticles, as represented by the locust tibial cuticle, which fails at a strain of 0.015. What should be clear, however, is that the arthrodial membranes are much softer, weaker, and more extensible than the hard cuticles. They are likely very viscoelastic as well, although we do not have information on this issue. These authors also tested the arthrodial membranes between the abdominal tergites of a beetle, *Pachynoda*. The beetle membrane was tested only at 90° to the body axis, and its stiffness and strength were considerably higher than those of the locust membrane, at levels that are comparable to the endocuticle from the locust (see tables 12.3 and 12.6), but with a failure strain of 0.85, or 85% extension. Thus, this very rigid arthrodial membrane is 20 to 40

Table 12.6. Material Properties of Selected Arthrodial Membranes and Soft Cuticles

Cuticle Source	Test Orientation Relative to Body Axis (Degrees)	Tangent Modulus (MPa)	Tensile Strength (MPa)	Failure Strain
Locusta abdominal	0	3.2	0.85	1.32
intersegmental	45	10	–	–
membrane	90	74	16	0.53
Pachynoda (beetle) abdominal intersegmental membrane	90	500	200	0.85
Bombyx mori (silk-moth	0	34	14	0.66
caterpillar) body-wall	45	42	13	0.5
cuticle	90	37	12	0.5

Data from Hepburn and Chandler (1976).

times more extensible than the rigid cuticles of the limbs and thorax. They also tested the body-wall cuticle of the caterpillar of a silk moth, *Bombyx*, and the properties appear very similar in all test directions. Thus, the silk-moth caterpillar's cuticle likely contains primarily helicoidal chitin microfibrils, which should reinforce equally in all directions.

More recently, Lin and colleagues (2009) have measured the mechanical properties of the soft cuticle that makes up the body wall of the caterpillar of the hawk moth, *Manduca sexta*. They carried out uniaxial load-cycle tests in the longitudinal and circumferential directions, and their results are shown in fig. 12.10B. Here we see that the tissue shows a J-shaped stress-strain curve, much like that shown for intersegmental membranes, but their test did not take the tissue to failure. Instead, we see a load cycle that provides us with information about the stiffness and viscoelasticity of these materials. The cuticle stiffness in the longitudinal direction is about one-quarter of that in the circumferential direction. The initial stiffness levels are about 5 and 20 MPa respectively, and the final stiffness levels are about 0.15 and 0.45 GPa. The load cycles are quite open, with hysteresis values of about 0.45, which indicate that this is a viscoelastic material, with properties that likely vary with temperature and strain rate. Clearly, this tissue does not function as an efficient energy-storage structure, as do the rigid cuticles of limbs and tendons. Rather, the viscoelasticity likely makes this a tough material that can withstand repeated high-strain load cycles.

The structural origins of these properties remain poorly understood. As discussed earlier, there are important differences in the amino-acid composition of hard and soft cuticles, which likely lead to the development of more structured and rigid chitin-protein complexes in the hard cuticles, and to more flexible proteins and higher water content in the soft cuticles. In addition, hard cuticles all appear to have well-developed exocuticles, where chemical cross-linking and dehydration increase stiffness and hardness. The arthrodial membranes and body-wall cuticles typically have a reduced or missing exocuticle layer, and these membranes are thus largely un-cross-linked and function at higher hydration levels. Beyond that, however, it is not possible to specify the structural origins of the properties described here for these soft-cuticle tissues.

There are, however, two very interesting examples of soft cuticles whose properties and design are sufficiently well documented to be worth further discussion. These are the abdominal cuticle of the assassin bug, *Rhodnius prolixus*, and the abdominal intersegmental membranes of the pregnant female locust. This particular assassin bug makes its living by feeding on the blood of unsuspecting vertebrate animals, including humans, and when it finds a suitable victim it takes a very large blood meal indeed. The animal's capacity for this large meal resides in its ability to expand its abdomen greatly as it feeds, and it has been estimated that the surface area of the abdominal cuticle increases about fourfold when the animal feeds. This expansion is made possible by two features of the cuticle. First, the exocuticle of the abdomen is highly folded, and as a result it does not restrain the deformation of the underlying endocuticle, which exists as a flat sheet of cuticle with a fully helicoidal organization of its chitin microfibrils. The whole-cuticle chitin content is 11% by dry weight, which suggests that the chitin volume fraction of this cuticle is very low, perhaps on the order of 0.05, or less. Second, the animal possesses a hormonally controlled process that triggers a dramatic change in the material properties of the abdominal endocuticle, and, fortunately, this hormone's action can be mimicked by the pharmacological agent 5-hydroxytryptamine. Stuart Reynolds (1975b) showed that the changes in the abdominal cuticle's properties arise from cellular mechanisms that lower the pH of the cuticle tissue to levels below pH 6. The result of this drop in pH is that the cuticle proteins are shifted further from their isoelectric pH, and hence they develop a greater net charge that causes the cuticle to absorb water from its internal tissues. The water plasticizes the cuticle matrix proteins, lowering the cuticle stiffness. Measurements indicate that the water content of the cuticle more than doubles, from about 50% to more than 100% of the dry weight, when the pH falls to 5.8. Interestingly, the animal can reverse this process by increasing cuticle pH, which causes the cuticle matrix protein to lose water and to stiffen.

We will now concentrate on the changes in mechanical properties of the cuticle that arise from this change in water content. Fig. 12.11A shows the results of constant-strain-rate tests on normal and plasticized cuticle. In this type of test it appears that the normal cuticle is about 12 times stiffer than the plasticized tissue, as indicated by the slopes of the linear portions of the curves. The strain rate for this tensile test was about $5 \times 10^{-4} \text{ s}^{-1}$, but because this tissue is very viscoelastic the stiffness varies dramatically with time. Reynolds (1975a) documented these viscoelastic effects in both creep and stress-relaxation tests, and it is clear that the time-dependent stiffness of the cuticle is much lower in the plasticized state than in the normal state. However, he presented quantitative data for only the stress-relaxation tests, which are shown in fig. 12.11B. The bold lines show the average trend of tests on five cuticle samples of the plasticized and the normal cuticle. The data presented for the plasticized tissue are restricted to about 10 minutes of stress-relaxation because the plasticized state begins to decline significantly after this time. The data for the normal cuticle are shown at long times, to emphasize the time shift in the data, which indicates that the plasticized cuticle relaxes in one-twelfth the time of the normal cuticle. However, an equally important shift is the vertical shift between the two data sets, which is best shown by the extrapolated

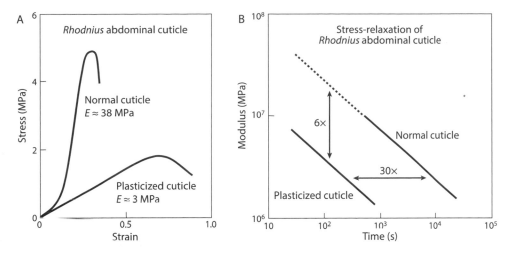

Figure 12.11. Properties of the abdominal cuticle of the assassin bug. **Panel A.** Tensile-stress-strain curves for the abdominal cuticle of the assassin bug, *Rhodnius*, in its normal and plasticized forms. **Panel B.** Stress-relaxation data for the abdominal cuticle of *Rhodnius* in the normal and plasticized states. *E*, elastic modulus.

dashed line for the normal cuticle. Here we see that the plasticized cuticle has one-sixth the stiffness of the normal cuticle at any time in the relaxation process. The broad, gradual fall in modulus is likely due to the interaction of an un-cross-linked polymer system that is reinforced by the short chitin microfibrils we have discussed previously, and it is not likely to exhibit elastic recovery. If that is the case, then the recovery described above, where the tissue regains its normal properties, means that the animal must have some mechanism other than elastic recoil to bring the abdominal cuticle back to its contracted starting state. Indeed, the creep tests with the plasticized cuticle caused rapid creep that ended in tissue failure after quite a short period of time, and the creep of the normal tissue was shown to continue at a low rate for up to 8 hours without failure. Thus, the mechanism for bringing the abdominal tissues back to their starting state is likely powered by the contraction of muscles inside the animal's abdomen.

The abdominal intersegmental membranes of the female locust provide a similar system, but in this case the hormonal signals are associated with reproductive development of the animal, and these signals cause dramatic changes in the material properties of a soft cuticle that is stretched when the female locust buries her eggs deep in the soil to protect them from desiccation. To do this the female must dig into the soil with her abdomen, which is about 2.5 cm long, to deposit the eggs 8–10 cm into the soil, and this requires a dramatic extension of the animal's abdomen. The locust abdominal wall is constructed from a series of rigid plates—dorsal tergites and ventral sternites. These rigid plates are connected by soft intersegmental membranes, and the three membranes between rigid segments 4 and 7 (out of 10 segments total) are the membranes that show the greatest change in properties as the female becomes sexually mature and ready to lay her eggs. As we will soon see, the stiffness of the stretchable membrane is nearly two orders of magnitude lower than that of the normal cuticle, so this is an exceptionally soft material system.

Julian Vincent and colleagues have investigated this system in some detail (e.g., Vincent, 1975), and they observed that the tissue has quite a high chitin content, with chitin making up about 50% of the dry weight of the "stiff" membrane and about 40% of the dry weight of the stretchable membrane. It is not clear how this translates into volume fraction of chitin in the hydrated composite, but it likely means that the chitin-microfibril volume fraction is of the order of 0.1, because the matrix of this cuticle is quite highly hydrated. More importantly, the organization of the chitin microfibrils is quite unusual because virtually all are oriented at right angles to the long axis of the abdomen. Since these intersegmental membranes are stretched in the direction of the abdomen's long axis, the cuticle composite is being loaded in series, which means that the chitin microfibrils provide very little reinforcement to the composite in the longitudinal direction. This likely explains the very low longitudinal modulus of both the stiff and stretchable forms of this membrane. If the fibers are not actually reinforcing this material in the direction of its deformation, why are they there? The answer lies in the effect of these transverse fibers on the biaxial behavior of this composite.

A study by Peter Tychsen and Julian Vincent (1976) provides very useful information on the properties of the intersegmental membranes in their initial inextensible form and in the extensible form seen in the sexually mature female. These authors presented their data as force-extension curves for the longitudinal extension of this membrane, but these have been converted into stress-strain curves by incorporating their initial modulus values for the membrane in the inextensible (170 kPa) and extensible (3 kPa) forms. These stress-strain curves are plotted in fig. 12.12A. The inextensible form of the cuticle reaches a peak stress of about 275 kPa at a strain of about 3, and failure, as indicated by the rapid drop in stress, occurs soon after the peak stress is reached. It is important to note that this curve for the inextensible cuticle is significantly different from the curve for the same material plotted in fig. 12.10A, with the current plot having significantly lower stiffness and strength. The reason for this difference is that the test sample for the curve in fig. 12.12A was strained in the longitudinal direction, and the sample in fig. 12.10A was tested in the transverse direction (90° to the long axis of the abdomen), and hence was strongly reinforced by the transverse chitin fibers in this tissue.

The curve for the extensible cuticle has a very low stiffness of 3 kPa to strains of about 6, and then it stiffens to about 36 kPa to a strain of about 10. After this, the material softens and reaches a plateau at a strain of about 15, which is about the maximal strain seen when the animal is extending its abdomen to lay eggs. It turns out that imposing a strain of 15 onto these three intersegmental membranes, which start out with a length of about 2 mm each, is more than sufficient to extend the abdomen to the required length of 8–10 cm. That is, 6 mm of membrane times 15 equals 9 cm of extension, which would make the abdomen about 11 cm long. There are two key questions we need to answer to understand how this system works. How does the animal extend its abdomen, and what material properties of the membrane are required for this process to work?

The mechanism for extension of the abdomen does not involve a hydrostatic expansion of this tubular structure; rather, it is powered by the motion of the ovipositor valves at the very posterior end of the abdomen. These valves function as "shovels" to

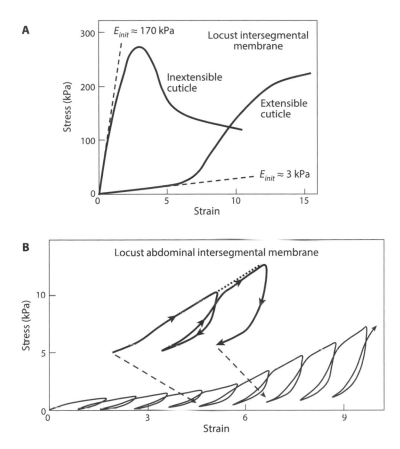

Figure12.12. Consequences of short-fiber composites in insect cuticle. **Panel A.** Tensile-stress-strain curves for the locust intersegmental membrane cuticle in its inextensible and extensible forms. The tissues in these tests were at in vivo hydration levels. **Panel B.** Load-cycling behavior of the plasticized locust abdominal intersegmental membrane from a fully mature female, in the state that exists when the female lays her eggs.

push soil aside and to pull the abdomen further into the hole it has created. Thus, during the digging cycle, there are periods when the valves actively rotate towards the head and thus act to extend the membranes, and there are times when valves lose contact with the sides of the hole, and thus opportunities for the membranes to recoil elastically, causing the abdomen to retract. The beauty of this material system is that the membranes do not spring back immediately, like a perfect spring, as we will soon see.

Before we look into this load cycling in more detail, it will be useful to consider the overall reversibility of stretching for this material. To do this, Vincent stretched membranes slowly to strains of up to 7 and then held them at this strain for 1 hour. The samples were then allowed to relax at zero stress for 1 hour, and the residual strain was measured. Under these conditions the test samples recoiled slowly to a strain that was about 10% of the initial strain imposed on the tissue, suggesting that the deformation of this tissue is quite reversible. The living animal extends its abdomen over a period of about 30 minutes to even higher strain levels, and the elastic recoil at the end of this process is essentially complete. So this is definitely an elastic material. How, then, does this material system work during the digging cycles used by the animal?

First, let's think about the role of the transverse chitin microfibrils in this composite. As this membrane is extended to strains of up to 15, we might expect that the lateral dimensions of the membrane would contract, just as a stretched rubber band gets thinner and narrower because rubber typically deforms without changes in volume. If this were the case for the locust, then the intersegmental membrane would taper down to a narrow tube, which it does not do. Vincent measured the Poisson's ratio for this material for extension in the longitudinal direction, which is the ratio of the lateral strain divided by the longitudinal strain of the test sample, as indicated by eq. (2.9). At low strains the ratio is about the expected value of 0.5, which is appropriate for a material that deforms at constant volume, but by strains of about 3 (i.e., 300% extension) the Poisson's ratio falls to very low values of about 0.03, which indicates that the lateral contraction of the membrane is negligible compared with the longitudinal extension. Thus, as abdominal extension proceeds the extended intersegmental membrane remains at relatively constant width. This behavior implies a relatively high compressive stiffness of the material in the transverse direction, and this lateral stiffness is the result of the transverse chitin microfibrils described above.

Next, let's think about the membrane's behavior in the longitudinal direction, and the role of its viscoelastic properties, as documented in Vincent (1975). This study shows that incremental extension of the abdomen, followed by periods of viscoelastic relaxation, allows the gradual extension of the abdomen in the rather complex digging cycle employed by the animal. Video analysis of the digging cycle indicated that the extension phase, when the valves pull the abdomen forward, lasts about 2 s and the recovery phase lasts about 1 s. The animal typically carried out three to five cycles of extension/recovery, separated by intervals of about 4 s when she held the membrane at an extended length and used the valves to tamp the sides of the hole that she was creating. During this "hold" period, the stress in the tissue relaxed quite considerably. Vincent built a test machine that allowed him to mimic these movements to measure the mechanical behavior of the membrane under continued cyclic loading, and he created a stress-strain profile that allowed the membrane to achieve the exceptional strains observed in the living animal. A simplified version of this load cycling is shown in fig. 12.12B, where we see the results of an experiment where the extension rate was equivalent to that created by the animal, but involved longer times for extension and retraction. In this case extension occurred for 60 s and retraction occurred for 30 s, making the extension-retraction cycle 30 times longer than the one employed by the animal. The process in this experiment caused the sample to reach a maximal strain of about 11 over 9 load cycles, whereas it would have required about 300 load cycles by the insect using the same strain rate. Note that there are no "hold" periods in this test protocol.

This diagram gives us a clear look at the shape of the extension-retraction cycle and allows us to understand the mechanics more clearly. The inset curve in this figure gives an expanded view of two load cycles, and you will see that the membrane material is very viscoelastic, because if the retraction had proceeded to the starting strain level, there would be a very large, open hysteresis loop indicating that much of the energy to stretch would be dissipated in each cycle. However, since Vincent only allowed half the time for the recovery phase of the first cycle, the second cycle started at low stress but did not follow the trajectory of the initial extension. Rather, the stress rose just slightly above the unload portion of the first cycle, and it did not reach the level of the extrapolated first cycle (dotted line) until the end of the second extension. This type of material behavior is called "stress softening," and is often called the "Mullins effect." It is not well understood, but for those interested in learning more about this effect, Diani and colleagues (2009) provide a good review. Stress softening is commonly seen in filled rubbers, such as the rubber used in tires, where rubber networks are filled with and reinforced by rigid particles, such as carbon-black particles. The key feature of this effect is that in incremental load cycles the loading curve to reextend the sample follows a stress level that is lower than the stress during the initial loading. This behavior likely provides an additional function for the transverse chitin microfibrils seen in this cuticle, which clearly function as the reinforcing filler particles.

The final piece of information to complete our understanding of this unusual type of cuticle comes from a study by Stephens and Vincent (1976), who used infrared spectroscopy to evaluate the secondary structure of the cuticle proteins in the mature, extensible intersegmental membrane. Their results show that the majority of the protein in this membrane exists in the form of random coils, with about a third in the form of α helices. There was no indication of β structure in the protein. These observations are entirely in agreement with the rubberlike properties observed for the membrane.

Clearly, our last two cuticle examples have strayed from the initial focus of this section on rigid materials, as they are definitely soft and extensible materials. They were included here because arthropod cuticle design is so radically different from most other material systems in animals. One of the wonderful aspects of the overall design of insect cuticle is that the low-aspect-ratio chitin microfibrils allow insects to modulate the stiffness and extensibility of these proteins through control of hydration and cross-linking, and this has made it possible for arthropods to produce an enormous range of structural materials with stiffness levels that span from 3 kPa to 20 GPa, a range of almost seven orders of magnitude. In the next section we will address the design of soft materials in general, and we will see that there are many designs that employ collagen-fiber-based reinforcement with extensible matrices that range from rubberlike proteins to hydrated polysaccharide gels.

10. THE FUNCTIONAL CONSEQUENCES OF SHORT-FIBER COMPOSITES IN INSECT CUTICLE. Before we leave our discussion of insect cuticle it will be interesting to revisit the basic hypothesis that we started with; namely, that insect cuticle's material properties are significantly limited by the requirement for this class of material to accommodate the complex developmental processes imposed by molting. We have seen that the reinforcing chitin microfibrils have low aspect ratios relative to those of similar composite systems made from cellulose, and that the low aspect ratios are necessary to achieve the low stiffness levels required for the animal to expand its newly synthesized cuticle when in molts and to create the soft cuticle structures that exist in the flexible parts of an insect's exoskeleton. Clearly, the short-fiber composites in insect cuticles allow the animals to produce soft, extensible cuticle structures as well as rigid structures, and this diversity of material properties makes it possible for insects to manage essentially all of their key biomechanical mechanisms without need for extensive use of collagen-based connective tissues. In fact, it has been established that

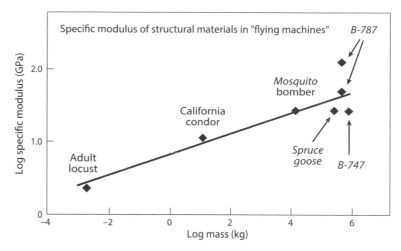

Figure 12.13. Specific modulus of structural materials in "flying machines." Scaling of the specific modulus of structural materials used in the construction of living and artificial flying machines. E_{init}, initial modulus.

some insects, specifically the fruit fly *Drosophila*, and presumably other flies, have lost the gene for collagen, which is present in the genomes of more primitive insects. This means that they have replaced all their soft-tissue functions with soft-cuticle alternatives. Are there functional consequences to the insect's commitment to chitin-based composite materials?

One possible consequence is that the short-fiber-composite design has set an upper limit to the stiffness and strength of insect exoskeletons, and this limitation may have imposed size limits on these wonderful flying machines. A comparison of chitin-based cuticle with cellulose-based wood and artificial materials should provide us with some interesting insights. Fig. 12.13 shows a log-log plot of the specific modulus (modulus divided by the specific gravity of the material) for a number of materials that have been used in the construction of flying machines, both artificial and living, against the mass of these machines. The examples plotted here were chosen to represent the upper end of the size range for flying machines made from five types of structural material: insect cuticle, bone, spruce wood, aircraft aluminum, and carbon-fiber composites. The specific modulus is used here, rather than the modulus alone, because this allows us to incorporate differences in the density of the materials in this analysis. This is important for an analysis of flying machines because flight requires aerodynamic forces to support the full mass of the airframe plus the "cargo," which in the case of living machines such as the locust and the condor includes all the nonlocomotor body components.

In this plot we see a very strong correlation between the mass of these flying machines and the specific modulus of the materials used to construct their various airframes. The artificial machines include two wood-based structures (the British *Mosquito* bomber from World War II, and Howard Hughes's *Spruce Goose*, which flew only once), an aluminum-based design (the *Boeing 747*), and a design based largely on carbon-fiber composite (the *Boeing 787*). The data for the specific modulus of the insects is based on the properties of locust tibial cuticle, and the data for the California condor are based on this bird's size and weight and on the modulus and density of bone from the sarus crane given in John Currey's book (Currey, 2002). A similar strong correlation is seen between wingspan and specific modulus. So, it seems plausible that the specific modulus of the structural materials used to create flying machines has played an important role in determining the maximal size that flying machines can achieve. That is, the threefold greater specific stiffness of bird bone relative to that of insect cuticle likely contributes to the roughly one-thousand-fold difference in maximal body mass between insects and birds. Interestingly, the correlation between the specific tensile strength and the mass or the wingspan is only marginally significant, although our data set for strength of insect cuticle is exceptionally limited.

You should be amazed to see where the wood-based aircraft lie on this graph because they are of similar size to modern aircraft made from high-strength aluminum alloys and carbon-fiber composites. The reason that wooden airframes can be so big is probably the exceptionally low specific gravity of wood, which is about 0.4, because in dry wood the cell walls are air filled. Aluminum and carbon-fiber composites have specific gravities of 2.7 and 1.6 respectively, much higher than the specific gravity of wood. Two data points are plotted for the *Boeing 787* because carbon-fiber composites come in many forms, and the two points indicate a reasonable range for a 0.6-volume-fraction uniaxial composite (the higher value) and for a 0.5-volume-fraction cross-ply.

We have seen that the cellulose microfibrils that reinforce the cell walls of plants are virtually identical to the chitin microfibrils that reinforce the cuticles of insects. But there is one crucial difference, the aspect ratios of the fibrils in the woody cell walls and in the cuticles of insects. The secondary cell walls in woody plants have cellulose microfibrils with aspect ratios in the range of 10,000 or more, and chitin microfibrils in insect cuticle have aspect ratios of the order of 100. Trees can grow to heights that approach 100 meters, and the largest insects in the world today have linear dimensions of the order of 0.2 meters. This discrepancy is due in part to the aspect ratios of their microfibrils, but the fiber volume fraction and the matrix shear modulus likely also contribute to the differences in stiffness and strength between these composites. I wonder what our world would be like if insects had developed the ability to give their cuticles material properties similar to those of plant cell walls.

Before we move to the collagen-based soft composites, it will be useful to mention several other kinds of rigid materials that have not been covered in this section. Two material systems that play key roles in animal designs are the highly mineralized ceramic materials seen in shell-like structures of many animal skeletal systems and the rigid protein composite known as keratin. The mineralized ceramics include structures such as calcium carbonate–based shells, seen in mollusks, echinoderms, coelenterates (corals), and others. The protein-based composite called keratin is a cellular structure made from epithelial cells that contain high densities of the intracellular protein fibers called intermediate filaments. When these cells die, they dry and stiffen to form structural materials that are found in the vertebrates as the surface layer of skins, and in reptilian scales, bird feathers, and mammalian hoof, horn, claws, and hair. For those interested in these kinds of materials, Vincent's book *Structural Biomaterials* (2012) will be a useful source, as well as the very extensive review article by Marc Meyers and colleagues (2008).

SECTION V

THE MECHANICAL DESIGN
OF PLIANT BIOMATERIALS

Chapter 13

The Evolutionary Origins
of Pliant Biomaterials

Pliant materials provide the structural basis for the bodies of the earliest metazoan organisms that evolved in the ancient oceans, and the coelenterates provide important insights into the designs of these early animals. The body walls of animals such as sea anemones and coral polyps are constructed from a collagen-based fiber-reinforced composite called mesoglea. In addition, the collagen fibers are embedded in a matrix that is made from protein-polysaccharide complexes of rather unique organization. We will start our analysis of pliant materials with these most primitive of pliant composites.

1. THE EVOLUTION OF COLLAGEN FIBRILS. Interestingly, the collagen molecules in the reinforcing fibers in these primitive composites are very similar to those of vertebrate collagens described previously in sections on tendon (chapter 9) and bone (chapter 11), as shown by Nordwig and Hayduk (1969). There is clear biochemical and structural evidence to support this contention. That is, there are very strong homologies between the gene sequences of sea anemone and mammalian collagens, and in the structural organization of the tropocollagen molecules into fibrils and fibers. What is interesting, however, is that the collagen fibrils in these ancient composites are considerably thinner than the fibrils seen in tendons and other collagen-based structural materials in vertebrate animals. Fig. 13.1 shows some of the evidence for the structure of this primitive collagen and its organization in mesoglea.

In fig. 13.1A we see a diagram that provides clear evidence of the strong sequence homology between coelenterate collagen and vertebrate collagen. The diagram shows the results of experiments in which isolated tropocollagen molecules are caused to aggregate into an unusual pattern in which individual molecules line up in perfect lateral packing. That is, there is no partial overlapping of the tropocollagen molecules as seen in collagen fibrils; rather, the molecules are lined up without overlap. This produces structures that can be observed in the electron microscope, and with appropriate staining they reveal a specific banding pattern that reveals the distribution of polar amino acids along the full length of the tropocollagen molecule's sequence. The top portion of this panel shows a series of arrows arranged in parallel to represent the organization of the tropocollagen molecules, and the banding pattern below shows the

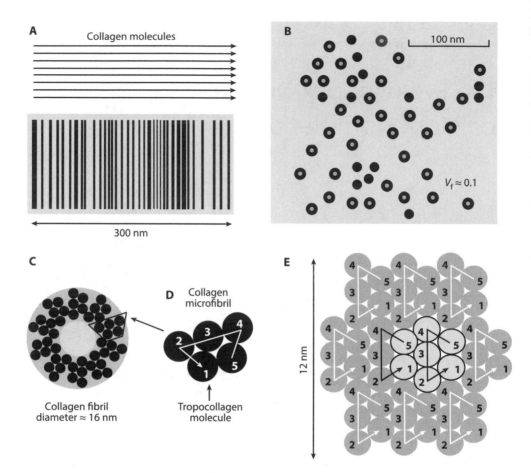

Figure 13.1. Structure of coelenterate collagen. **Panel A.** Segment-long spacing crystals of collagen, where individual collagen molecules line up in register, creating a banded structure that indicates the distribution of polar (*black*) and nonpolar regions in the collagen molecule's amino-acid sequence. This banding pattern is identical in vertebrate collagen and collagen from sea anemones. **Panel B.** This diagram shows the appearance of collagen fibrils in electron micrographs of *Metridium* mesoglea. The fibrils appear in two forms: small, solid fibrils (diameter ca. 12 nm), and larger, hollow fibrils (diameter ca. 16 nm). V_f, fiber volume fraction. **Panel C.** This diagram provides a plausible structure for the 16 nm collagen fibril, where 10 collagen microfibrils are organized into a hollow ring. **Panel D.** The structure of an individual collagen microfibril that is made from an overlapping array of five individual tropocollagen molecules. **Panel E.** The collagen microfibrils are packed into a solid crystalline array in the small-diameter (12 nm) collagen fibril. As shown here by the set of seven lightly shaded collagen molecules, we see that the collagen molecules form a hexagonal crystal structure in the fibril.

characteristic distribution of polar amino acids along the collagen molecules isolated from sea-anemone collagen. A single banding pattern is drawn because the pattern in the sea-anemone collagen is identical to that seen in vertebrate collagens. That is, the length of the triple-helical tropocollagen molecule and the sequence design of the peptide chains described in detail in chapter 9 (3. Tropocollagen, the Collagen

Molecule, and 4. The Assembly of Collagen Fibrils) also applies to the collagen fibrils in sea-anemone mesoglea.

What is different about mesogleal collagen, however, is the lack of discrete collagen fibers that are made of tightly packed collagen fibrils. Rather, the fibrils are distributed individually within an "amorphous" matrix. In addition, the mesogleal fibrils are much thinner than the fibrils seen in vertebrate tissues. The diameter of vertebrate fibrils typically ranges from 50 to 250 nm; in the sea anemone *Metridium senile*, the collagen fibrils have a diameter of about 12–16 nm, as measured from the high-resolution image provided by Grimstone and colleagues (1958), and many of the fibrils appear as hollow structures and others appear as solid structures. Typically, the hollow fibrils have diameters of about 16 nm, and the solid fibrils have diameters of about 12 nm. It is not clear, however, if these two cross-sectional shapes indicate two distinct classes of fibrils, or if mesogleal collagen fibrils taper at their ends into a compact structure, but are hollow in central regions of the fibril. Fig. 13.1B provides a diagram of the structure of *Metridium* mesoglea obtained with electron microscopy at high magnification. The solid, dark circles and the hollow circles represent the two forms of collagen fibrils, and their distribution in this diagram has a fiber volume fraction $V_f \approx 0.1$. This is a reasonable estimate of the volume fraction seen in the EM images, but it is likely that this is an overestimate for the fully hydrated material because there was probably considerable shrinkage of the tissue during preparation for electron microscopy. Indeed, it is likely that the fiber volume fraction in the native tissue varies between 5% and 10%. We will take 5% volume fraction as a plausible average value for the whole-tissue collagen content when we create composite models to analyze the mechanical design of this tissue. At this stage, however, we will use this structure to introduce an additional level of the structural hierarchy that exists in collagen; namely, the collagen microfibril.

Fig. 13.1C shows a diagram of the possible substructure of one of the hollow fibrils. Note that it appears as a ring of subunits, which we will identify as microfibrils, the smallest collagen substructure made from tropocollagen molecules, each of which has a diameter of 1.5 nm. The microfibrils are drawn as five dark circles, where each circle represents a single tropocollagen molecule, cut to reveal its essentially circular cross-section. Fig. 13.1D shows a larger view of the microfibril cross-section. This microfibril, which extends out of the page toward you, is a thin filament assembled from five overlapping tropocollagen molecules. To understand this structure it will be useful to refer back to fig. 9.4C that shows the overlap pattern of adjacent tropocollagen molecules, which are indicated by long arrows. Recall that the tropocollagen molecules overlap with their neighbors and that there are five overlap zones along the length of each tropocollagen molecule. Fig. 9.4C presents the overlap pattern as a flat sheet, and it explains the nature of the overlap between adjacent molecules, but it does not indicate how this overlap pattern is achieved in planes above and below the plane of the diagram. It turns out that collagen molecules do not interact as sheets; rather, they are organized as microfibrils that contain just five tropocollagens arranged in a loop pattern, as shown in fig. 13.1D. Thus, the white arrow in fig. 13.1D indicates the packing of the five overlapping tropocollagens within this microfibril. That is, the interface between tropocollagens 4 and 5 indicates a pair of tropocollagens in which repeat-sequence blocks 4 and 5 are adjacent and bonded together by secondary bonding forces (hydrogen bonds, ionic bonds, and hydrophobic interactions). Thus, this ring of five tropocollagens is equivalent to the five-tropocollagen flat-sheet pattern shown in fig. 9.4C, but we now see how this overlap pattern is organized in three dimensions. The adjacent microfibrils also interact laterally with their neighbors, but the primary structural integrity of the collagen fibril is established by the interactions within this microfibril. To fully bring this microfibrillar structure into the structural hierarchy of collagen, fig. 13.1E shows an assemblage of 10 microfibrils packed tightly together as they exist within the fibrils of vertebrate collagen. Here we see that the tight packing of the microfibrils produces hexagonal packing of the tropocollagen molecules, revealed clearly by the seven lightly shaded circles at the center of this diagram. This arrangement of the tropocollagens in adjacent microfibrils creates the hexagonal crystal lattice that provides the basis for the X-ray diffraction patterns that have been used extensively to document the structural organization of collagen fibrils (Orgel et al., 2006; Wess et al., 1998). In addition, this diagram, as a whole, provides a plausible structure for the smaller, ca. 12 nm diameter, collagen fibrils seen in electron micrographs of *Metridium* mesoglea that do not show a hollow center.

2. SEA ANEMONE ANATOMY AND FUNCTION. With this background on the structure of collagen fibrils in mesoglea, we can begin our analysis of the mechanical design of mesoglea, and to do this we need to consider the function of this material in the living animal. The early studies by Batham and Pantin (1950) documented the behavior of *Metridium* during its daily schedule, which is dominated by the tidal cycles in its nearshore marine environment. This animal is a filter feeder; it gathers its food with a crown of tentacles located at the top of its body column. Its preferred habitat is in the subtidal waters of temperate oceans. Typically, it feeds during periods of tidal flow, which bring microscopic prey into contact with its tentacles. The mouth of this animal is located at the center of the tentacle disc, and extending into the fluid-filled center of its body column from the mouth is a cilia-covered groove (called the siphonoglyph) that pumps seawater into the cylindrical body cavity to inflate the animal (fig. 13.2A). During feeding the body is inflated by the siphonoglyph, which generates internal pressures of the order of 10 Pa (this is about one ten-thousandth of atmospheric pressure). Figs. 13.2A and 13.2B show the animal's transition from a resting state, where the column height is about 1.5 times its width, to the extended state, where the column height is 2–3 times the resting height. Fig. 13.2B shows the typical appearance of a feeding animal, in which the column is greatly extended to allow the animal to move its tentacles up into the tidal flow to capture prey. Note that the column diameter is reduced, particularly in the upper half of the column, and this reduction in diameter allows the column to bend in the current and expose the maximal area of the tentacle ring to the oncoming flow. The thickness of the mesoglea is greatest in the bottom half of the column; its thickness decreases with height above this level and reaches a thickness about one-half that in the lower half of the column, as described by Mimi Koehl (1977c). Following feeding, the animal contracts its body column to a minimal shape (fig. 13.2C) in the process of defecation, which releases the remnants of its meal after digestion.

These animals work on a time scale that is very different from that of most other animals, with the behavioral cycle described above extending over many hours, typically determined by local tidal cycles. The animal has at its disposal three important design features that allow it to carry out the processes described above; namely, a viscoelastic body-wall material, the mesoglea; the siphonoglyph pump that brings water into the

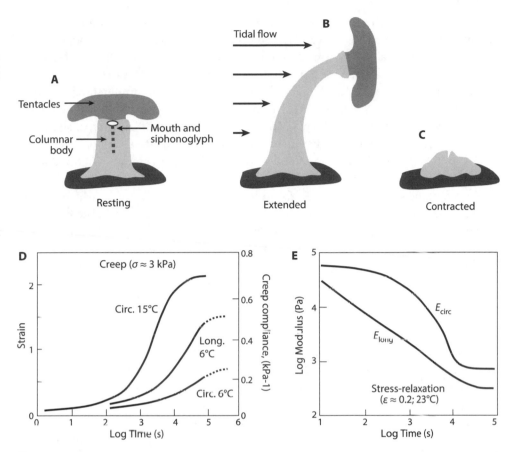

Figure 13.2. The sea anemone *Metridium* and the properties of its mesoglea. **Panel A.** Diagram of the sea anemone *Metridium senile* in its resting state. **Panel B.** During feeding the anemone extends its body column, raising the ring of tentacles into the tidal flow where it can capture plankton. **Panel C.** After feeding, in the process of defecation, the anemone contracts its body into a minimal shape. **Panel D.** The results of long-term creep tests carried out with samples of *Metridium* mesoglea. Circ., circumferential; long., longitudinal; σ, stress. **Panel E.** The results of long-term stress-relaxation tests on *Metridium* mesoglea. E_{circ}, modulus, circumferential test sample; E_{long}, modulus, longitudinal sample; ε, strain.

internal cavity to inflate this cylindrical structure; and longitudinal muscles that can actively contract to shorten the column as needed.

3. MESOGLEA: VISCOSELASTIC PROPERTIES. Alexander (1959) documented the viscoelastic behavior of the mesoglea in creep tests. Tissue samples were obtained from rings of tissue, which were loaded in the circumferential direction of the column. The experiments documented the time-dependent creep of the mesoglea when loaded at constant stress for periods of the order of 24 hours. Because the tissue samples were strained to nearly three times their initial length, the tissue cross-sectional area was reduced dramatically to one-third of its initial level (assuming deformation at constant

volume), and Alexander employed a hyperboloid weight that sank into water as the tissue stretched to reduce the load and hence maintain a constant stress throughout the experiment. A typical result from a circumferential test is shown in fig. 13.2D, where we see strain plotted against the log of time in seconds. Alexander's data are identified by the notation "circ. 15°C," as the sample was strained in the circumferential direction and the experiment was run at 15°C. The graph shows the progression of a creep test starting several seconds after the application of stress, with an initial strain of about 0.1. Strain then rises in a sigmoid curve over about 10 hours, to a final engineering strain of just over 2. That is, the sample extended to a length that was more than three times its initial length.

The constant stress level used in this test was approximately 3 kPa, which, as calculated by Koehl (1977c), is the approximate tensile stress developed in the body wall because of the bending of the animal's body column in the tidal flow. The scale on the right side of the panel gives the creep compliance of this material in units of kPa^{-1}. If you are confused by this measure of tissue properties, recall from eq. (3.9) that compliance is defined as:

$$D(t) = \frac{\varepsilon(t)}{\sigma_0},$$ (13.1)

where σ_0 is the constant stress applied to the test sample and $\varepsilon(t)$ is the time-varying strain that occurs in the sample. It is related to the time-dependent stiffness, and if we choose to invert the compliance, we obtain a reasonable estimate of the time-dependent stiffness (however, it will not be exactly the same as the actual stiffness, $E(t)$, measured in a stress-relaxation test). That said, we see that the initial compliance is about 0.03 kPa^{-1}, making the "stiffness" about 33 kPa. Over a period of about a day the compliance increases about 25-fold to about 0.7 kPa^{-1}, or the "stiffness" falls to about 1.4 kPa. Alexander showed that this curve could be fitted with a simple viscoelastic model, which we called a standard linear model (see fig. 3.5A). This model contains a spring in series with a Voigt element, and it has a single time constant that Alexander estimated at about 3000 s. The fit of this single-element model, however, is not perfect, and a much better fit is obtained with a model that contains two Voigt elements in parallel, having time constants of 1600 and 10,000 s. Unfortunately, Alexander did not provide plots for longitudinal creep, and it is therefore not possible to make comparisons of the stress-strain behavior between circumferential and longitudinal properties from his study.

Koehl (1977b) measured both the longitudinal and circumferential creep profiles for *Metridium* body-wall tissue; the data presented were obtained from a single animal and are also plotted in fig. 13.2D. In this case, however, the experiments were carried out at 6°C. It turns out that the 6°C–15°C temperature range used in these tests is a reasonable approximation of the seawater temperatures where this animal lives, so they likely give us a good indication of the biologically relevant material properties for the mesoglea. Note that the longitudinal compliance of the mesoglea is about three times greater than the circumferential compliance, which means that the mesoglea is stiffer in the circumferential direction than in the longitudinal direction, by a factor of about three. This difference plays a key role in the overall structural design of this cylindrical animal, as we will soon see.

Gosline (1971) used stress-relaxation tests to measure the time-dependent properties of mesoglea in the longitudinal and circumferential directions, as shown in fig. 13.2E. The top line shows the stress-relaxation behavior of a circumferential test sample, and the bottom line shows the stress-relaxation of a longitudinal sample; both were strained initially by about 20%. The data confirm the mechanical anisotropy described above from creep tests, and they suggest that the difference in stiffness may be even larger than the threefold value described earlier, particularly at low strains. In this case, the data are plotted as log of elastic modulus, $E(t)$, against log time, and at the middle of the relaxation profile the ratio of circumferential stiffness to longitudinal stiffness approaches eight. These data, obtained at a higher temperature of 23°C, provide additional insights into the time-dependent behavior of this unusual material. It is difficult to apply these data directly to the functional design of mesoglea, as described above, because the sea anemone's body wall actually functions in a creep process and not in a stress-relaxation process. However, the creep and stress-relaxation data both support the conclusion that the mesogleal material must be reinforced in some manner to create greater stiffness in the circumferential direction. The next issue to consider is the structural basis for this mechanical anisotropy.

Fig. 13.3 shows how the collagen fibers in mesoglea are organized in the fiber-reinforced-composite structure that creates the mechanical properties just described. The diagrams are based on polarized-light microscopy of mesogleal tissue sections that were cut in three different orientations in the body wall. The data come from my own study (Gosline, 1971) and one by Koehl (1977b). Starting at the left of this figure, we see a hollow cylinder, which represents the cylindrical body wall of the animal. To its right, a rectangular block of mesogleal tissue is shown, and four diagrams are provided to indicate the pattern of collagen-fiber reinforcement that exists in the body-wall material. The diagram labeled "cross-section" reveals the pattern of collagen fibers viewed from the top of the tissue block, the diagram labeled "radial section" shows the collagen fibers as they would appear on the side of the tissue block, and the diagrams labeled "tangential sections" show how the collagen fibers would appear at different levels through the thickness of the tissue. There are three different fiber systems running in two distinct tissue layers in the body-wall composite. Look first at the radial section, where we see on the right a row of vertically oriented solid lines that represent vertically oriented collagen fibers. However, this section reveals only the vertical component of this fiber system, because we can see in the tangential sections that these collagen fibers in the outer 60% of the tissue form a cross-lattice structure. In addition, we see a set of horizontal dotted lines running in the radial direction that represent radially oriented collagen fibers spanning the full thickness of the mesoglea.

Next, look at the diagram labeled "cross-section," and here again we see the inner and outer layers in the mesoglea. In this case, to simplify the drawing, the radial fibers are not shown. The inner layer reveals a second set of tightly packed collagen fibers that all run in the circumferential direction and occupy about 40% of the body-wall thickness. These fibers are not arranged in a cross-lattice; rather they run at 90° to the long axis of the mesogleal column. The outer portion of the cross-section shows the cut ends of the cross-lattice of collagen fibers described above. The actual organization of this layer will be best seen in tangential section.

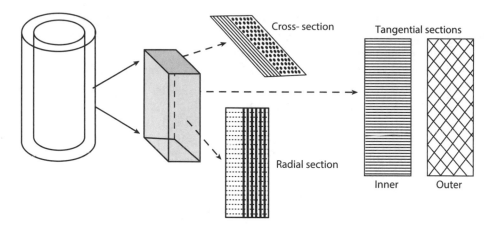

Figure 13.3. The organization of the collagen fiber lattice in *Metridium* mesoglea. The cylinder at the left represents the shape of the whole cylindrical animal, and the rectangular block to its right represents a sample of the animal's body-wall mesoglea. The three arrows extending from this block lead to diagrams that show the organization of collagen fibers that would be seen with polarized-light microscopy of tissue sections cut from the three faces of this tissue block. The radial section reveals the longitudinal orientation of a set of fibers in the outer two-thirds of the tissue, along with a small amount of radially oriented collagen that spans the full thickness of the material. The cross-section shows a layer of circumferential collagen fibers in the inner third of the tissue, and the outer two-thirds of the tissue shows the cut ends of the longitudinally oriented collagen fibers seen in the radial section. The tangential sections clarify the relative orientations of the inner and outer fiber systems. As seen in a tangential section from the inner third of the body wall, the circumferential fibers do indeed run in a plane that is transverse to the body axis of this cylindrical animal. The tangential section from the outer two-thirds of the body wall shows that the collagen in this region forms a cross-lattice of fibers that are oriented at about 45° to the body axis in resting animals.

The tangential sections are shown separately for the inner and outer layers that were described in the radial and cross-sections. Again, the radial fibers are not shown. The tangential section through the inner layer clearly reveals the densely packed circumferential collagen fibers. Images of this layer look similar to those of tendons, thus they are likely to produce higher mesogleal stiffness in the circumferential direction. The tangential section of the outer layer clearly shows the collagen-fiber cross-lattice that was mentioned above. However, the cross-lattice layers are separated by nonbirefringent layers, which means that the collagen-fiber content in the outer layer is considerably less than that in the inner layer. In sections taken from relaxed animals, these fibers are oriented at about 45° to the vertical axis of the column. However, as the animal inflates itself and extends its body column, these fibers will reorient to become more aligned with the vertical axis of the body.

4. MESOGLEA: A COMPOSITE MODEL. We will now attempt to explain the animal's repertoire of body shapes in terms of the reinforcement pattern of the collagen

fibrils, as described above. This is a rather complex composite structure, but it should be possible to gain insights with discontinuous-composite models and with some reasonable simplifications of this structure. We will assume that the collagen fibrils have mechanical properties similar to those of vertebrate collagen, as discussed previously. That is, fibril stiffness is about 3 GPa. We need, however, to estimate a number of structural features that are not well established. Specifically, we need to know the aspect ratio of the collagen fibrils and their approximate volume fraction (V_f) in the material. In an early study (Gosline, 1971), I measured the salt-free dry weight of collagen as about 7% of tissue wet weight, which suggests that the average collagen volume fraction $V_f \approx 0.05$, based on a dry collagen specific gravity of 1.35. This value is in reasonable agreement with the volume-fraction estimates taken from electron micrographs, as discussed above.

The available electron micrographs do not, however, provide any direct indication of fibril length, as the collagen fibrils do not run for long distances in available EM images. However, the very low stiffness of this tissue suggests that the collagen fibrils likely have quite low aspect ratios. We will assume that the aspect ratios fall in the range of 50–250, and with these values we can create discontinuous-composite models for the inner circumferential-fiber layer that can be compared to the initial circumferential stiffness, $E_{c,0}$, as shown in fig. 13.2E, at a strain of about 0.2 in a stress-relaxation test. The initial stiffness $E_{c,0} \approx 60$ kPa at 10 s, and we start our analysis with models that explain this level of circumferential stiffness.

Since the inner layer of circumferential collagen fibers occupies about 40% of the tissue's thickness and likely dominates the circumferential properties of the body wall, we assume that the inner layer's stiffness is actually 2.5 times greater than that based on the full thickness of the tissue. This means that the inner layer's circumferential stiffness is actually about 150 kPa. Composite models based on a fiber volume fraction of 0.05 suggest that composites made with fibrils having aspect ratios of 100 ($AR = 100$) would require a matrix shear modulus of $G_m \approx 0.75$ kPa to create the observed tissue stiffness of 150 kPa. However, the inner layer has more densely packed collagen fibers than the outer layer, and thus its collagen content is likely to be higher than the average. Models based on $V_f = 0.1$ and $AR = 100$ suggest that a matrix shear modulus $G_m \approx 0.25$ kPa would explain the observed tissue stiffness. However, if the collagen-fibril aspect ratio is smaller, at $AR = 50$, then the matrix shear modulus must be higher, at $G_m \approx 1$ kPa, and if aspect ratio is higher, at $AR = 250$, the matrix shear modulus must be lower, at $G_m \approx 0.05$ kPa. These calculations provide matrix-shear-modulus values in the 0.05–1 kPa range, which seem plausible, although it is possible that aspect ratios may be larger, and hence G_m may be even smaller. Thus, the initial matrix shear stiffness is about one-thousandth that of a rubber band.

As time proceeds through the circumferential stress-relaxation test in fig. 13.2E, the tissue stiffness falls by a factor of 60. That is, at long times the whole-tissue circumferential stiffness levels off at about 1 kPa, making the inner layer's stiffness about 2.5 kPa. This stiffness level is achieved because the matrix has "relaxed" to a much softer state, to a shear modulus of the order of $G_m \approx 0.005$ kPa at aspect ratio $AR = 100$. Note that the circumferential stiffness remains constant for a considerable period at long times, and this indicates that the "relaxed" matrix contains a cross-linked elastic network that reaches its elastic equilibrium after 3–5 hours of stress-relaxation at 23°C. Thus, the matrix shear stiffness at long times is five orders of magnitude lower than the shear stiffness of a rubber band. Although this analysis does not produce a definitive model for the design of the mesogleal composite, it clearly documents the crucial role that an extremely soft matrix plays in this system.

The longitudinal properties follow similar trends, again shown in fig. 13.2E, but also in the creep tests at 6°C in fig. 13.2D. In both cases, however, the longitudinal stiffness of the tissue is considerably less than its circumferential stiffness. In the creep tests the lower stiffness is indicated by the greater strain in the longitudinal direction, and in the stress-relaxation test the lower longitudinal stiffness is indicated by its lower modulus. The low longitudinal stiffness arises from two factors. First, in longitudinal tests the radial and the circumferential collagen fibers of the inner layer are loaded in series (i.e., the stress is applied perpendicularly to these reinforcing fibers), and this series composite's properties are dominated by the tensile modulus of the matrix, E_m. Our estimate above for the matrix shear modulus at between 0.1 and 1 kPa at short times means that the matrix tensile modulus should be in the range of 0.3–3 kPa (because $E_m \approx 3G_m$). Thus, the tensile modulus of the inner layer in the longitudinal direction should be close to the tensile modulus of the matrix, making its longitudinal stiffness of the order of 1 kPa at short times. This means that the circumferential stiffness of the inner layer is about 100 times greater than its longitudinal stiffness, and if the inner layer of the mesoglea dominates the inflation mechanics of this animal, its body column should extend greatly in height over the several hours of inflation, and there should be relatively little, if any, increase in diameter. This is a somewhat correct description of the animal's body-inflation mechanics, but, in addition, the mechanical contribution of the outer mesogleal layer increases the vertical stiffness of the mesogleal system over time, and this likely sets limits on the extent of the animal's vertical expansion.

Without going into detail, the stress levels that develop in the walls of a pressurized cylinder differ by a factor of two in the circumferential and longitudinal directions, with the circumferential stress being twice that in the longitudinal direction. Thus, if an elastic cylinder made from an isotropic material is inflated by an internal pressure, the cylinder's circumference will increase more than its length. The three- to eightfold greater stiffness of the mesoglea in the circumferential direction seen in the creep and stress-relaxation tests is key to the pattern of inflation seen in the normal behaviors of this animal, where the animal increases its height with essentially no increase in diameter. In fact, the diameter declines with expansion—particularly in the top half of the column, where the column diameter is reduced by half (see fig. 13.2B), and this is a key design feature of mesoglea that arises from the structural organization of the collagen-fiber lattice in its outer layer. The outer tangential section in fig. 13.3 shows the cross-lattice organization of the collagen fibers in the outer mesogleal layer. The fiber angle of this lattice is thought to be at about 45° in the "resting" state, and in this diagram the fiber angle of the lattice in the outer section is drawn at 38° relative to the vertical axis to indicate the organization of this fiber lattice as it might exist during the stress-relaxation test run at a longitudinal strain of 0.2. It is this shift in angle with inflation that provides the rising longitudinal stiffness of the column that may ultimately limit the height of the column's extension. In addition, as the fiber angle decreases with extension, the width of the lattice will also decrease, and this decrease in width is what causes the diameter of the upper portion of the body column to decrease. How does this system work?

The trigonometry of a continuous-fiber lattice is dominated by the fixed length of the filaments, and a lattice that started at rest at a 45° angle could only extend by about 40% before the fixed-length fibers would rotate to a vertical orientation and stop further extension. In addition, the rotation of these fixed-length fibers would cause a dramatic decrease in the diameter of the body column. Clearly, a continuous-fiber lattice is not present in mesoglea. A discontinuous-fiber composite in mesoglea, however, would allow the cross-lattice structure in the outer body wall to achieve much higher longitudinal strains because the collagen fibrils can slide relative to one another, and in this process the viscoelastic matrix would be sheared, allowing the tissue to achieve much greater vertical strains. In addition, the cross-lattice of the fibers will reorient toward the vertical axis as the tissue is strained in the vertical direction, and this will increase the vertical stiffness of the body wall. Thus, the sliding and reorienting of the overlapping collagen fibers in the outer fiber system will limit the reduction in diameter due to the fiber rotation and may explain the maximal vertical strain of the body column. These are the key features of the mesogleal design that explain the animal's ability to extend its body longitudinally threefold but only reduce diameter by a modest amount as the column extends.

Because we do not have a definitive model for the composite structure of this material, it is not clear what actually limits the final height of the column. There are, however, several possibilities. First, the elongation of the collagen cross-lattice in the outer layer may rotate sufficiently toward a vertical orientation and increase the system's vertical stiffness to a level that prevents further extension at the very small pressures that are generated by the siphonoglyph pump. Alternatively, the very long time constants for the creep processes, shown by the samples in fig. 13.2D, may simply be large enough that the column does not have time to creep to greater strains during the duration of a tidal cycle, removing the requirement for elastic limits to the longitudinal expansion. This certainly appears to be the case for the creep curves at 6°C, but does not seem to be the case in the 15°C creep test. Perhaps the animal simply has a sensory system that monitors column height, and it turns off its siphonoglyph pump at appropriate extensions. One thing that is clear, however, is that the cross-lattice in the outer layer does indeed cause the mesoglea to contract laterally, and hence make the body column thinner as it extends, as shown in fig. 13.2B.

5. MATRIX CHEMISTRY. This brings us to the final question about mesogleal design, the biochemical nature of the matrix in this unusual structural biomaterial. Unfortunately, little is known about the chemical structure of the matrix material, and we are forced so speculate on possible designs. As a starting point, recall the structure of decorin, the proteoglycan found at the interface between the tightly packed collagen fibrils in the fibers of vertebrate tendon. This molecule has a small, compact core protein that binds to the surface of collagen fibrils, and a polysaccharide chain that extends between fibrils to form a matrix that binds collagen fibrils together into fibers. In decorin the polysaccharide contains about 50 disaccharide units, each of which contains a uronic acid that has a negatively charged carboxyl group and a sulfated sugar that also carries a negative charge. These negatively charged polysaccharides take on expanded conformations and likely form entangled gels that hold the collagen fibrils together in collagen fibers. The situation with mesogleal collagen and matrix is quite different.

Analysis of the sugar content of mesoglea from *Metridium dianthus* by Katzman and Jeanloz (1970) shows that the mesoglea is devoid of the negatively charged sugars that are typical of the matrix proteins in vertebrate collagens and many other soft tissues, but that mesoglea contains large amounts of the uncharged sugars: glucose, galactose, fucose, mannose, xylose, and hexosamine. Together, the content of these uncharged sugar polymers can range from 8%–10% of the dry weight of the tissue. The biophysical form of these sugar polymers remains to be determined.

In an early study (Gosline, 1971), I analyzed the collagen content of *Metridium* mesoglea by autoclaving the tissue to solubilize the collagen and its associated polysaccharides. The insoluble material contained precipitated matrix proteins and their associated polysaccharides. Starting with a 100 g sample of mesogleal tissue, the salt-free dry weight of macromolecules in mesoglea was 8.8 g, the salt content about 5 g, and the total water was thus about 86 g. Autoclaving produced an autoclave-soluble fraction that was 7.4 g per 100 g of tissue. This fraction contained all of the collagen and about two-thirds of the polysaccharide. The autoclave-insoluble fraction contained the remainder of the matrix material, which included both matrix protein and matrix polysaccharide.

Based on these data, it is possible to estimate the total collagen content as 6.7% of tissue wet weight and the matrix content as 2% of tissue wet weight. Finally, partitioning the water between collagen and matrix provides an estimate of the matrix concentration in mesoglea as about 2.5%. It is worth noting that a variety of biopolymer systems form soft gels at polymer concentrations of this level. For example, agar, a structural polysaccharide from seaweeds, will form stable soft gels from 0.5% solutions and forms rigid gels at 2%. Gelatin, derived from heat-processed collagen, forms stable gels from 1% solutions. So the matrix-forming macromolecules in mesoglea are certainly present in sufficient quantity to account for the soft-composite matrix that is inferred from materials testing. It remains for future research to establish the biochemical and biophysical properties of the matrix materials that transform the collagen-fiber lattice into this remarkable time-dependent material. As a start, it is possible that the abundant polysaccharide polymers provide an entangled network that accounts for the short-term stiffness and the time-dependent tissue extension/relaxation over three log decades of time to the elastic equilibrium of this system. It is also possible that the matrix contains an autoclave-insoluble rubberlike protein that provides the very soft elasticity that limits tissue extension at long times.

For those readers who are not familiar with invertebrate zoology, it is important to note that not all sea anemones live in subtidal habitats where flows are steady and vary over periods of hours. Other sea anemones live in the wave-swept intertidal zone, and they experience velocity oscillations on time scales of the order of 10 s. Koehl (1977a) also documented the mechanics of the body wall of an intertidal anemone, *Anthopleura xanthogrammica*. These animals experience drag forces with each wave, but Koehl discovered that the animals typically occupy sheltered locations in surge channels where they experience peak-flow velocities that were essentially the same as those seen by *Metridium* in subtidal currents. However, the flow velocities in these channels can be an order of magnitude greater a few tens of centimeters above their sheltered attachment sites, and they have developed body forms and mesogleal-material designs that suit their habitat. The "resting" height of their body column is less than half that of

Metridium, and the thickness of their mesoglea is roughly twice that of *Metridium*. In addition, the viscoelastic time constant for the extension of this column is at least an order of magnitude greater than that of *Metridium*, so they simply do not increase in height very much during their normal activities. Together, these properties allow *Anthopleura* to thrive in its turbulent, wave-swept environment. Readers should consult Koehl's papers (1977a, b, c) to learn about this animal's mesoglea in greater detail. The next chapter will consider the functional design of several rubberlike proteins produced by animals to provide high-efficiency, long-range elastic materials that have been studied in considerable detail.

Chapter 14

Rubberlike Proteins

The mesogleal system discussed in chapter 13 represents a body-wall design that is rather unusual in the many different kinds of animals that exist in nature. That is, composite structures made only from collagen fibers and a highly-extensible matrix of protein-polysaccharide complexes are relatively unusual. More frequently, we see material systems in which collagen-proteoglycan systems contain rubberlike proteins that provide long-range elastic properties to these composites. We will discuss these systems below, but first it will be useful to consider the range of designs seen for the rubberlike proteins that exist in nature. We will start with the most primitive of these materials, the fibrillin microfibrils that exist in essentially all invertebrate phyla and in vertebrates as well.

1. MICROFIBRILLAR ELASTOMERS. Evolution has produced a number of proteins that exhibit long-range soft elasticity, and several of these proteins meet all of the criteria established for this class of materials. That is, the proteins are known to take on dynamically mobile, random coil conformations, and they are cross-linked into three-dimensional random networks, as described in chapter 6. However, we will start with an unusual protein, fibrillin, that forms filamentous rubberlike structures called microfibrils. Please note that the term "microfibril" is also used to describe the basic substructure of collagen fibrils; collagen microfibrils are bundles of five tropocollagen molecules, as described in fig. 13.1C. Interestingly, fibrillin microfibrils exhibit elastomeric properties that are similar to those of true random-coil rubber networks. Fibrillin protein molecules, however, are quite rigid, folded structures, and they assemble into fibrillin microfibrils with diameters of about 16 nm. These microfibrils have a "beads-on-a-string" morphology with an axial periodicity of about 56 nm, as illustrated in fig. 14.1A, but their length is considerably greater than that shown in this diagram. Sherratt and colleagues (2003) report that bovine aorta fibrillin microfibrils have an average length of about 2000 nm, making the aspect ratio of microfibrils about 125. These microfibrils are bundled into elastic fibers that have diameters of the order of 2–5 μm. The fibers exhibit soft elasticity like that seen for cross-linked networks of random-coil molecules, in spite of the fact that they are built from quite rigid macromolecular subunits. The biophysics of this interesting material remains somewhat uncertain, but, given the material properties, it is likely that an entropic mechanism is involved. The structure and assembly of microfibrils, however, have received a great

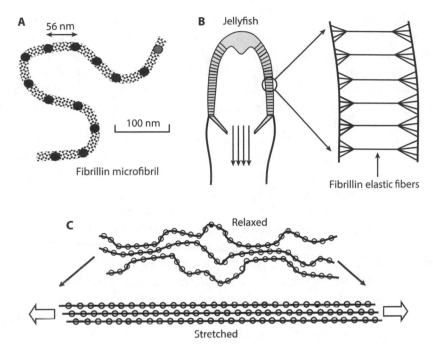

Fig. 14.1. The structure of microfibrillar elastomers. **Panel A.** The structure of a fibrillin microfibril as viewed by high-resolution electron microscopy. Microfibrils show a "beads-on-a-string" morphology, with a spacing of the "beads" of about 56 nm and a diameter of about 22 nm. **Panel B.** The organization of fibrillin microfibrils in the mesoglea of the jellyfish *Polyorchis penicillatus*. **Panel C.** A plausible structure for the organization of microfibrils in the zonule fibers that form the ciliary ligaments of the vertebrate eye.

deal of attention recently, largely because genetic defects in this material are known to cause a large range of human genetic diseases, the most prominent of which is Marfan syndrome (Jensen et al., 2012, 2013).

The first appearance of fibrillin microfibrils occurred in the earliest metazoan animals, as shown by Reber-Muller and colleagues (1995), and their presence is well documented in the mesogleal "bell" of the hydromedusa *Polyorchis penicillatus*, which is reinforced in the radial direction by discrete fibrillin-based elastic fibers, with diameters of about 2.5 μm, as illustrated in fig. 14.1B. The jellyfish diagram shows a surface cut through the center of the animal's bell-shaped body, which reveals an array of radially oriented elastic fibers that span the thickness of its body wall. The expanded diagram on the right indicates the organization of these radial fibers, which fan out near the mesogleal inner and outer surfaces to distribute their tension more uniformly across the structure. These elastic fibers function in the elastic recoil of the mesogleal bell during swimming. DeMont (Demont and Gosline, 1988) documented the dynamic strains that occur in the bell mesoglea during the animal's swimming contractions, and he observed maximal circumferential strains of about −0.35 during the power stroke, which is created by the contraction of circumferential muscles that line the inner layer of the bell mesoglea. Their contraction causes the expulsion of seawater from the bell-shaped body through a funnel-like structure at the base, which creates a pulsed jet of fluid that provides vertical thrust for the animal (downward-pointing arrows in fig. 14.1B). The maximal radial strains that occur in the mesoglea during this contraction are about 0.6. That is, the mesoglea thickens radially by about 60%, and thus the microfibrils undergo extensions of the order of 60% with each contraction. The elastic recoil of these microfibrils plays a major role in powering the refilling of the mesogleal bell in preparation for the next contraction.

Megill and colleagues (2005) carried out mechanical tests of the intact mesoglea in compression and tension to obtain estimates of the properties of these fibrils. The logic of their analysis is that the stiffness of the mesogleal matrix, which is composed of a collagen-fibril-reinforced proteoglycan gel, can be measured in compression tests without any contribution from the microfibrils because the microfibrils will buckle in compression and thus will not contribute to tissue stiffness. Tensile tests, however, show greater stiffness because the microfibrils are stretched, and hence the fibers will contribute to the tissue's tensile stiffness. The compressive stiffness of the mesogleal composite was about 0.35 kPa, and the tensile stiffness was about 1.2 kPa. Thus, the difference of about 0.9 kPa can be attributed to microfibrillar fibers. These authors then used a continuous-fiber-composite model (eq. [8.1]) to analyze this system and calculate the stiffness of the mesogleal microfibrils:

$$E_{c,0} = E_f V_f + E_m (1 - V_f). \tag{14.1}$$

Rearranging this equation to solve for E_f produces:

$$E_f = \frac{E_{c,0} - E_m (1 - V_f)}{V_f}. \tag{14.2}$$

Megill et al. used the tissue's compressive stiffness as E_m, and they measured microfibril diameter and microfibril density to calculate the fiber volume fraction, which turned out to be $V_f = 0.09\%$, giving a predicted fiber modulus of $E_f \approx 1 \pm 0.5$ MPa. Mesogleal microfibrils do indeed have rubberlike properties. However, it is not clear if this soft elasticity continues to high strains, as it does in rubber networks.

In addition to these invertebrate fibrillin-based elastic tissues, the vertebrate eye contains an array of fibrillin-based elastic fibers that form the ciliary ligament that attaches the lens of the eye to the muscle that controls the focal length of the lens. Research on the molecular origins of microfibrillar elasticity indicate that the fibrillin microfibrils are organized into bundles to form the micrometer-scale zonule fibers, and their organization within these fibers changes dramatically with extension, as shown in fig. 14.1C. This diagram, which is based on a study by Wright and colleagues (1999), provides a plausible organization for several of the many microfibrils present in a relaxed, unloaded zonule fiber. Note that the individual microfibrils are tightly clustered in parallel arrays in some areas, but they are quite disordered in adjacent areas. With extension of this filament bundle, the disordered—and presumably mobile—areas become extended and, at some point, they become tightly packed into a semicrystalline structure, as shown in the lower part of fig. 14.1C where they are extended by about 30%. This transition indicates how the early stages of microfibrillar extension may involve rubberlike elastic mechanisms associated with the extension of the disordered and mobile segments in the microfibril structure.

Mechanical testing of isolated bovine zonule fibers by Wright et al. (1999) provides a stress-strain curve for this interesting material, as shown in fig. 14.2A. The solid line in this figure shows the results of their mechanical test on the isolated zonule fibers, which reveals a J-shaped stress-strain curve that becomes nonlinear at strains of about 0.3. Note that once taut, these fibers show a stiffness of about 0.7 MPa, which is similar to the stiffness observed for the radial mesogleal microfibrils in the jellyfish. Thus, the zonule fibers show long-range elasticity with a stiffness that is consistent with a rubberlike entropic mechanism. However, the shape of this stress-strain curve is quite different from the stress-strain curve for a typical entropy-elastic polymer. Recall from fig. 6.6 that the stress-strain curves for rubberlike materials show declining stiffness in tension, leading to a constant stiffness at extension ratios above about 2 that is one-third that of the initial modulus. This is the Gaussian behavior of a long-chain rubber network. At high strains, however, rubberlike materials show increased stiffness as they enter into the range of non-Gaussian behavior, where the molecular chains take on highly extended conformations and stiffness rises rapidly. Fig. 6.7 shows how the number of random segments in the polymer chains between cross-links determines the strain at which rubber networks enter their non-Gaussian behavior. Note the curve for network chains that contain 12 random segments between cross-links. In this case, the non-Gaussian curve tracks the Gaussian curve for the initial 50% of extension, and then stiffness rises rapidly with further extension. Note also that the slope of the stress-strain curve for the zonule fibers rises continuously from the very start of their extension. This suggests that the zonule fibers contain an exceptionally short chain network. The dotted and dashed curves in fig. 14.2A show the results of network models that allow us to understand more about the structural design of microfibrils. The dotted line labeled "Gaussian" shows the behavior of a long-chain rubber network with a stiffness $G = 15$ kPa. This stiffness level was chosen because the initial stiffness of the curve matches that of the zonule fiber. However, short-chain versions of this rubber network will show increasing stiffness with extension, and non-Gaussian network models may be sufficient to explain the properties of the zonule fibers. The two dashed lines labeled "non-Gaussian" show the result of calculations for short-chain rubber networks containing 2 and 3.4 random segments between cross-links. The curve labeled "$n = 2$" has essentially the same initial stiffness as the initial portion of the zonule-fiber sample, and its stiffness rises significantly as strain increases, but not to the level seen for the zonule fibers. Network models based on even shorter chains come closer to matching the high-strain stiffness of the zonule fibers, but their initial stiffness exceeds that of the zonule fibers. Thus, it seems that simple non-Gaussian rubber-network models are not able to explain the mechanical behavior of microfibrils.

The model for relaxed zonule fibers in fig. 14.1C provides us with a structure that more accurately explains the properties of microfibrils. There are regions of disordered rubberlike chains in series with blocks of tightly packed extended chains. Network models based on this type of structure can explain the molecular origins of the microfibrillar stress-strain curve. The bold dashed line labeled "non-Gaussian hybrid" in fig. 14.2A was calculated for short network chains with a shear modulus of 15 kPa and containing 3.4 random segments. These short rubber chains are placed in series with more rigid, straight segments with a stiffness of 250 kPa, and the relative lengths of these regions were assumed to be equal. It is possible to get reasonable fits for models

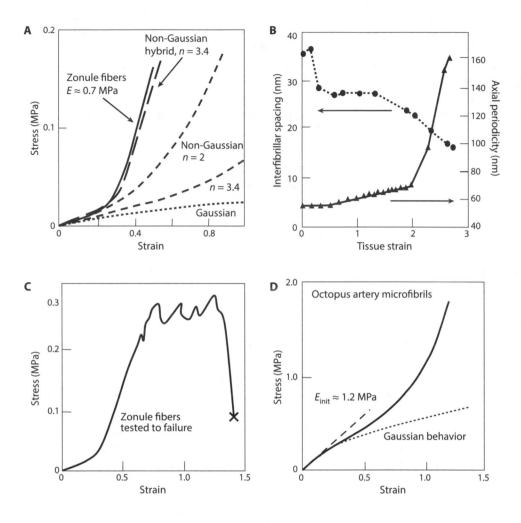

with longer rubberlike chains, but not with shorter rubberlike chains. The calculated stress-strain curve (bold dashed line) is almost a perfect fit for the zonule-fiber stress-strain curve (solid line), and it provides a clear indication that the extension of kinetically free segments is responsible for the initial portion of the stress-strain curve, and that the elastic deformation of fully extended microfibrils is responsible for the linear elastic behavior (at stiffness = 0.7 MPa) of the zonule fibers at strains above about 0.35. X-ray diffraction studies provide strong support for this model.

Synchrotron X-ray diffraction studies by Sherratt and colleagues (2003) and by Glab and Wess (2008) document the changes in molecular dimensions that occur with the stretching of zonule microfibrils. They documented changes in the axial periodicity (the interbead spacing) of the microfibrils and the average lateral spacing of the microfibrils (fig. 14.2B). They also measured the diameter of the microfibrils with extension, and observed that the diameter remained constant, at 16 nm, up to a strain of 2, at which point the diameter began to decline. At starting length, the interbead spacing was about 56 nm, as expected from electron micrographs, and the average interfibrillar spacing of the microfibrils was about 36 nm, which is much larger than the 16 nm

Figure 14.2. (*left*) The mechanics of microfibrillar structures. **Panel A.** The *solid line* indicates the stress-strain behavior of zonule fibers, showing a J-shaped curve with a final stiffness (*E*) of about 0.7 MPa. The *dotted* and *dashed lines* indicate the properties of polymer-network models that are used in the analysis of network structures that account for the mechanical properties of the microfibril-based zonule fibers. See the text for details of this analysis. **Panel B.** The results of X-ray diffraction experiments on microfibrils obtained from zonule fibers. The *solid line with triangles* indicates the change in the axial spacing of the microfibrils as the zonule fibers are strained. The *dotted line with circles* indicates the change in interfibrillar spacing of the individual microfibrils that make up the fiber bundles in the zonule fibers. Note that the axial periodicity of the tissue remains constant for an initial strain of about 0.4, which corresponds to the low-stiffness region of the zonule-fiber stress-strain curve in panel A. Above a strain of 0.4 the axial periodicity of the fibers increases, indicating that the disordered microfibrils, as shown in the "relaxed" structure in fig. 14.1C, are now straight and are being stretched. **Panel C.** Stress-strain curve for a zonule-fiber sample tested to failure, where we see a transition from a linear-elastic behavior to a behavior that indicates structural failure within the microfibrils at strains above approximately 0.7. Beyond this "failure" point stress remains relatively constant but oscillations in stress suggest sequential unfolding of structural domains within the microfibrils. **Panel D.** Tensile-stress-strain curve for microfibrils isolated from the aorta of an octopus. The *dotted line* indicates the Gaussian behavior of a rubber network with an initial modulus E_{init} = 1.2 MPa, and the *solid line* shows the properties of octopus microfibrils.

diameter of the microfibrils. This large interfibrillar spacing arises from the disordered structure of segments of the relaxed filaments, as shown in fig. 14.1C.

With extension of this tissue, the interbead spacing of the microfibrils remains constant up to a strain of about 0.4, but there is a dramatic drop in the interfibrillar spacing from 36 to 27 nm. This indicates that extension of the disordered regions allows the now-straightened microfibrils to pack more closely together. Thus, at a strain of about 0.4, the microfibrils are fully straightened, as they appear in the "stretched" diagram in fig. 14.1C. At strains above this level the interbead spacing begins to increase gradually from 56 nm, reaching about 70 nm at a strain of 2. This indicates that the "chain of beads" in the straightened microfibrils is being stretched, and in this process the roughly 100% increase in sample length resulted in a 30% increase in the axial periodicity observed by X-ray diffraction. This is a rather puzzling relationship, and it suggests that at strains above about 0.4 there are major reorganizations within the microfibrillar proteins that are not understood at this time. It also suggests that the high-stiffness region of the microfibrillar stress-strain curve results from the elastic deformation of the straightened microfibrils. That is, the rapidly rising stress at strains of 0.4 and above is not due to the non-Gaussian behavior of the relaxed filaments, but rather to the structural reorganization and unfolding of protein domains within the fibrillin molecules that make up the microfibrils. It remains for future research to identify the precise structural basis for the high-stiffness behavior of the microfibrils.

The dramatic changes in microfibril dimensions that were observed at strains above 2 suggest a process that results in the full unfolding of the microfibril structure. The axial periodicity and interfibrillar spacing begin to change rapidly, with the axial periodicity increasing more than twofold to about 160 nm and the interfibrillar spacing decreasing by about 40% to 16 nm as strain approaches 3. Thus, fibrillin's molecular structure and its packing into microfibrils change dramatically at high strains in ways that are not currently understood. Unfortunately, this X-ray diffraction study did not include data for tissue stress, so we do not know if the large-scale changes in morphology indicate increasing stiffness at high strain, or if the data simply document the mechanical disintegration of the microfibril, as it is pulled out into a thinner structure. Jensen and colleagues (2013) present a recent review that summarizes several models for the elastic deformation of fibrillin filaments, and much remains to be learned about this interesting system. However, is seems plausible that the elasticity of microfibrils is due to the straightening of the short-chain random-coil structures in relaxed fibrils, as shown in fig. 14.1C, as well as the elastic deformation of protein domains within the bead structure of the straightened filament.

Finally, the stress-strain curve in fig. 14.2C, which comes from the study by Wright et al. (1999), gives some insights into how the mechanics of the material match with the X-ray diffraction data discussed above. This figure shows a stress-strain test for an isolated-zonule-fiber sample, and the plot shows the full test result up to the point of sample failure. Interestingly, the form of the stress-strain curve changes dramatically at just above the stress and strain levels reached in fig. 14.2A. In fig. 14.2C we see that the test sample appears to fail, but it does not break immediately. Rather, the system yields, and then the stress rises and falls in a sawtooth pattern at an average stress of about 0.25 MPa over about another 80% of extension before the sample actually breaks at a final strain of 1.4. The authors mention that this failure pattern extended to a total strain of 3 in one test sample, and it is tempting to apply these stress-strain data to the X-ray diffraction data in fig. 14.2B, which also extended to a strain of about 3. There are several different theories about the structural organization of the proteins that make up the fibrillin filaments, but it is clear that the molecules are made from a large number of structural domains and are likely assembled to form an elongate structure (see Jensen et al., 2013, for details). Many of the structural domains are cross-linked by disulfide bonds, making them quite rigid, but others have inherent flexibility and may explain the apparent soft elasticity of the zonule fibers. Thus, the high-stiffness region of the stress-strain curve, at strains from 0.4 to 0.6, likely arises from the extension of softer domains in the fibrillin molecule. Then, at a stress level of about 0.25 MPa, the stress-strain curve suggests that some of the more rigid structural domains in fibrillin begin to pop open in a sequential process that creates the sawtooth stress pattern shown in fig. 14.2C. It seems likely that the functional range for this material is well below the stress where the structural domains in fibrillin begin to fail, making the zonule stress-strain curve in fig. 14.2A the most likely representation of the functional properties of this interesting "rubberlike" material.

To this point we have discussed microfibrillar materials from coelenterates, from echinoderms (in sea-urchin catch ligaments), and from chordates, but microfibrils have been documented in many other animal phyla, including annelids, arthropods, and mollusks. Indeed, it is likely that they exist in all animal phyla. However, there are relatively few additional studies on the properties of elastic structures made from microfibrils in invertebrate animals. McConnell and colleagues (1996) have documented the role of microfibrils in the arteries of an arthropod (lobster), and Shadwick and Gosline (1981) documented the role of microfibrils in the arteries of a mollusk (octopus). These soft arterial tissues are rich in microfibrils and, interestingly, they exhibit J-shaped stress-strain curves at low strains that are similar to those shown for isolated microfibrils, as in fig. 14.2A. That is, the intact tissues, which typically contain circumferentially organized microfibrils, show properties that reflect the J-shaped stress-strain curve of the individual microfibrils. Fig. 14.2D shows the stress-strain curve for a single bundle (ca. 2 μm in diameter) of octopus artery microfibrils that shows very clearly the rubber-elastic behavior of this material. The properties are quite similar to those in fig. 14.2A for zonule fibers, as we see that the sample enters into its non-Gaussian regime at a strain of about 0.35. The complex shape of this stress-strain curve makes it difficult to define a single stiffness value for these fibers, as is the case with the zonule fibers, but the initial tensile modulus of $E_{init} \approx 1.2$ MPa indicates a shear modulus of about 0.4 MPa. Above this strain level the tensile stiffness rises to values of about 3 MPa. The range of stiffness values reported for zonule fibers, octopus arterial fibrils, and similar materials from other animals is likely due to difficulties in establishing accurate cross-sectional areas for the respective test samples. So, it is likely that the tensile stiffness of microfibrils is somewhere in the range of 0.5–1.5 MPa. It is clear, however, that microfibril bundles in zonule fibrils and octopus arterial fibrils both exhibit a J-shaped stress-strain behavior that is consistent with the mechanical behavior of all elastic arteries.

It is interesting that the elastic arteries in higher chordates—such as fish, amphibians, reptiles, birds, and mammals—all show similar J-shaped stress-strain curves, and it is generally agreed that this mechanical behavior plays a key role in the hemodynamics of animals. However, in these higher chordates the elastic properties arise from a more complex structural arrangement of another elastic protein, elastin, that shows linear elastic behavior to quite high strains, in combination with wavy collagen fibers that straighten with extension to increase stiffness. Together, these two materials create J-shaped stress-strain curves for vertebrate arteries that are essentially identical to the stress-strain curves of the microfibril-based systems that exist in invertebrate arteries. However, the composite structure of vertebrate arteries allows them to achieve greater stiffness and strength, with similar extensibility, and this improvement in properties allows vertebrate animals to run their circulatory systems at higher pressures, thus facilitating more active lifestyles. We will consider the design of vertebrate arteries in greater detail in chapter 15.

Finally, there is one additional aspect of microfibril function that remains to be considered; namely, that microfibrils in vertebrate animals play a key role in the assembly of elastin. That is, the microfibrils in elastin-based elastic tissues serve as a structural scaffold that directs the assembly of elastin fibers, and this leads us to the molecular structure and mechanical design of elastin, the rubberlike protein found in the elastic tissues of vertebrate animals.

We learned a bit about the dynamic properties of elastin in chapter 3. Specifically, Figs. 3.8 and 3.10 show data for the small-amplitude dynamic properties of elastin as a function of loading frequency (fig. 3.8) and the effect of temperature and strain rate on the failure behavior of elastin (fig. 3.10). It will be useful to review these sections before proceeding with this chapter, in which we will analyze the polymer-sequence designs of protein rubbers and then the functional properties that arise from these designs.

2. THE AMINO-ACID COMPOSITION OF PROTEIN RUBBERS. The key features of these rubberlike proteins that are responsible for the mechanical properties of elastin, resilin, and abductin (an elastic protein found in bivalve mollusks) are found in their amino-acid sequences, which contain repeat motifs that encode long, random chain segments that are interspersed with sequence blocks that encode the cross-linking domains that enable the formation of rubber networks. We can begin to see aspects of the overall sequence designs in the amino-acid composition of the proteins, but, more importantly, they are clearly revealed in the repeating sequence motifs that exist in the full-length protein chains. Rauscher and colleagues (2006) provide a very interesting analysis of the protein dynamics of rubberlike proteins that is based on the abundance of two key amino acids, glycine and proline. Glycine is the smallest amino acid, with a single hydrogen occupying its side chain. Proline is an imino acid, where the three-carbon side chain folds around and forms a ring structure with its amino group. The arrangement of proline in a polypeptide is shown in fig. 14.3A, where we see a tripeptide structure with the sequence alanine-proline-glycine. The proline ring can be seen to form a rigid structure that limits the backbone flexibility of this small peptide. Glycines arranged in an extended tripeptide repeat with prolines and hydroxyprolines that extend for more than 1000 amino acids result in the formation of an extended triple-helical structure called tropocollagen, as described in chapter 9, but prolines linked to glycines and alanines in longer but less repetitive repeat motifs create a "kink" in the peptide backbone that prevents the formation of stable α-helical or β-sheet structures. Thus, glycine- and proline-rich sequences can facilitate the formation of polypeptide chains that have considerable conformational freedom, allowing them to take on dynamically mobile random-coil structures. These are exactly the types of protein sequences that give rise to rubberlike proteins. Table 14.1 provides data for the amino-acid composition of mammalian elastin; two different insect resilins, from dragonfly and locust; and scallop abductin. They show

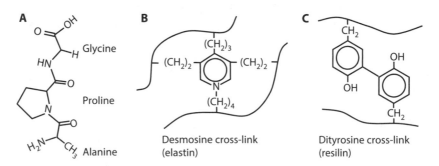

Fig. 14.3. Amino-acid composition of protein rubbers. **Panel A.** The tripeptide alanine-proline-glycine is shown here to represent a common sequence feature of rubberlike proteins. That is, proteins can achieve rubberlike properties if they are rich in relatively disorganized sequences that contain the amino acids proline, glycine, and alanine. **Panel B.** Desmosine is the dominant cross-link found in the protein rubber elastin. It is formed from four lysine side chains, as described in the text. **Panel C.** Dityrosine is the dominant cross-link found in the insect rubber called resilin. It is formed from two tyrosine side chains.

Table 14.1. Amino-Acid Composition of Random-Network Elastic Proteins

Amino Acid	Elastin (Bovine)	Resilin (Dragonfly)	Resilin (Locust)	Abductin (Scallop)	Fibrillin (Bovine)	Horseradish Peroxidase
Aspartic acid/ asparagine	6	94	99	39	111	145
Glutamic acid/ glutamine	15	42	46	21	105	59
Threonine	9	20	31	7	55	70
Serine	10	128	79	51	66	72
Glycine	*324*	*422*	*391*	*584*	*116*	*45*
Proline	**120**	**75**	**76**	**8**	**64**	**66**
Alanine	**223**	**70**	**112**	**67**	**37**	**110**
Valine	**135**	**12**	**26**	**6**	**38**	**64**
Methionine	–	–	–	**125**	**14**	**9**
Isoleucine	**26**	**9**	**17**	**3**	**50**	**45**
Leucine	**61**	**30**	**23**	**4**	**41**	**111**
Tyrosine	**7**	**12**	**25**	**1**	**31**	**7**
Phenylalanine	**30**	**21**	**26**	**69**	**30**	**57**
Lysine	7	8	5	9	37	19
Histidine	1	12	9	1	16	13
Arginine	5	46	35	4	45	72
Half cystine	–	–	–	2	143	35
Hydroxyproline	11	–	–	–	–	–
Desmosine	8	–	–	–	–	–
Total amino acids	**998**	**1001**	**1000**	**1001**	**1000**	**999**
% glycine	33	42	39	58	12	5
% proline	12	7	8	1	6	6
% glycine + proline	45	47	47	59	18	11
% nonpolar	60	23	31	28	31	47
% polar and charged	7	35	30	13	58	45
% half cystine	–	–	–	–	14	4

Amino acids carrying negative charges and polar side chains are listed at the top. Glycine, next, is the smallest amino acid and shown in italics. The nonpolar amino acids are shown in bold, and the positively charged basic amino acids—lysine, histidine, and arginine—come next, followed by half cystine, which can form disulfide cross-links in a protein. Empty cells indicate the lack of an amino acid in the protein.

some interesting similarities, but each illustrates a different biophysical strategy for a random-coil protein and for the formation of a cross-linked random network. In addition, amino-acid composition data are provided for fibrillin and a typical intracellular enzyme, horseradish peroxidase, as examples of typical nonelastomeric proteins. The individual amino acids in this table are arranged in a sequence that groups the amino acids with similar types of side chains. That is, the first four amino acids have negatively charged or polar side chains, and the next amino acid, glycine (in italics), is the smallest amino acid, having only a single hydrogen as its side chain. The eight amino acids below glycine (in bold) have nonpolar hydrophobic side chains, and the next three have positively charged side chains. The last of the normal amino acids, half cystine, has a thiol (-SH) side chain and can form covalent disulfide bonds within and between protein molecules. Data are not provided for tryptophan because this amino acid is degraded when proteins are hydrolyzed to form amino acids. Hydroxyproline (OH-proline) is an unusual amino acid that was described in chapter 9, on collagen. It is not one of the 20 amino acids coded in the gene sequence; rather, it is formed by hydroxylation of proline after a protein is synthesized and released into the extracellular space. Desmosine is the covalent cross-link found in elastin that is formed from four lysine side chains.

Below the list of individual amino acids in table 14.1 is a listing of the relative proportions of the various classes of amino acids in these proteins, and this allows a more convenient way to compare the chemical compositions of these three rubberlike proteins. These data will allow us to see how well these proteins meet the conclusions of Rauscher et al., that high levels of glycine and proline are key to the formation of rubberlike proteins. Note first that glycine levels are very high in all three of the rubberlike proteins: elastin contains 33% glycine, the two resilin contain 39%–42% glycine, and abductin contains 58% glycine. For comparison, fibrillin contains only 12% glycine, and the soluble enzyme protein, horseradish peroxidase, contains only 5% glycine. The importance of glycine in rubberlike proteins is that its very small side chain, a single hydrogen, contributes to the overall flexibility of the peptide backbone. Next, we see that proline content can vary dramatically between the three rubberlike proteins, ranging from 12% in elastin to only 1% in abductin, with the resilins having about two-thirds the level found in elastin. The key feature, however, is that the sum of glycine + proline in these proteins is dramatically higher than that seen for other proteins. That is, the glycine + proline levels for elastin, resilin, and abductin are 45%, 47%, and 59%, which are very high levels indeed.

Another interesting feature of the amino-acid composition of the rubberlike proteins is that there are very large differences in the nonpolar and the polar and charged amino acids. Elastin contains 60% nonpolar amino acids, and only 7% polar and charged amino acids, making it one of the most hydrophobic of all known proteins. This high hydrophobic content must certainly reduce the overall hydration of elastin. The two resilins, on the other hand, have much lower nonpolar content, 23%–31%, which is lower than most typical globular proteins, but their polar + charged amino-acid contents, at 30% and 35%, are four to five times greater than that of elastin. Thus, the resilin proteins will be more fully hydrated than elastin, and this may explain differences in the viscoelastic properties of elastin and resilin. It is also important to note that there are quite large differences in the amino-acid compositions of these two resilins, with dragonfly resilin having a much lower nonpolar-amino-acid content, and hence a higher polar-amino-acid content than locust resilin. This difference in polar/nonpolar-amino-acid content likely creates significant differences in network hydration, and it likely explains differences observed in the material properties for these two resilins, as we shall soon see. Abductin shows an even more extreme composition pattern, largely based on its remarkably high glycine content of 58%. Its nonpolar content of 28% is similar to that of fibrillin and is not much lower than that of typical globular proteins. Abductin's polar + charged amino-acid content, at 13%, is somewhat higher than that of elastin but is considerably lower than that of the resilins, at 30% and 35%. Overall, this composition suggests that the hydration of abductin is lower than that of resilin and likely similar to that of elastin.

3. THE AMINO-ACID SEQUENCE AND CROSS-LINKING OF PROTEIN RUBBERS. Amino-acid composition is only part of the story. Clearly, amino-acid sequence motifs play the dominant role in the creation of rubberlike proteins. Table 14.2 shows some typical sequence motifs that are found in elastin, resilin, and abductin. It is important to note, however, that these motifs are not repeated as exact copies through the full length of the polypeptide chain, as they are for the glycine-X-Y tripeptide in the amino-acid sequence of collagen. Rather, there are significant variations in length and amino-acid content of the sequence motifs that prevent the formation of permanent, stable secondary structures. Thus, the inherent flexibility of glycine-proline-rich motifs, combined with variation in these motifs, ensures that these proteins maintain dynamically mobile random-coil conformations that are consistent with rubberlike elasticity. More detailed information about these repeat-sequence blocks can be found in the references that are given in the legend to table 14.2.

Typical amino-acid sequence motifs shown—e.g., PGVGV and PGVGVA for elastin—are written in the single-letter format. That is, P = proline, G = glycine, V = valine, etc. A more complete list of these abbreviations is given in the legend to table 14.2. Note the presence of the proline-glycine (PG) unit in these motifs for elastin, as it creates a tendency for β turns in the peptide backbone. However, these elements are separated

Table 14.2. Amino-Acid Sequence Motifs in Elastin, Resilin, and Abductin

Elastin

Elastomeric motifs: PGVGV, PGVGVA

Cross-linking motifs: AAAA**K**AA**K**A, GPGG**K**PL**K**PV

Resilin

Exon-1 repeat motif: GG**RPSDS**Y*GAP*G**QGN**

Exon-3 repeat motif: G**Y**SGGRPGG**QD**LG

Abductin

Repeat motif: GGFGG**M**GGGX

Elastin sequence motifs are from Muiznieks et al. (2010), resilin sequence motifs are from Andersen (2010), and abductin sequence motifs are from Cao et al. (1997).

Large bold letters indicate the amino acids that are used to form the network cross-links. Small bold letters indicate polar and charged amino acids. A, alanine; D, aspartic acid; F, phenylalanine; G, glycine; K, lysine; L, leucine; M, methionine; N, asparagine; P, proline; Q, glutamine; R, arginine; S, serine; V, valine; X, variable amino acid; Y, tyrosine. The underlined letters are a highly conserved repeat motif in resilins.

by hydrophobic glycine-valine (GV)-rich motifs, and it is these and similar elements that establish the extreme hydrophobicity of elastin. The elastomeric segments of the elastin protein in mammals and birds are about 40 amino acids long on average, but the individual segments range in length from about 25 to 75 amino acids. In the full human protein, there are 15 hydrophobic network-chain segments, with 15 interspersed cross-link domains.

The cross-link sequence motifs are shown to the right of the elastomeric motifs, and we see two quite different sequence motifs that contain a pair of closely spaced lysine residues, which are indicated by the large bold K. The top motif contains a distinctive array of alanines before the first lysine, with two, or occasionally three, alanines between the first and second lysine, and this is followed by a small number of additional alanines. The second cross-link motif contains a pair of lysines that are separated by a pair of other amino acids, but in this case the lysines appear to be inserted into a typical, PGV-rich, elastomeric sequence. The cross-linking motifs shown in table 14.2 are about 10 amino acids long. However, the initial polyalanine block in the top motif can be longer than the 4 alanines shown, and in some cross-link domains the initial alanine block may be up to 10 alanines long. Because polyalanine polymers are known to form α-helical structures, it is thought that these alanine-rich cross-link domains form short α-helical segments in which the pair of lysine side chains emerges on one side of the helix, and this facilitates the assembly of the desmosine cross-link from two pairs of lysine side chains on neighboring peptide chains. That is, the pair of lysines on one peptide react together with another pair of lysines on an adjacent peptide to form the ring structure called desmosine, as shown in fig. 14.3B. This four-lysine ring structure is the predominant cross-link in elastin, but, in addition, a small number of straight-chain cross-links, called lysino-norleucine, are formed from two single lysines, one each from two chain segments. Two elastomeric sequence motifs are shown for resilin from the fruit fly *Drosophila*, and are labeled exon-1 and exon-3 repeat motifs. The resilin gene encodes three distinct protein-sequence domains, where the central domain (exon 2) contains a typical R-R consensus element that forms a chitin-binding domain, as described previously in chapter 12 (6. The Protein Matrix and Its Sclerotization). The two elastomeric domains contain similar but distinct sequence motifs. However, both contain a single tyrosine, indicated by the large, bold "Y" in the sequence. The phenolic side chains of the tyrosine pair react to form the dityrosine (see fig. 14.3C) and a smaller number of trityrosine cross-links that together create the rubber network. Both motifs are rich in glycine, and both contain proline, but not always in PG sequences. As we saw in table 14.1, glycine makes up about 40% of the composition of dragonfly resilin, but the level of proline is quite modest, at 7%. Interestingly, the polar and charged amino-acid content of resilin is about five times that of elastin, and this high polar content will substantially raise the hydration level of resilin relative to that of elastin. As a consequence, the peptide chains in the resilin network should be more mobile than those in elastin, and, as we will soon see, this enhanced mobility will give resilin a higher resilience than elastin.

The abductin sequence is the most extreme of the three proteins, with a glycine content that approaches 60% of the total composition, as discussed above. However, the proline content is very low, at about one-tenth the proline contents of elastin and resilin. So, the prevalence of glycine, with its small, hydrogen "side chain," likely plays a dominant role in establishing the random-coil behavior of this protein. However, it is clear that other features of this sequence are essential to this behavior, because pure polyglycine peptides take on stable extended structures. That is, polyglycine can exist in solution as a helical structure or as an antiparallel β-sheet structure, neither of which can explain the elastomeric behavior of abductin. Clearly, the other 40% of the composition is responsible for the random-coil structure of this protein. Methionine, phenylalanine, alanine, and serine are the other major amino acids, and presumably they prevent the formation of the stable polyglycine-based structures. The amino-acid sequence of the scallop abductin protein, as shown by Cao and colleagues (1997), contains many, di- and triglycine blocks that are typically separated by a single other amino acid. The consensus repeat motif for this protein is shown in table 14.2, where we see a 10-amino-acid peptide with a central methionine, indicated by the large, bold "M." As a consequence of the methionine in this short repeat motif, the methionine content of abductin is remarkably high, at 12.5%. Interestingly, this amino acid is completely absent from both elastin and resilin and is typically found in smaller quantities in most other proteins. Methionine has a nonpolar sulfur-containing side chain, which means that it should not contribute to the hydration of the protein network in this material. Consequently, the many short glycine repeats, which should impart great mobility to the network chains, are not going to contribute to the hydration of this material. As with elastin, the protein chains in the abductin network should be less mobile than those of resilin.

The cross-linking of abductin is currently not known, although there are several possibilities. Andersen (1967) suggested a tyrosine-based cross-link, but this is unlikely because the cross-link he suggested could not explain the fluorescence of abductin, as dityrosine does in resilin. However, the amino acid sequence obtained by Cao and colleagues indicates the presence of two tyrosines, one at position 20 in the full amino acid sequence and one at the end of the molecule, at position 135. Since the first 19 amino acids in the full protein most likely form the signal peptide that directs the secretion of this protein into the extracellular space and are removed from the protein during transport through the cell, the protein that forms abductin in the extracellular space is only 116 amino acids long. It appears, therefore, that there is a tyrosine located at both ends of this protein. Thus, it seems plausible that tyrosine may contribute to cross-linking in this material. However, with only two tyrosines at the chain ends, tyrosine-based cross-linking alone is not sufficient to explain the cross-linking of the abductin network, because as we will soon see, multiple cross-links are required to explain the relatively high stiffness of abductin. In addition, the sequence also contains three lysines that are interspersed in the sequence and which might form straight-chain cross-links like those found in elastin, but this type of cross-link does not explain the fluorescence of abductin. Dityrosine cross-links in resilin produce fluorescence; abductin is also fluorescent, so the most likely source of fluorescence would be a ring-based cross-link. Thus, the search for cross-link structures is linked with this feature. Another possibility is a sulfilimine cross-link, which has been observed by Vanacore and colleagues (2009) in the basement membranes of most animal groups, and involves the reaction of lysine and methionine. Given the presence of 3 lysines and 14 methionines in this short peptide, there is potential for this type of cross-link to explain the mechanical properties of abductin, although this remains to be confirmed.

4. THE NETWORK STRUCTURE AND MECHANICAL PROPERTIES OF PROTEIN RUBBERS. As described in chapter 6, the mechanical properties of random-network elastomers are defined by the number of random-network polymer chains per unit volume. Specifically, the shear modulus of a rubber network, G, can be calculated as:

$$G = NkT = \frac{\rho RT}{M_c}, \tag{14.3}$$

where k is the Boltzmann constant, M_c is the average molecular weight of network chains between cross-links, N is the number of chains per unit volume, R is the molar gas constant, T is temperature, and ρ is density. In addition, we can calculate the tensile- and compressive-stress-strain curves for a rubber network as:

$$\sigma_{eng} = G\left(\lambda - \frac{1}{\lambda^2}\right) = \frac{\rho RT}{M_c}\left(\lambda - \frac{1}{\lambda^2}\right), \tag{14.4}$$

where λ is the extension ratio (see fig. 6.6 for a plot of this equation). The initial tensile and compressive modulus of a rubber network is three times the shear modulus, and thus it should be possible to estimate the initial stiffness of protein rubbers. However, this equation is based on the assumption that the rubber network is made entirely from polymer chains, and that all of the protein rubbers are hydrated so that their network chains can function as kinetically free random-coil chains. Thus, we need to include the volume fraction of polymer in the hydrated random-coil network, v_2, as shown in eq. (14.5). In this equation, G is the shear modulus of the swollen network, G_0 is the shear modulus of the unswollen network, and ρ is the density of the unswollen network:

$$G = G_0 v_2^{1/3} = \frac{\rho RT}{M_c} v_2^{1/3}. \tag{14.5}$$

Now, we can predict the shape of the stress-extension-ratio curve for a swollen rubber, as:

$$\sigma = G\left(\lambda - \frac{1}{\lambda^2}\right) = \frac{\rho RT}{M_c} v_2^{1/3}\left(\lambda - \frac{1}{\lambda^2}\right). \tag{14.6}$$

Thus, if we know the volume fraction of polymer in the hydrated protein rubbers, we can estimate the average molecular weight of chains between cross-links, M_c, from stress-strain curves, and we can compare these estimates with the sequence organization of the proteins, as discussed above. Table 14.3 provides the information needed for this analysis.

The analysis in table 14.3 starts with an estimate of the extent of cross-link formation in each protein. The initial calculation for each protein assumes 100% cross-linking, which means that all of the cross-linking sites in the polymer, as described in table 14.1, are used in the formation of cross-links. Then calculations are made for reduced cross-link levels, in an attempt to estimate the extent of cross-link formation that is required for the development of network stiffness that matches the properties observed in mechanical tests of these materials. Looking first at elastin, we start with the assumption that the elastin protein contains 840 amino acids organized into 15 cross-link and

**Table 14.3. Analysis of Network Structure and Mechanical
Properties of Elastin, Resilin, and Abductin**

Cross-Linking Level	Cross-Links per Molecule	Amino Acids per Chain	M_c (kg/mole)	v_2	G_{calc} (MPa)	E_{calc} (MPa)	E_{meas} (MPa)
Elastin							
100%	15	56	4.8	0.65	0.59	1.8	1.2
80%	12	70	6.0	0.65	0.40	1.2	1.2
Resilin							
100%	33	16	1.4	0.45	1.8	5.5	1.9
50%	17	33	2.9	0.45	0.9	2.7	1.9
35%	12	45	3.9	0.45	0.65	1.9	1.9
Abductin							
3 cross-links	3	38	3.1	0.5	0.8	2.5	4.0
4 cross-links	4	29	2.3	0.5	1.1	3.3	4.0
5 cross-links	5	23	1.9	0.5	1.4	4.1	4.0

The rows shown in bold give the results of calculations that closely match measured values for these rubberlike materials. E_{calc}, calculated tensile stiffness; E_{meas}, measured tensile stiffness; G_{calc}, calculated shear stiffness; M_c, average molecular weight of chains between cross-links; v_2, polymer volume fraction.

15 network domains. Next, we will estimate the effect that 100% cross-linking would have on the stiffness of elastin. Full 100% cross-linking should produce a network that contains random chains that are, on average, 56 amino acids long, giving a molecular weight of network chains between the cross-link domains of about $M_c = 4.8$ kg mole^{-1}. Note that this calculation is based on a polymer volume fraction $v_2 = 0.65$, which is a reflection of the network's low hydration level due to the extreme hydrophobicity of the elastin protein. With these parameters, the shear stiffness predicted by eq. (14.5) for elastin with 100% cross-linking is $G \approx 0.6$ MPa, making the predicted initial tensile modulus $E \approx 1.8$ MPa.

Aaron (Aaron and Gosline, 1981) measured the tensile elastic behavior of single, approximately 10 μm diameter, elastin fibers obtained from the elastic neck ligament of a cow, and he was able to obtain a value for the elastic modulus, E, that is not affected by the complex organization of elastin fibers and lamellae that are found in most vertebrate elastic tissues. His analysis indicated that the shear modulus of bovine elastin is $G \approx 0.4$ MPa, making its initial tensile stiffness $E_{init} \approx 1.2$ MPa. Thus, it appears that cross-link formation is not 100% complete in this material. However, if we assume that cross-link formation is 80% complete, the calculations predict that network chains contain about 70 amino acids, producing a shear modulus of 0.4 MPa and a tensile modulus of 1.2 MPa, which are essentially identical to the values observed for bovine elastin. Thus, it appears that cross-link formation in elastin makes use of almost all the potential cross-link sites present in the protein's amino-acid sequence.

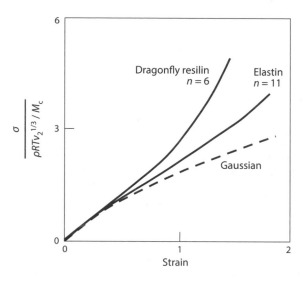

Figure 14.4. Properties of protein rubbers. Non-Gaussian stress-strain curves for resilin and elastin indicate that the average number of random links, *n*, in the random chains of these random-coil elastomers is quite different, with dragonfly resilin's network chains containing 6 random links, and elastin's network chains containing 11 random links. However, the individual random links in the protein chains of these two rubberlike proteins comprise about seven amino acids. For the symbols comprising the equation for normalized stress (*y* axis) see eq. (14.6) and surrounding text.

In addition, Aaron documented the non-Gaussian behavior of these elastin fibers by following the stress-extension relationship to high strains, and he observed that the stress-extension-ratio curve matches a theoretical non-Gaussian network model that contains network chains with 11 random links. These data are shown in fig. 14.4, but they are plotted in an unusual way. Rather than plotting stress against strain, this figure plots the normalized stress, which is σ divided by the modulus G of the swollen network. This transformation allows us to plot data for several different materials on a single axis, and allows us to compare their non-Gaussian curves. In this figure we see the actual data for elastin and for resilin, along with a theoretical Gaussian network curve. That is, the line labeled "Elastin, $n = 11$" shows the actual stress-strain curve for elastin, but the theoretical non-Gaussian curve for $n = 11$ is not shown because it almost perfectly matches the elastin data. With about 70 amino acids per chain and 11 random links in each chain, each random link in the elastin random chain must be about 7 amino acids long. Since each amino acid provides two single bonds (except proline, which contains only one single bond) that give partial rotational freedom, it appears that 10 to 12 of these single bonds are required to provide sufficient flexibility to form an effective random link in this protein.

The resilin analysis in table 14.3 is based on the amino-acid sequence data provided by Andersen for *Drosophila* (fruit fly) resilin. It indicates that if all potential cross-linking sites (33) in this protein are utilized in network formation (100% cross-linking), the protein chains between cross-links would be 16 amino acids long, on average. These chains would have an average molecular weight of 1.4 kg/mole, and the resulting network would have a shear modulus $G_{init} \approx 1.8$ MPa and hence an initial tensile modulus $E_{init} = 5.5$ MPa. We do not have good mechanical test data for the properties of fruit-fly resilin, but Torkel Weis-Fogh (1961) provides us with a stress-strain curve for a chitin-free resilin tendon from a dragonfly that provides data for both the initial tensile modulus as well as the volume fraction of resilin at physiological pH levels, $\nu_2 \approx 0.45$. The observed initial tensile modulus is $E_{init} = 1.9$ MPa, making the initial shear modulus $G_{init} \approx 0.66$ MPa. Thus, it is clear that the cross-linking level must be well below 100%. At 50% cross-linking the predicted stiffness remains above the observed stiffness, but at 35% cross-linking (12 cross-links per molecule) the predicted stiffness of resilin is essentially identical to that measured by Weis-Fogh. Further, the calculations indicate that the network chains in resilin have an average molecular weight between cross-links (M_c) of 3.9 kg/mole and hence contain about 45 amino acids on average. Weis-Fogh also documented the high-strain non-Gaussian properties of the dragonfly tendon, and the results are shown in fig. 14.4, where we see that the protein chains in dragonfly resilin appear to contain only six random links between cross-links. Again, the actual data so closely match the theoretical curve for $n = 6$ that the theoretical curve is not shown. The resilin curve shows greater departure from the Gaussian network curve because the resilin network chains are considerably shorter than those in elastin. With about 45 amino acids per chain, there are about 7 amino acids in each random link in the resilin protein, which is essentially identical to that observed for elastin.

The situation for scallop abductin is quite different. At present, we do not know what the actual cross-linking sites are, and therefore it is not possible to make firm predictions. However, we can estimate the consequence of introducing various levels of cross-linking in the abductin sequence. In table 14.3 we calculate the consequence of introducing three, four, or five cross-links into the 116-amino-acid protein we discussed above, giving us estimates of the average chain length at different cross-link levels. With three cross-links we expect an average chain length of 38 amino acids, etc. Molecular weight between cross-links, M_c, estimated for each cross-link level provides us with estimates of the shear and tensile stiffness of this material, as shown in table 14.3. There are two estimates in the literature for the compression stiffness of scallop abductin, one at 3.5 MPa and one at 4.5 MPa, and we will take a value of 4 MPa as a reasonable estimate for scallop abductin. The analysis in table 14.3 indicates that an abductin network with three or four cross-links would produce too low a stiffness to match the 4 MPa stiffness measured for the material, but at five cross-links per protein chain, the predicted network stiffness, 4.1 MPa, is essentially identical to the observed stiffness. So, we can conclude that some combination of the potential cross-linking systems discussed above is present, sufficient to create the high-stiffness rubberlike hinge ligament of the scallop.

There is one final issue about the extent of cross-link formation in these rubberlike proteins. Why do elastin and abductin appear to use almost all of their potential cross-linking sites, whereas dragonfly resilin appears to use only a small fraction of its cross-linking sites? Here are two possible answers, but others may apply. First, there are significant differences in the time available for these three systems to grow and to add cross-links, with those systems that are cross-linked over long time periods achieving a higher cross-link density than those cross-linked rapidly. Elastin and abductin are formed over many years through the continuous addition of new material, and in the case of vertebrate elastin, through physical reshaping of old material. As a consequence, they have sufficient time to achieve high cross-linking efficiency simply because the cross-linking reactions may take place over months or years. Elastin deposition starts in the final months of gestation in mammals, and it continues from early development through to adulthood. For large mammals such as horses and humans, elastin deposition may continue for many years or even decades. Similarly, the hinge ligament and shell complex of bivalve mollusks (like the scallop) will grow continuously as the animal increases in size over a number of years, and abductin synthesis and cross-linking must continue over these long time periods as well. The case for dragonfly resilin, and all insects in general, is that the developmental time and the full lifetime are remarkably short. Recall our discussion of the synthesis of locust leg cuticle in fig. 12.6, where we saw that the during final molt from the fifth instar juvenile to the adult animal, it took about 1 day for cuticle cross-linking reactions to bring the exocuticle to its maximal stiffness. We do not have sufficient information to specify time limits for all insects, but most insects live for only a few months, and in that time they go through a number of discrete developmental stages, each of which requires the formation of a new expandable cuticle layer to allow for their increase in body size. Cuticle maturation, and hence cross-linking, must therefore occur quickly, and it is likely that this limits the extent of cross-link formation. A second possible benefit of this "limitation" is that with many available cross-linking sites, it is possible that different species may have evolved developmental processes that allow them to adjust the extent of cross-linking and hence to control the stiffness of resilin (and other cuticle proteins) to match specific functional requirements. At present, there is no direct information about this kind of variation in the properties of resilin or other cuticle proteins, but it is something that is worth looking into.

5. THE DYNAMIC MECHANICAL PROPERTIES AND FATIGUE OF RUBBER-LIKE PROTEINS. Rubber-network stiffness is not the only mechanical property of significance for these materials. In all three cases considered here, the protein rubbers typically function in strain-energy storage and release mechanisms. That is, they function as dynamic springs, and their dynamic mechanical properties are of considerable importance. It is crucial that they have high resilience at the load-cycle frequencies that occur in the living animal, so that energy that is input by an applied deformation can be efficiently recovered in elastic recoil. Dynamic mechanical testing documents the behavior of these materials, as illustrated in fig. 14.5A. This graph compares the storage and loss moduli of elastin with those of two forms of resilin. The elastin test sample was taken from cow neck-ligament elastin (solid lines; Gosline and French, 1979). The locust resilin test sample was the prealar arm ligament (dotted lines with diamonds; Jensen and Weis-Fogh, 1962), which is a composite structure made from resilin that is reinforced by chitin lamellae. The dragonfly resilin test sample (dashed lines; Dudek and Gosline, unpublished data) was a resilin tendon that is entirely free of chitin. The dynamic properties of elastin are shown for frequencies that span from about 0.5 to about 10^5 Hz, and were constructed using the time-temperature superposition principle, as described in figs. 3.6C and 3.6D.

In all three polymer systems we see that the storage modulus, E', is at a level of approximately 1 MPa and remains fairly constant over the frequency range of 0.5 to about 100 Hz. The loss modulus E'', however, shows a strong frequency-dependent rise, particularly in elastin and the locust prealar-arm resilin. Thus, in all cases we see behavior indicating entry of these polymers into their glass transition, but the extent of this transition varies between test samples. Look first at the elastin data, and we see that E''

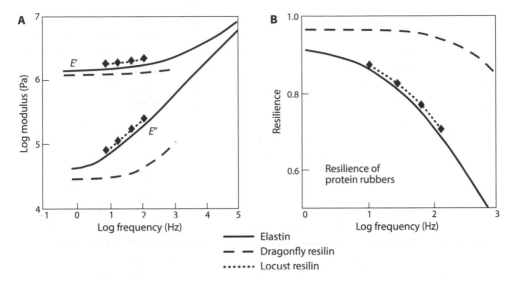

Figure 14.5. Dynamic behavior of resilin and elastin. **Panel A.** Dynamic mechanical properties of resilin and elastin. Note that elastin and locust resilin have very similar properties, whereas dragonfly resilin has a much lower loss modulus, E'', at high frequencies, which indicates that dragonfly resilin has significantly higher resilience. E', storage modulus. **Panel B.** The resilience of elastin and resilin, derived from the data in panel A. Note that dragonfly resilin has a resilience of about 95% at frequencies up to about 100 Hz, whereas elastin and locust resilin have somewhat lower resilience.

is low at frequencies of the order of 1 Hz, but it rises rapidly above about 3 Hz. At the highest frequencies (ca. 10^5 Hz) the storage modulus and loss modulus are nearly equal, indicating that the material is nearing the midpoint of its glass transition. Interestingly, the locust resilin data, which span only a small frequency range, parallel the elastin data closely, indicating that the dynamic properties of the chitin-filled resilin sample are very similar to those of elastin. The dragonfly resilin tendon, however, shows quite different behavior. At low frequency, in the range of 1 Hz, the loss modulus (E'') is significantly lower than that of elastin and the locust resilin sample, and as frequency rises, E'' rises more slowly than that for elastin and locust resilin. As a consequence of these differences in dynamic properties, there will be significant differences in the resilience of these materials, as shown in fig. 14.5B. This figure shows the resilience, calculated as $R = e^{-\pi\tan\delta}$, as a function of test frequency for the three elastomeric proteins, and again we see that elastin and the locust resilin sample show similar properties over the range of test frequencies from 1 to about 600 Hz, above which this equation is not able to accurately predict resilience. Resilience for elastin at 1 Hz is somewhat above 90%, but it falls to 60% at 300 Hz. The locust resilin data do not cover the full frequency range, but the data essentially parallel those of elastin, and thus it seems reasonable that the chitin-filled resilin sample from the locust has resilience levels that are similar to those of elastin. The situation for the dragonfly resilin, again, is quite different. This chitin-free test sample has a much higher resilience over the full frequency span, with resilience at 1 Hz of about 96%, and this remains above 80% at 1000 Hz. This exceptional performance likely makes the dragonfly resilin the most resilient rubberlike material amongst all natural and artificial rubberlike materials. It is not clear why dragonfly resilin is more resilient than the locust resilin, but it is likely because there are considerable differences in the amino-acid content, and hence the amino-acid sequences of these two proteins. The exceptionally high content of polar and charged amino acids in the dragonfly resilin likely increases hydration levels that enhance resilience. In addition, the presence of chitin fibrils in the locust resilin might reduce its resilience.

How do these dynamic properties relate to the functions of elastin and resilin in live animals? Clearly, for slow movements, at frequencies of 1 Hz and below, both proteins have resilience levels of the order of 0.90 (90% efficiency) and above, but at frequencies above 1 Hz the resilience of elastin and locust resilin falls quickly. How high are the functional frequencies for elastin- and resilin-containing tissues in living animals? For humans, the elastin in the arterial system will experience dynamic oscillations caused by the contractions of the heart, and maximal heart rate should provide an indication of the upper range of dynamic loading in elastin-containing tissues. Resting heart rate in humans is typically around 60 beats per minute (1 Hz), and maximal heart rate in young athletes is about 200 beats per minute (3.3 Hz). Clearly, elastin resilience remains well above 90% at resting heart rates, and even in intense activity elastin resilience remains at about 90%. Hummingbirds, however, have heart rates that are much higher, in the range of 1250 beats per minute (about 20 Hz), and at this frequency the resilience of their elastin likely falls to about 84%. However, it is important to note that no one has actually measured the frequency-dependent properties of hummingbird elastin. Similarly, for locust resilin, the wing-beat frequency of locusts is in the range of 20 Hz, and hence resilin in these animals should be functioning at well above 80% elastic efficiency. Dragonflies have wing-beat frequencies of the order of 30–40 Hz, and with the properties shown in fig. 14.5B it appears that the resilience in these animals remains at about 95% throughout the full range of their wing-beat frequency. At present, we do not understand the specific requirement for such exceptional resilience in the dragonfly tendon.

We have not included scallop-shell abductin in our discussion of dynamic mechanical properties to this point, because there are no studies that have measured its dynamic properties over a broad range of frequencies. Rather, experiments have focused on measuring the resilience of abductin at a frequency that matches the swimming "frequency" for this animal, which is approximately 4 Hz. For those who are not familiar with scallop swimming, these animals are clam-like bivalve mollusks that surround their soft tissues with a shell that is made from two mineralized, domed discs that are attached by a flexible hinge along one edge, as shown in the diagram in fig. 14.6. Just inside this hinge there is a triangular rubberlike pad, called (appropriately) the resilium, that is made from the protein abductin. This pad is located between angled surfaces of the two shell halves, and the pad is compressed when the shell's muscle (M) contracts to close the shell. This type of organization is found in many bivalve mollusks, but it is particularly interesting in scallops because they can use this muscle-spring-hinge system to forcefully expel water from the shell interior as a high-velocity stream that allows the scallop to "swim" by jet propulsion. The elastic recoil of the abductin pad then opens the shell, drawing in water that will be expelled in a subsequent muscular contraction. What material properties are required for the resilium in this propulsive

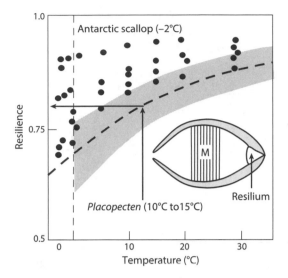

Figure 14.6. Resilience of scallop abductin. At any given temperature, the resilience of Antarctic scallop hinges (*black dots* are measurements taken by Denny and Miller (2006)) is higher than that of scallops found in temperate waters (*gray band* shows the range for the temperate-water scallop *Placopecten magellanicus,* from Bowle et al. (1993)). M, muscle that contracts to close the mollusk's shell.

mechanism? They are twofold. The resilium must be rubberlike and capable of storing a large amount of energy, which is clearly provided by its properties, as described in table 14.3. In addition, the material must have a high resilience so that muscle energy that is applied to the ligament can be recovered in elastic recoil to power the opening of the shell in the recovery phase of the animal's swimming cycle.

Fig. 14.6 provides data for the resilience of abductin from two different scallop species, and they indicate an interesting adaptation of this material for function in different temperature regimes in the ocean. *Placopecten magellanicus* is a temperate-water animal that lives at temperatures in the range of 10°C–15°C. Bowie and colleagues (1993) measured the resilience of the abductin hinge ligament in damped free-oscillation experiments over a range of temperatures from 0°C to 30°C. Their full data set is indicated in fig. 14.6 by the shaded area labeled *Placopecten,* and the midline-trend average resilience at 12.5°C indicates a functional resilience for this material of 0.8 (80% efficiency). The black dots indicate individual measurements taken by Denny and Miller (2006) on abductin from the Antarctic scallop *Adamussium colbecki,* which lives at water temperatures of −2°C. Regardless of the scatter in this data, it is clear that the Antarctic animal produces an abductin that has higher resilience at all temperatures than that of the temperate animal, and it appears that on average the resilience of Antarctic abductin at −2°C is essentially the same as that of *Placopecten* at its environmental temperature of about 12.5°C. Denny and Miller looked at the amino-acid composition for clues to the origin of the enhanced resilience of the Antarctic abductin, but did not find any compelling answers. However, comparisons between the Antarctic abductin and abductins from three species of temperate-ocean scallops revealed some small changes that may account for the differences in the temperature-dependent resilience. Specifically, they saw that the glycine content of the Antarctic abductin was about 4.5% higher than the average for the three temperate species, and they saw compensating reductions in the content of alanine, serine, and aspartic acid. It is not clear at this point how these changes explain the differences in temperature-dependent resilience. However, the additional 4.5% glycine brought the total glycine content of the Antarctic abductin to

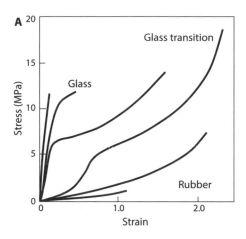
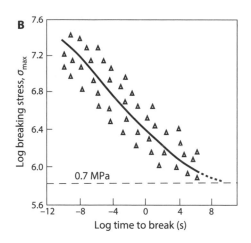

Figure 14.7. Fatigue behavior of mammalian elastin. **Panel A.** The fatigue lifetime of elastin was obtained from tensile tests to failure that were carried out over a broad range of strain rate. Time, temperature, and solvent shifting of these properties produced the data shown in **panel B**, which plots the log of breaking stress versus log of time to break. When this data set is extrapolated to a breaking stress (0.7 MPa) that is equivalent to the in vivo stress experienced by elastin in living animals, the fatigue lifetime can be extrapolated to a time of 10^9–10^{10} s, which corresponds to fatigue lifetimes of 30–300 years.

71%, which may have increased the inherent flexibility of the Antarctic protein's backbone, and through this increased the inherent resilience of the material.

Unfortunately, there is relatively little information on the fatigue behavior of protein rubbers, with data available only for mammalian elastin. However, the data on elastin are quite interesting, as they indicate that the expected lifetime of elastin is roughly equal to or greater than the lifetime of the organism that produces the elastin. The stress-strain curves presented in fig. 14.7A are from tensile tests to failure of arterial elastin ring samples taken from pig aortas by Lillie and Gosline (2002). Here we see the full range of behaviors that are observed as elastin is tested through its transition from rubber to polymer glass. Fig. 14.7B shows the results of experiments to measure the fatigue lifetime of elastin. This graph is a plot of log failure stress, σ_{max}, against log time to break, t_{max}, in seconds, for tensile tests to failure of pig aortic elastin. The tests were carried out over a broad range of strain rates, from 4×10^{-5} s^{-1} to about 1 s^{-1}. That is, the range of strain rate was about 25,000-fold. In addition, shifts in temperature, ranging from $-15°C$ to $+75°C$, and reduction in hydration level increased the overall time range of the shifted data sets to create time-to-break values that ranged from about 10^{-10} s to 10^6 s (a 10^{16}-fold time range). The triangles in this figure represent the results for individual test samples, but the actual data set contained 298 test values in this study. So, the triangles do not accurately represent the full data set, although they reasonably indicate the scatter of the data. The time axis on the graph is not simply the time for the individual test samples, but an adjusted time to break that is determined by applying two shift factors that account for the effect of test temperature and sample hydration on the effective time scale of the test. These shift factors were determined by dynamic testing of elastin, as shown in figs. 3.6C and 3.6D, and they were applied to the actual failure time to create the adjusted time-to-break data in this figure.

The data at very short times correspond to the behavior of elastin in its glassy state, as represented by the two curves labeled "glass" in fig. 14.7A. At the mid time range (10^{-6}–1 s) in fig. 14.7B the samples are in their glass transition, and at times longer than 1 s the samples are in a rubberlike state. The solid line represents the general trend of this data set, up to an adjusted time of about 10^6 s, and the dotted line extends the trend of the data to an intersection with a dashed line that indicates the approximate average stress that elastin experiences in the arterial system of a living animal, in this case a pig. This extrapolation suggests that the failure time for elastin in a living animal is in the range of 10^9–10^{10} s, which corresponds with fatigue lifetimes of the order of 30–300 years. Interestingly, 30 years is about equal to the lifetime of a pig, and 300 years is well beyond the lifetime of humans, so we can conclude that the fatigue lifetime of elastin at normal blood pressures in humans and animals is likely as long or longer than the actual lifetime of the animal or human. It is interesting that experiments to determine the biochemical turnover of elastin in the mouse aorta by Elaine Davis (1993) and in human lung elastin by Shapiro and colleagues (1991) have shown that the turnover time for elastin is longer than the animal or human life-span. Thus, the long-term biomechanical stability exhibited by pig elastin, as shown in fig. 14.7B, is consistent with our understanding of the biochemical stability of this important rubberlike protein.

Chapter 15

Pliant Matrix Materials and the Design of Pliant Composites

With this information on rubberlike proteins, and an understanding of the structural design of collagen fibers from chapter 9, we have most of the background needed to understand the design of pliant composite materials. What remains to complete our analysis is an understanding of the soft matrix materials that surround the collagen fibers and rubberlike-protein fibers in pliant composites. Generically, these matrix materials are referred to as proteoglycans, which are typically made from a core protein with attached polysaccharide chains called glycosaminoglycans. Glycosaminoglycan chemistry was introduced in chapter 9, where we saw that aligned collagen fibrils are embedded in and interconnected by a matrix of a small proteoglycan called decorin. This is one of a range of proteoglycans that form the matrix for a wide variety of pliant composite materials. Generally, the proteoglycans in pliant composites are much larger than the decorin molecules found within collagen fibers, although their structural components are quite similar. However, before we discuss the matrix molecules of pliant composites, we will consider a matrix material that is made entirely of proteoglycans. This material is mucus, a soft protein-polysaccharide-gel material that functions on its own, and it will provide us with interesting insights into the mechanical behavior of the soft gel-like matrix materials found in pliant composites.

1. THE MECHANICAL DESIGN OF MUCUS. Much of our knowledge of mucus structure is based on experiments with mammalian material, and we will assume that the mucus produced by other animals is similar. The diagrams in Fig. 15.1 provide a reasonable indication of the molecular organization of the mucin molecule, the proteoglycan that forms the material we call mucus. Fig. 15.1A shows the organization of an individual mucin molecule, which contains a single protein chain, indicated by the solid line, and a number of polysaccharide side chains, indicated by the dotted lines. In general, there are three structural domains to mucin proteins. Two of these are the globular terminal domains, and the third is the central extended chain. The numerous polysaccharide side chains are attached to the central extended protein domain. These side chains are quite short, typically containing from 3 to 15 sugar rings, and the larger side chains are branched. The side-chain composition includes a number of sugars,

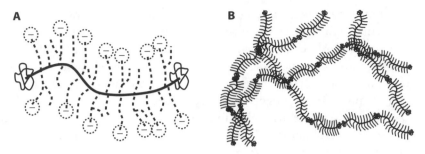

Figure 15.1. The structural organization of mucus. **Panel A.** The structural design of an individual mucin molecule, which contains a central protein chain (*solid line*) that is decorated by short polysaccharide side chains (*dotted lines*), many of which contain a terminal sialic acid group that carries a negative charge. The mucin molecules contain globular terminal domains that interact to form long linear chains, which can interact by noncovalent bonding forces to form semistable gel networks. **Panel B.** The structure of a mucin gel network.

but it is dominated by N-acetylglucosamine and N-acetylgalactosamine. The hydroxyl groups and the N-acetyl groups on these sugars make them very polar, even though they do not carry a full charge. In addition, the "bubbles" with minus signs located at the ends of the polysaccharide chains indicate the presence of a terminal sialic acid, which is a sugar derivative that carries a negatively charged carboxyl group. Together, these aspects of the side chains favor expanded, highly hydrated chain conformations, and this results in an open bottle-brush structure, as is shown in this diagram. The tightly coiled terminal domains of the protein are rich in the amino acid cysteine, and these amino acids form disulfide bonds with neighboring molecules, creating long chains of mucin molecules, which can form entangled networks with other mucin chains, as shown in fig. 15.1B. The four mucin chains in this figure do not indicate the full length of each chain, as they can be made from dozens of mucin molecules. Typically, individual mucin molecules have molecular weights of the order of 200–500 kilodaltons (kDa), and the assembled mucin chains can have molecular weights up to 50 megadaltons (MDa), suggesting that a single mucin chain may contain up to about 100 or more mucin molecules (Bansil and Turner, 2006). The diagram in fig. 15.1B does not indicate what kind of interactions might exist between the long mucin chains to create the gel-like properties of mucus, but there are two possibilities. The mucus chains can interact by transient entanglements, where the chains loop around each other but do not form intermolecular bonds, or they can interact through noncovalent bonding forces to create semistable cross-linked networks. The experiments described below will strongly support the concept that the mucin chains interact through noncovalent bonding forces to create transient polymer networks, and that these transient networks are key to the function of this interesting material.

What are the functional properties of these interesting molecules? Studies with isolated mucus from mammalian systems, including gastric mucus and tracheal mucus, show interesting sol-to-gel transitions that occur with dynamic loading and with shifts in pH. However, it is not clear how these transitions relate to the function of the mucus in the living animal because most test data on mucus are based on low-strain dynamic mechanical testing of the material in its semisolid state (Celli et al., 2007). However, the study by Denny and Gosline (1980) on the mechanics of the pedal mucus of the banana slug, *Ariolimax columbianus*, provides us with very different insights into the functional properties of mucus; namely, the material's behavior under high-strain shear deformation. To understand the properties of the slug's pedal mucus we need first to understand the mechanical role of the mucus in the animal's locomotion. Fig. 15.2A shows a drawing of the slug, which is about 10–15 cm long in life, and fig. 15.2B shows a diagram of the bottom of the animal's foot as documented in videos taken as the animal "walks" on a glass surface. In this diagram we see that there are three regions of the foot that make contact with the glass through a layer of pedal mucus. The mucus is released at the front of the animal from a gland that opens just behind its head, and the forward motion of the animal creates a mucus layer that is about $10\,\mu$m thick under the entire foot. The diagram in fig. 15.2B shows three functional zones of the foot. In the center of the foot we see light and dark gray bands that indicate areas of foot movement and foot attachment to the substrate. The light gray bands indicate areas where the foot is sliding forward relative to the surface of the glass at a velocity indicated by the length of the arrow on the second-to-last light gray band. The dark gray areas indicate areas where the foot is stationary relative to the glass. Outside this central region there is a "rim" of white, which is an area of the foot that moves forwards at a constant speed that is equal to the speed of the whole animal. Note that the arrow on the rim is shorter than that on the moving segment in the center of the foot. This is because the faster movements of the wave occur only transiently, and the slug as a whole moves at a constant but slower rate.

The important movements required for locomotion occur in the central region of the foot. Forward motion is created when muscles located at the back end of a dark, fixed band contract, and this contraction causes the lighter band behind the dark band to be pulled forward, at a velocity indicated by the arrow. This local contraction first occurs at the tail end of the animal, and it progresses forward as a moving wave that delivers a small forward extension of the foot at the front of the animal and ultimately moves the whole animal forward by a small "step." A continuous succession of these waves creates continuous forward motion of the whole animal. How does this system actually work? The answer lies in the properties of the mucus. The waves pass forward at a rate of about one wave every 2–3 s, and as each wave passes over its underlying mucus layer the mucus is sheared to high strains for a period of about 1 s. Each wave moves about 1 mm, and in this process it will shear the $10\,\mu$m mucus layer to a strain of the order of 100, resulting in a shear strain rate of about $100\ \text{s}^{-1}$. These high strains transform the soft mucus gel into a viscous fluid, which essentially lubricates the sliding of the moving wave in the forward direction. But the only way that this can happen is if the fixed, dark gray region ahead of this moving band is somehow able to remain fixed to the substrate and not slide backward. The key is that the mucus under the fixed bands exists in a solid elastic state, and the mucus under the moving band exists in a fluid state. Large-strain shear testing of isolated slug pedal mucus shows exactly how this mucus-based system works.

Fig. 15.2C provides a diagram of a cone-and-plate viscometer, which is the shear-testing apparatus that can be used to determine the high-strain material properties of the slug's slime. Fig. 15.2D shows the results of tests from this device that mimic the

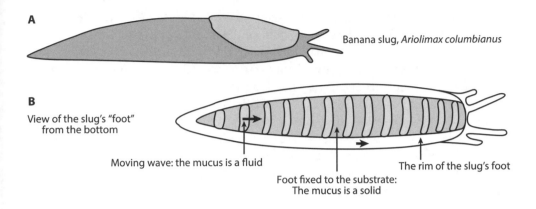

B

View of the slug's "foot"
from the bottom

Moving wave: the mucus is a fluid

The rim of the slug's foot

Foot fixed to the substrate:
The mucus is a solid

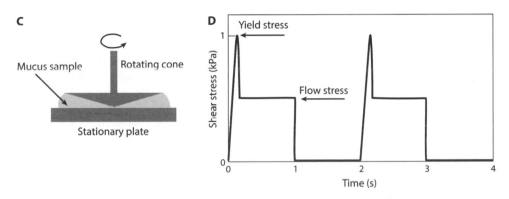

C

Mucus sample   Rotating cone

Stationary plate

D

Yield stress

Shear stress (kPa)

Flow stress

Time (s)

Figure 15.2. The mechanics of pedal locomotion in a terrestrial slug. **Panel A.** Side view of the banana slug, *Ariolimax columbianus*. **Panel B.** The banana slug's foot viewed through glass to reveal the movements of the foot during locomotion. Local contractions of muscles in the foot create waves of forward movement that constitute individual, local "steps" in the locomotor cycle, as described in the text. **Panel C.** Diagram of the cone-and-plate apparatus used to carry out shear testing on isolated slug mucus. **Panel D.** Results of shear tests on slug mucus. Note the initial linear rise in stress at the very start of the test when the cone rotates to shear the mucus sample. This indicates the elastic behavior of the slime. However, the slime yields after about 0.2 s at shear strains of about 10, transforming the slime gel into a viscous fluid. The shearing of the fluid slime lubricates the forward sliding of the moving wave on the slug foot. After a 1 s pause, however, the fluid slime has transformed back into its soft gel form, as indicated by the second linear rise in shear stress that starts a second "step" for the animal.

shearing of the pedal mucus with each forward movement of the wave on the animal's foot. The cone-and-plate system used in these tests produced a shear strain of 250 for each revolution of the cone, so to mimic the high-strain deformation that exists under the moving portions of the slug's foot the cone must be rotated by 0.4 of a full rotation of the cone. The results of shear testing slug pedal mucus at strain rates of the order of 100 s^{-1} reveal the remarkable dynamic behavior of this material that provides the basis for the animal's locomotion. Fig. 15.2D shows a typical result for the slime tested in this system. Rotation of the cone is turned on for about 1 s and then turned off for 1 s to mimic the pattern of movement seen in the animal, and this pattern is repeated. We see a series of abrupt transitions that occur when the mucus is sheared to high strains and then allowed to recover. Starting from a resting state, the shear stress rises linearly, indicating that at the start of shearing the slime exhibits elastic behavior, with an elastic shear modulus $G \approx 200$ Pa. Thus, the solid form of the mucus is very soft indeed. However, after the mucus is sheared to a strain of about 10, the mucus yields, and it is transformed into a viscous fluid, where stress remains constant, at about 60% of the yield stress, until the rotation of the cone is stopped. In this process the elastic gel is transformed into a fluid with a viscosity of about 40 poise, which is about 4000 times greater than the viscosity of water.

The sample is then allowed to recover for 1 s, and it is then sheared again. Remarkably, we see that over a period of 1 s the sample has regained its elastic properties, and that it then goes through the transition to a viscous fluid again. This cycling of properties from elastic solid to viscous fluid and back on a 2 s time scale can be repeated 20–30 times without significant changes in the properties of the mucus, and this remarkable set of properties is what enables slugs to move forward over the slime layer that they create. The key features of this system are the yielding of the elastic mucus at the start of movement that creates a viscous fluid to lubricate the sliding of the moving wave, and then the rapid recovery of the fluid mucus back to its elastic state that allows the previously mobile section of the foot to become fixed to the substrate, where it can then act to pull adjacent mobile sections of the foot forward. It is remarkable how sophisticated the materials design of mucus can be.

More recently Taylor and colleagues (2003) have demonstrated similar behavior in pig gastric mucus, although with a very different test protocol. They employed dynamic mechanical testing, as described in chapter 3, which involved sinusoidal strains applied to test samples. Taylor et al., however, applied a broad range of shear-stress levels to the mucus samples that ranged over three orders of magnitude, and they observed major shifts in properties as stress amplitude (and hence strain amplitude) was changed. At low stress amplitude the mucus behaved as an elastic solid, where the storage modulus, G', was about an order of magnitude greater than the loss modulus, G'', indicating the presence of an elastic network in the mucus gel. However, as stress amplitude was increased, G' fell and G'' increased, indicating that there was a disruption of the forces holding the molecular chains into a polymer network, and thus the mucus was transformed into a viscoelastic fluid. Interestingly, when stress amplitude was then decreased, this shift in properties was reversed, and stable interactions between the network chains were reestablished. The authors found that this process could be repeated over a number of cycles, and the behaviors remained the same, much as occurs in the slug mucus, as shown in fig. 15.2D. They concluded that this behavior arises from the formation of labile secondary cross-linking interactions between the network chains in the gastric mucus, and not from chain entanglements. Thus, the locomotion of the slug can likely be explained by the breaking and rapid reformation of weak bonding forces that stabilize the mucin gel that is present in the slug's pedal mucus.

Slugs employ a variety of other mucin-based materials with quite different properties in their daily lives. These animals are in the phylum Mollusca, which includes a wide range of shelled animals, such as snails, clams, and oysters. The terrestrial slugs, however, do not have an external shell for protection, but most are able to release a very firm and sticky defensive slime from the skin on their backs, which is known to deter predators. In addition, the leopard slug, *Limax maximus*, a close relative of the banana slug, is renowned for its acrobatic mating behavior, in which two mating animals entwine and descend from an elastic slime cord as they copulate. Clearly, mucus can play many interesting and diverse roles in the lives of animals.

2. THE MECHANICAL DESIGN OF CARTILAGE. Cartilage is an interesting pliant composite that functions in a variety of structures in the bodies of vertebrate animals. Perhaps the most important form of cartilage is the articular cartilage that is found at essentially all junctions between the bones in vertebrate skeletons. Joint cartilage is loaded in compression during load-bearing activities, and it is also loaded in shear as cartilage-lined bone ends rotate and slide during limb rotation. In addition, there are elastin-containing elastic cartilage structures, such as the ears and the nose, that function largely in bending modes. We will focus on the structural organization of articular cartilage.

Articular cartilage is always found in intimate contact with mineralized bone, as indicated in fig. 15.3, where we see a diagram of a synovial joint between the head of a thigh bone (femur) and its socket in the pelvis. The light gray band lying between these two bones indicates the location of the articular cartilage in this hip joint. Note the dotted black line in this band, as it represents the interface between the two cartilage layers, one attached to the femoral head and one attached to the joint socket in the pelvis. Static and dynamic loading of this joint imposes quite large stresses on the bone and the cartilage, and these stresses can be transient, as in the ground-contact phases of terrestrial locomotion, or static, as in weight bearing of stationary animals. The cartilage serves as a soft interface between the two bones that allows the transfer of force between them without imposing breaking stress levels onto the rigid bones. This is achieved by the relatively large strains that can occur in the much softer cartilage, which allows the cartilage to distribute stresses laterally and thus can minimize local stress concentrations that would exist at a bone-bone interface. This is possible because of the essentially "fluid" nature of the material we call cartilage.

Cartilage is a composite constructed from collagen fibers and a matrix of mucus-like proteoglycan molecules. The expanded portion to the right in fig. 15.3 provides an indication of the organization of the collagen (black lines) at the interface between bone (dark gray) and cartilage (light gray), as well as the organization of the collagen within the cartilage itself. However, these lines are not meant to represent individual collagen fibers; rather they represent an edge view of a layer of collagen fibers, and in the real tissue these layers are packed quite tightly together. There are two basic structural arrangements of the collagen layers in cartilage. One set of the collagen layers emerges vertically from the bone-cartilage interface, and they rise through the cartilage thickness in an arching pattern, ultimately curving into a plane that is parallel to the external surface of the cartilage, as described by Jeffery and colleagues (1991), who called these layers "leaves." The leaves are made from a cross weave of very fine collagen fibers, and they are interconnected by fine collagen fibers that run perpendicularly to the surface of the leaves. The vertical orientation of the collagen leaves at the bone-cartilage interface plays a crucial role in anchoring the cartilage to the bone. The arching of the

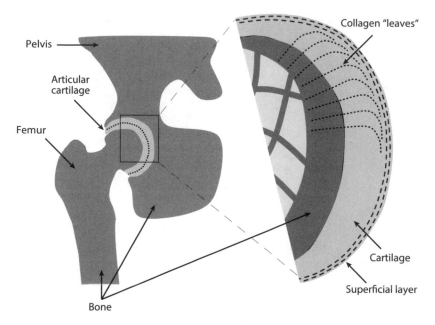

Figure 15.3. The structure of a synovial joint, the hip. The diagram to the left shows the overall structure of the hip joint between the head of the femur and its socket in the pelvis. The light gray area between the two bones is the articular cartilage, and it is important to note that the dotted line indicates the interface between two separate cartilage layers, one attached to the femur and one attached to the pelvis. The expanded diagram to the right shows the structure of the cartilage on the femoral head. Again, the light gray represents the cartilage and the dark gray represents bone. The dotted and dashed black lines represent the collagen fibers that are present in this structure. There are two important sets of collagen fibers. The collagen "leaves" (*dotted lines*) are thin sheets of collagen fibers that emerge from the underlying bone and extend vertically into the cartilage, but as they approach the cartilage surface they arch over to run parallel to it. In addition, there is a superficial layer of more densely packed collagen fibers (*dashed lines*) that run parallel to the cartilage surface.

collagen fibers at the cartilage surface creates densely packed collagen layers that run parallel to the cartilage surface, a design that contributes to the reinforcement of the surface, where high compression and shear stresses occur between the two cartilage surfaces in a typical synovial joint. The second set of collagen layers runs parallel to the cartilage surface at the very top of the cartilage, forming what is called the superficial layer, which occupies about 10% of the cartilage thickness. This layer contains relatively more collagen and less proteoglycan than are found in the lower layer of the cartilage, and it provides the majority of the structural reinforcement of the cartilage surface. The actual architecture of cartilage collagen is considerably more complex than indicated in this diagram, as the orientation of these arching collagen leaves varies with location in the joint, in patterns that match the direction and intensity of loading. Interestingly, when the surface of a cartilage sample is punctured with a round pin, the cartilage splits open, forming sharp-ended elliptical cuts, and the orientation of such an ellipse

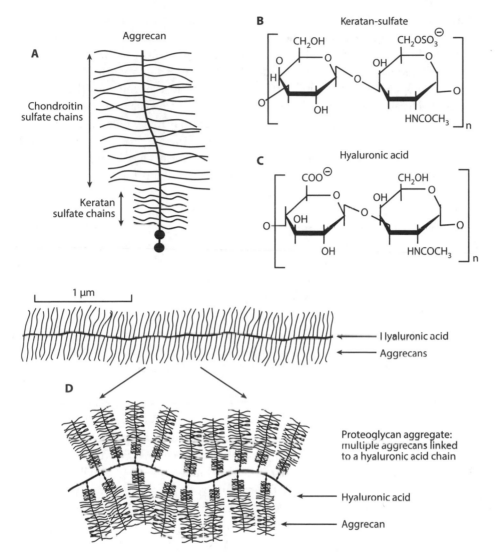

Figure 15.4. The structure of proteoglycan aggregates. **Panel A.** The structure of aggrecan, the cartilage proteoglycan. **Panel B.** The disaccharide repeat unit of keratan sulfate, **Panel C.** The disaccharide repeat unit of hyaluronic acid, the polymer that links multiple aggrecans into a proteoglycan aggregate. **Panel D.** The structure of the proteoglycan aggregate in cartilage.

provides a clear indication of the orientation of the collagen leaves at any particular location on the cartilage surface. That is, the long axis of the ellipse runs parallel to the plane of the collagen leaves at the location of the pinprick. Thus, it is quite easy to reveal the large-scale pattern of collagen orientation of the cartilage in a synovial joint.

The other major component of the joint cartilage is the proteoglycan "matrix" that exists within the dense collagen-fiber mesh described above. Fig. 15.4 shows the major structural features of these complex macromolecules. The predominant macromolecule in the cartilage matrix is the proteoglycan called aggrecan, which has a structure as shown in fig. 15.4A. Detailed information on the chemical structure of this proteoglycan can be found in the article by Kiani and colleagues (2002). Here we see a "bottle-brush" structure similar to that described previously in the section on slug mucus (1. The Mechanical Design of Mucus), but the overall dimensions of aggrecan and aggrecan aggregates are significantly larger than those of mucins. The aggrecan core protein has a molecular weight of about 260 kDa, which is about the same size as a fully decorated mucin molecule. Aggrecan, however, has many more and longer glycosaminoglycan chains than those attached to the mucin core protein. As indicated in this diagram, there are two regions where glycosaminoglycan chains are attached to aggrecan: a chondroitin sulfate domain that typically carries about 100 chains, and a keratan sulfate domain that typically carries about 30 chains. We have seen the structure of chondroitin sulfate previously in chapter 9 on collagen tendon, and as a reminder, chondroitin sulfate chains are built from a disaccharide repeat that contains a uronic acid, which carries a negatively charged carboxyl group, and a N-acetylglucosamine that carries a negatively charged sulfate group. Keratan sulfate is yet another glycosaminoglycan, and its disaccharide repeat structure is shown in fig. 15.4B, where we see that each disaccharide unit contains a single negatively charged N-acetylglucosamine sulfate ring. The chondroitin sulfate chains in aggrecan typically contain about 40–50 chondroitin sulfate units, and the keratan sulfate chains each contain 15–20 keratan sulfate units. This brings the total molecular mass of a single aggrecan molecule to a level of the order of 2.5 MDa, or roughly 10 times the molecular mass of an individual mucin molecule.

Unlike mucins, however, aggrecan does not form end-linked aggregates. Rather, the individual aggrecan molecules attach to another, very long, glycosaminoglycan chain called hyaluronic acid, which is a linear polysaccharide with a repeating disaccharide structure as shown in fig. 15.4C. This repeat contains a negatively charged uronic acid ring and an uncharged but highly polar N-acetylglucosamine ring. Finally, fig. 15.4D shows the structural organization of a full proteoglycan aggregate, which is clearly an enormous macromolecular assemblage, with a molecular mass of the order of 100–200 MDa. The top diagram shows the organization of a typical proteoglycan aggregate, with a full complement of aggrecans, which may reach levels of 100 aggrecans per aggregate. The expanded view below shows the structure in greater detail. The key feature of this aggregate is that it is so large that it is fully immobilized within the meshwork of fine collagen fibers, creating the composite material we call cartilage.

What are the mechanical properties of this unusual material? Perhaps the most useful set of properties that explains the mechanical function of cartilage comes from studies in which a cartilage sample is subjected to a static compressive stress, at a level that approximates the stresses in normal function in animal joints. Fig. 3.6 shows a long-term creep curve for articular cartilage, which was redrawn from Setton and colleagues (1993). This figure is also seen in fig. 15.5A; however, in this figure the curve for the Voigt element has been omitted. Here we see the results of a compression creep test that spans a time scale of 10^{-1} to 10^4 s. The material reaches its long-term equilibrium strain at about 3×10^3 s (about 1 hour). Note that the creep strain is normalized so that the equilibrium strain is assigned a value of 1.0. As a consequence, we do not have specific information on the magnitude of the tissue's strain or its stiffness through this creep process. However, other studies provide information on the equilibrium compressive

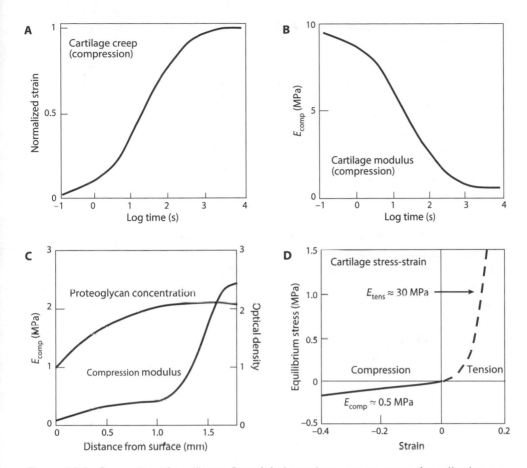

Figure 15.5. Properties of cartilage. **Panel A.** Long-term creep curve for articular cartilage. **Panel B.** Time-dependent cartilage compressive creep modulus, $E_{comp}(t)$. **Panel C.** The compression modulus and proteoglycan concentration at different levels in cartilage. **Panel D.** The equilibrium properties of cartilage in compression and in tension. E_{tens}, tensile modulus.

modulus of cartilage and the short-term dynamic stiffness of cartilage. The long-term equilibrium modulus is typically of the order of 0.5 MPa, and the short-term dynamic modulus of cartilage, on a time scale of about 1 s, is typically about 10 MPa. Thus, during the process of static loading a cartilage sample—or, for example, when a horse stands still overnight on straight legs—the effective cartilage stiffness should fall by a factor of about 20. Fig. 15.5B shows a plot of the time-dependent compressive modulus, $E_{comp}(t)$, that is based on the normalized creep data of fig. 15.5A and reasonable estimates of the dynamic stiffness and equilibrium stiffness of cartilage.

Table 15.1 provides more specific information for the equilibrium and dynamic stiffness levels. These data come from a study by Laasanen and colleagues (2003), who determined the equilibrium modulus from a long-term compression test, as described above, and dynamic properties that were determined by compressing a cartilage sample by about 10% over about 0.1 s, a deformation rate that is roughly comparable to the

Table 15.1. Variation in Cartilage Properties at Various Locations in the Bovine Knee

Location of the Cartilage Sample	Equilibrium Modulus (MPa)	Dynamic Modulus (MPa)
Patella	0.4	9
Tibial plateau	0.2	1.5
Medial femoral condyle	0.25	6
Medial patellar groove	0.4	15
Lateral patellar groove	0.6	15

Data from Laasanen et al. (2003).

shortest times reported in the cartilage creep data shown in Figs. 15.5A and 15.5B. The tests were carried out at a number of different locations in the bovine knee joint, and we see that there is about a threefold variation in the equilibrium and dynamic stiffness levels within this joint. Indeed, it is likely that there is similar variation in the stiffness at different locations in all joints. In addition, this study measured the variation in cartilage stiffness through the thickness of the knee cartilage by taking 10 slices, each about 0.15 mm thick, through the thickness of cartilage. The equilibrium modulus of individual slices was determined in compression tests. Although the process of slicing the tissue into thin layers likely disrupts the structural integrity of the cartilage to some degree, particularly the collagen-fiber lattice, the results are quite interesting. In addition, these authors used a stain that specifically labeled the cartilage proteoglycans in the tissue slices. It is known that the intensity of staining (optical density) is directly proportional to the proteoglycan content in normal cartilage. The results of this study are plotted in fig. 15.5C, where see that there are significant changes in both compression modulus and proteoglycan concentration through the thickness of the cartilage. That is, the compressive stiffness of the cartilage layers that are closest to the bone is about an order of magnitude greater than that of layers at or near the surface of the cartilage. The roughly twofold rise in proteoglycan concentration seen through the cartilage thickness likely accounts for part of this rise in stiffness, although the study also showed that there is an increase in collagen-fiber density toward the base of the cartilage.

What are the structural origins of the mechanical properties of cartilage described above? Why do we see the 20-fold reduction in cartilage stiffness through an hour of constant compression? The answer lies in the largely fluid nature of the cartilage composite. This fluid nature arises from the hydration of the polar glycosaminoglycans of the aggrecan molecules that are bound together in the aggrecan aggregates. The individual sugar rings in the glycosaminoglycans carry a number of polar hydroxyl groups and N-acetyl groups that hydrogen bond to water molecules, but more importantly the sugar rings carry negatively charged sulfate and carboxyl side groups that develop extensive hydration shells. Thus, the cartilage matrix is extremely hygroscopic, and the hydration and hence swelling of the tightly packed proteoglycan aggregates within the constraining meshwork of collagen fibers creates a "pressurized-fluid" structure to the cartilage matrix. At short times after a compression load is applied to a cartilage sample the measured compression strain is very low because the permeability coefficient, k, of the cartilage matrix is extremely low ($k \approx 10^{-14} \, \mathrm{m^4 \, N^{-1} \, s^{-1}}$). That is, water is not able to move away from the region under the compression load because it can not easily flow through the very dense cartilage matrix. This creates a high internal fluid pressure within the cartilage matrix that resists deformation, making the initial compression stiffness of cartilage quite high, as shown in fig. 15.5B. However, with time, water will move through the matrix away from the point of compression, allowing the tissue strain to increase (fig. 15.5A) and hence the internal pressure to fall.

There is one final material property of joint cartilage that is important for the structural design of this system; namely, the tensile stiffness of the cartilage surface layer in the plane parallel to the cartilage surface. Fig. 15.5D provides information on the relative stiffness of joint cartilage in tension and in compression. The data are from a study by Huang and colleagues (2005), and it is important to note that this graph actually shows the results of two very different mechanical tests. One is a compression test, where the compressive load is applied normal to the cartilage surface (indicated by the solid line), and one is a tensile test where the load is applied parallel to the surface of the cartilage (indicated by the dashed line). We will call this the transverse tensile stiffness. There are clearly two very different sets of properties. The equilibrium stiffness of cartilage in compression is, as discussed above, about 0.5 MPa, whereas the equilibrium transverse tensile stiffness starts out at a level of about 5 MPa, which is about 10 times greater than the compression stiffness. With increased strain, however, the tensile stiffness rises dramatically to levels in the range of about 10–40 MPa at different locations on the cartilage surface. That is, the cartilage-surface tensile stiffness is of the order of 20–80 times greater than the equilibrium compression stiffness. This high transverse tensile stiffness provides another important structural-design feature that is crucial to the function of articular cartilage; namely, the creation of a robust superficial layer. This layer has several important functions. First, it functions as a fluid-containment layer that restricts the movement of the pressurized cartilage fluid out of the cartilage matrix during periods of compression loading. Recall the description of the arrangement of collagen fibers in cartilage, as shown in fig. 15.3, where we saw that collagen fibers emerge vertically from the mineralized bone and rise vertically through the lower levels of the cartilage layer, but they then arch over as they approach the cartilage surface. In addition to these arched fibers, which form the "leaves" described above, there is a proteoglycan-deficient collagen surface layer that lies on top of the arched collagen fibers. This superficial layer provides a thin layer of collagen fibers that run parallel to the cartilage surface. In the experiment described in fig. 15.5D the tensile test sample was 0.25 mm thick, and thus it likely contains all of the superficial layer's collagen fibers plus a portion of the tops of the arched collagen leaves. As a result, the stiffness values given here likely represent a combination of the properties of the tops of the arched collagen-proteoglycan matrix and the more densely packed transverse collagen fibers at the very top of the cartilage. Thus, the true stiffness of the superficial collagen layer is likely somewhat higher than the values shown in this figure, and a guess of 20–80 MPa seems plausible. Based on the properties of parallel collagen fibers in tendons, with a tensile stiffness of the order of 1.5 GPa, a fully compact surface layer of multidirectional collagen fibers at the cartilage surface could have a stiffness of the order of 500 MPa. Clearly, the superficial layer is not simply a compact multidirectional layer of collagen fibers. Rather, the collagen fibers are surrounded by cartilage cells and proteoglycans, but the proportion of collagen is significantly higher in this layer, with the collagen likely occupying about 10% of the superficial layer's volume.

Van Mow and his colleagues have made many important contributions to our understanding of the structural basis for the mechanical properties of cartilage. Their initial models for cartilage mechanics were based on what they called the "biphasic poroviscoelastic behavior" of cartilage. The term "biphasic" relates to the presence of two discrete structural phases. Phase 1 is the proteoglycan-bound water, and this phase is immobilized within the tissue by the presence of phase 2, which is the collagen-fiber mesh (Setton et al., 1993). More recently, they have shown that there is a third contribution to the mechanics of cartilage; namely, that of the very large fixed charge density (FCD) that exists within the cartilage matrix (Lu et al., 2007). The large FCD arises from the numerous negatively charged carboxyl and sulfate groups that are attached to the cartilage glycosaminoglycan chains. The FCD forms the basis for the triphasic theory of cartilage biomechanics. The key feature of this theory is that the electrostatic repulsion forces between the numerous negative charges contribute significantly to the compression stiffness of the cartilage system. This contribution can be easily demonstrated by a simple experiment in which a cartilage test sample is placed in a solution with a high salt concentration. These authors conducted experiments where the cartilage test samples were immersed in PBS solutions of different concentrations. The control PBS solution had a 0.15 M concentration, which provides a reasonable approximation for the ionic concentration seen in normal tissue fluids. The experimental PBS solution had a 2.0 M concentration. PBS contains a mixture of sodium- and potassium-based salts, but the sodium salts dominate, so we will discuss the results in terms of the effect of sodium ions on the fixed charges present in the cartilage matrix. The disaccharide repeat unit of chondroitin sulfate contains two sugar derivatives, a uronic acid and a glucose-6-phosphate, each of which carries a single negative charge. If we assume that each sugar unit exists in vivo in a 1 nm cube of fluid, we can estimate the number of sodium ions associated with each charge on the chondroitin sulfate chain. Simple calculations allow us to estimate that at 0.15 M NaCl each charged sugar unit will have associated with it about 0.1 Na^+ ions, which means that there should be very little shielding of the fixed negative charge on the sugar's sulfate and carboxyl groups to reduce the FCD of the cartilage under normal conditions. However, at 2 M NaCl each charged sugar unit will have associated with it about 1.2 Na^+ ions, which should be sufficient to shield most of the fixed negative charge on the chondroitin sulfate chains. In effect, 2 M NaCl should largely eliminate the FCD of the cartilage matrix, and this should reveal the contribution of the fixed charge density to the overall stiffness of cartilage. The result of this experiment was that the equilibrium stiffness of cartilage in 2 M NaCl is reduced to about half that seen for cartilage in normal tissue ionic conditions. That is, roughly half of the equilibrium stiffness of cartilage arises from repulsive forces created by the fixed charge density of the glycosaminoglycans.

In summary, the structural design of cartilage employs a combination of collagen fibers, in highly organized arrays, with high concentrations of huge negatively charged proteoglycan aggregates that bind water to create a fluid-based system. Together, these constituents produce a constrained-fluid, fiber-reinforced composite material that is capable of transferring large compressive loads between bones in vertebrate skeletons without applying high local stresses, which would occur if rigid bone surfaces were placed in direct contact. Our next examples of soft-composite materials will deal with the design of tissues that function largely under tensile loading, those we find in animal skins and arteries.

3. THE MECHANICAL DESIGN OF VERTEBRATE SKIN. Skin is a very different material from cartilage; in fact it is a complex structure in vertebrate animals that typically contains two discrete layers—a collagen- and elastin-based dermis, which is covered by a keratin-based epidermis. We will focus on the dermis, which is largely responsible for the mechanical properties of skins. In vertebrate animals the skin functions as an extensible covering for the primary structural support system that is provided by the bone- and muscle-based internal skeleton. Invertebrate animals, however, generally lack rigid internal skeletons, and their structural support systems are based on a hydrostatic skeleton that is composed of a deformable external connective-tissue body wall that surrounds a muscle-lined, fluid- and tissue-filled chamber, much like that seen in the sea anemone described in chapter 13. That is, many wormlike invertebrate animals have collagen-based skins that provide the primary skeletal structure for the entire animal. In some cases, this "skin" layer is called a cuticle because it is an external noncellular structure much like insect cuticle. Nematode cuticle provides a good example of such a "cuticular" skin. For simplicity, we will use the term "skin" in our discussion of these connective-tissue-based skeletal systems.

We will start with the skins of mammals, as they likely are the most familiar kinds of skins for readers. Typically, mammalian skins are soft, pliable materials because they contain a collagen-based connective tissue in which the collagen fibers form a multidirectional network of wavy collagen fibers. In addition, interspersed within this collagen network is a multidirectional mesh of elastin fibers, and these two fiber systems are surrounded by a matrix "gel" that is formed primarily by the proteoglycan versican. Versican is similar to aggrecan, although it is somewhat smaller, and its properties are likely similar, although it plays a very different structural role as the matrix material in skin. That is, rather than functioning as a hydrated compression element that is constrained within a very fine mesh of collagen fibers, as is the case in cartilage, versican functions as a viscoelastic gel that surrounds the collagen and elastin fibers and provides a fluid environment that facilitates the extension and reorientation of these two reinforcing fibers. Details about the biochemical structure of versican can be found in a review by Wu and colleagues (2005), but the basics will be described here. The versican core protein has a similar structural organization and size to those of aggrecan, and it has a chondroitin sulfate–binding domain that binds about 20 chondroitin sulfate chains, each of which has a length similar to that found in aggrecan. As a consequence, the molecular weight of versican is about one-tenth that of aggrecan, with an approximate value of 0.75 MDa. Versican also has a hyaluronic acid–binding domain, and thus it forms proteoglycan aggregates similar to those shown in fig. 15.4 for aggrecan, but with molecular weights of about 10 MDa. Thus, versican is about one-tenth the size of aggrecan. It is known to bind to elastin fibers, and its primary mechanical role in skin is to provide a soft lubricating gel that facilitates the extension and reorientation of the reinforcing fibers, collagen and elastin, that form the structural integrity of skin.

Mammalian skins exhibit a characteristic J-shaped stress-strain curve, as illustrated in fig. 15.6A. This graph is based on a study by Veronda and Westman (1970), who

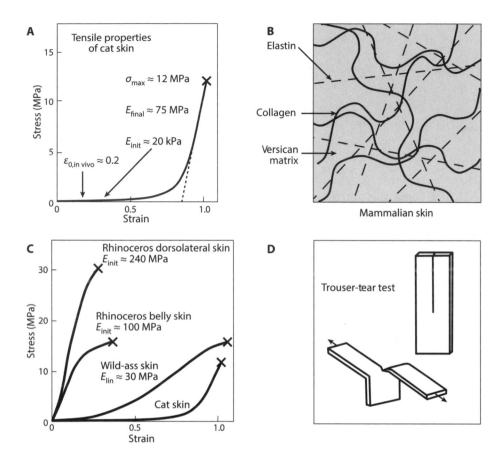

Figure 15.6. Properties and structure of mammalian skin. **Panel A.** The tensile properties of cat skin. E_{init}, initial modulus; E_{final}, final modulus; $\varepsilon_{0,in\ vivo}$, in vivo prestrain; σ_{max}, breaking stress. **Panel B.** General features of the structure of mammalian skin. This diagram provides a reasonable indication of cat skin, which is particularly extensible. The wavy collagen fibers run in many directions, but typically skin has anisotropic properties where stiffness levels may differ by about fourfold in orthogonal directions. The rapid rise in stiffness seen in panel A starts when the collagen fibers become extended and begin to carry significant levels of stress. **Panel C.** The mechanical properties of rhinoceros skin and skin from the wild ass. Data for cat skin are included for comparison. E_{lin}, linear portion of stress-strain curve. **Panel D.** The trouser-tear test that is used in measurement of skin toughness involves the measurement of the stress required to extend a precut notch in a skin test sample.

measured the properties of cat skin, specifically the lateral body skin, tested in a direction parallel to the animal's long axis. The test samples were taken in a procedure that allowed them to measure the in vivo prestrain on the tissue, $\varepsilon_{0,in\ vivo}$, which turned out to be about 0.2. That is, the skin on the animal existed at a longitudinal strain of about 0.2, as indicated in this figure. The tensile test, however, started from the fully relaxed state; i.e., at zero strain. The stresses measured at small strains were very low, on the order of 10 kPa, and as a result the initial tensile modulus at strains of 0.2–0.5 was very low, at about 20 kPa. That means the initial modulus of the skin is about 2% of the stiffness of a rubber band, or of a solid block of elastin. Above a strain of 0.5 stiffness rises rapidly, reaching values of about 75 MPa in the linear zone at strains above about 0.8. The sample broke at a strain of about 1.0, with a tensile strength of about 12 MPa. These material properties give us insights into the relative proportions of the three major components of skin: collagen fibers, elastin fibers, and the versican matrix. However, before we attempt to dissect out these contributions, we need to consider the distribution and orientation of these two fiber systems.

As shown in fig. 15.6B, mammalian skin, in general, is composed of multidirectional arrays of collagen and elastin fibers. Typically, collagen fibers in mammalian skins have a wavy appearance in the in vivo resting state, and the collagen fibers are oriented at a wide range of angles. Elastin fibers are similarly organized in a multidirectional array, but at resting dimensions they are straight. In most skins, these two fiber systems run almost entirely in the plane of the skin, with few or no fibers running through the thickness of the skin. Thus, the dermal layer in most skin is essentially a two-dimensional material. The linear elastin fibers become strained from the very start of extension; indeed, they are strained in the in vivo resting state, because, as shown in fig. 15.6A, the fully relaxed tissue contracts by about 20% when it is cut from the animal. With extension, however, the wavy collagen fibers become increasingly oriented toward the direction of strain, and their waves become extended, causing the collagen fibers to take on increasing levels of the applied load. With the material properties we have established for these two fiber systems, we can estimate the fraction of the tissue's cross-sectional area that is occupied by the collagen and the elastin. However, before we attempt this calculation, it is necessary to deal with the orientation of these fibers in the tissue composite. Fig. 15.6B is drawn in a way that suggests the collagen fibers are organized into a random composite structure, but this is generally not the case. Indeed, it has long been recognized that there is a degree of preferred orientation in the collagen-fiber lattice in skin. Ridge and Right (1966) showed that this preferred orientation can be demonstrated by the same method that was used with cartilage: a puncture made with a round pin created an elliptical cut, and the long axis of the cut indicates the direction of preferred orientation of the collagen fibers in the skin. As a result, these authors were able to document the origin of the mechanical anisotropy of skin as arising from this preferred orientation of collagen fibers, a phenomenon referred to as "Langer's lines." Typically, in human skin the extensibility of a skin test sample that is loaded in the direction of the ellipse's long axis is considerably less than a sample loaded at right angles to the ellipse. An interesting article by Carmichael (2014) provides insights into the concept of Langer's lines and the role that the properties of human skin play in surgical procedures.

Returning to the tensile properties of cat skin, we can now attempt to explain the magnitudes of the initial and final moduli of this skin and its tensile strength. First, we will consider the origins of the roughly 20 kPa initial stiffness, which we will attribute to the elastin-fiber mesh. Given the multidirectional nature of elastin fibers, we will base the calculation on the random-fiber-composite model given in eqs. (8.4) and (8.5), where the reinforcement efficiency of a random composite is three-eighths of the stiffness of a parallel-fiber composite. Since elastin has a tensile stiffness of about 1 MPa, a random composite of pure elastin should have a stiffness of about 0.4 MPa. With an initial skin stiffness of 20 kPa, we can infer that elastin occupies about 5% of the skin volume. Similarly, with collagen, which has a tensile stiffness of 1.5 GPa and a unidirectional fiber architecture that is created at high strains in a skin tensile test, we can infer that it also occupies about 5% of the skin volume. These calculations represent minimal levels for collagen and elastin in skin because they assume a simplified fiber structure that likely does not reflect the true structure of skin. In reality, the fiber architecture of cat skin is more complex than we have assumed, and there are likely many collagen and elastin fibers that do not carry the stress levels assumed in this calculation. Regardless, it is clear that the versican matrix and cells must occupy a large fraction, likely in the range of 75% of total skin volume. However, these values for skin do not hold for all mammalian skins, as there are large variations in the functional roles of skin in different mammals. Indeed, cat skin is particularly soft and extensible, and we should consider the skins of other animals.

Skin has a variety of functions in mammals, but generally it is a relatively soft and extensible material that allows the skin to deform easily during body movements. In some animals the skin also functions as a form of body armor, and Robert Shadwick and colleagues (1992) investigated the mechanical properties of the skin of the white rhinoceros, an animal that is known for its large, sharp horn, which it uses for defense and also in mating combat between males. This is a skin that requires quite different properties from those seen in cats and humans. In particular, the dorsolateral flank skin of the rhinoceros functions as a form of body armor, and these animals have evolved a quite specialized version of dermal skin to form this armor. Fig. 15.6C shows stress-strain curves for two kinds of rhinoceros skin, along with curves for the skins of a wild ass (a horselike animal) and the cat. Clearly, there are quite substantial differences amongst the skins of mammals. Interestingly, the tensile strengths of three of the four skins are quite similar, with values of the order of 15 MPa, but one, the rhinoceros dorsolateral skin, has about twice the strength of the others. This reflects the dorsolateral skin's functional role as body armor. However, the most significant difference in properties between these four skins is in their stiffness.

There are very large differences in initial modulus, E_{init}, in these skins, ranging from 20 kPa for cat skin, to about 240 MPa for rhinoceros dorsolateral skin. This is a 10,000-fold range of initial stiffness that arises from two factors. Normal skin (e.g., cat skin) has wavy collagen fibers that carry essentially no load at low strains, and thus the initial stiffness is determined entirely by the orientation and content of elastin fibers, as discussed above. Clearly, the collagen of rhinoceros skin plays a dominant role right from the start. This is because in rhinoceros skin the collagen mesh, which is much denser than that in cat skin, also contains nearly straight fibers. Indeed, the low-stiffness "toe zone" of stress-strain curves for both rhinoceros dorsolateral and rhinoceros belly skin is essentially nonexistent. The rhinoceros belly skin, however, is somewhat less stiff than the dorsolateral skin, with an initial linear modulus of about 100 MPa, but both types of rhinoceros skin are much stiffer than the skin of the wild ass. Interestingly, the shape of the stress-strain curve for wild ass skin is much closer to that of cat skin in that there is a clear low-stiffness zone that extends to strains of about 0.3 and then rises to a linear zone with a stiffness of about 30 MPa.

How can we explain this range of material properties? In cats the skin represents one extreme of material behavior that is dominated by a dilute mesh of wavy collagen fibers that do not contribute to the stiffness of this skin until the skin has been extended to strains of about 0.7. The wild-ass skin shows a similar behavior, but the recruitment of collagen likely is present even at initial strains, and the reorientation of the collagen fibers at strains of about 0.3 brings the stiffness to about 30 MPa. It is not clear, however, how much the elastin fibers may contribute to the initial properties of the wild-ass skin, but horse skin is known to contain significant levels of elastin. So the initial low-stiffness zone of this skin is likely due to both the stretching of elastin fibers and the straightening of collagen fibers that are much less "wavy" than those present in cat skin.

Finally, Shadwick and colleagues document the fracture toughness of rhinoceros and wild-ass skin, and there are some interesting differences. Specifically, they measured the work of fracture, kJ m^{-2}, in an experimental protocol called a "trouser-tear test," which is useful in studies of soft fabric materials. This test used a rectangular strip of skin that was cut to form a long notch that runs about half the length of the strip, and then the two free ends of this strip are pulled apart, as shown in fig. 15.6D, extending the notch by tearing the material. The work of fracture can be calculated as the strain-energy release rate, G (i.e., fracture toughness), which can be determined by quite simple calculations. Briefly, G is calculated as:

$$G = \frac{W}{A_p},$$
(15.1)

where W is the total work required to extent the crack, and A_p is the total area that is created as the crack is propagated. The work of fracture for the rhinoceros dorsolateral skin was measured as 78 kJ m^{-2} for the external layers of the skin, and 43 kJ m^{-2} for the deep layers of the skin. For comparison, the work of fracture for the wild-ass skin was 33 kJ m^{-2}. So the rhinoceros dorsolateral skin is not only stiffer and stronger than the wild-ass skin, but it is also tougher by a factor of about two. For comparison with other materials, table 5.1 gives toughness values for several structural materials, and you will see that the toughness of about 80 kJ m^{-2} for external layers of rhinoceros dorsolateral skin places it amongst the toughest of biomaterials. It is tougher than wood and horse-hoof keratin, and in this table its toughness is only exceeded by that of mild steel.

Peter Purslow (1983) measured the fracture toughness of two very different "skins," the mesoglea of the sea anemone *Metridium senile*, and the skin of rats. He also used the trouser-tear test protocol, described above. Tests with *Metridium* mesoglea were conducted in both the longitudinal and circumferential directions, and the toughness was essentially identical, at levels of approximately 1.1 kJ m^{-2}. This is rather surprising because the dominant circumferential orientation of collagen fibers in the inner layer of *Metridium* mesoglea should create greater toughness in longitudinal tests than in circumferential tests. However, the very large strains that can occur in this discontinuous-fiber composite likely result in huge strains at the tear surface, and these high strains may randomize the fiber organization at this interface.

Purslow also carried out more extensive testing on rat skin. Both fracture toughness (G) and stiffness (E) were measured in samples that were taken from anterior and posterior locations and from dorsal and ventral locations on the body. In addition, tests were carried out in both longitudinal and transverse directions. There were relatively small differences between anterior and posterior locations, but there were quite large differences between dorsal (i.e., back) skin and ventral (i.e., stomach/chest) skin. So we will concentrate on differences between dorsal and ventral skin properties. For the dorsal skin, the circumferential stiffness (ca. 8 MPa) was about 60% greater than the longitudinal stiffness (ca. 5 MPa), and for the ventral skin the circumferential stiffness (ca. 20 MPa) was about three times greater than the longitudinal stiffness (ca. 6 MPa). Thus, there must be significant differences in the collagen-fiber content and orientation between dorsal and ventral skin, with the ventral skin showing both greater collagen content and greater reinforcement in the circumferential direction. The pattern of fracture toughness followed a similar but essentially inverse pattern relative to that seen for stiffness. That is, the direction of greatest fracture toughness occurred in the longitudinal direction ($G \approx 22$ kJ m^{-2}) in both dorsal and ventral areas, and the fracture toughness in the circumferential direction ($G \approx 14$ kJ m^{-2}) is smaller by about one-third. Overall, these data indicate that the collagen mesh in rat skin is preferentially oriented in the circumferential direction of the animal's body, and this preferential fiber orientation creates greater stiffness in the circumferential direction and greater toughness in the longitudinal direction. Table 5.1 indicates that, in comparison to other materials, rat skin is as tough as horse-hoof keratin and tougher than wood.

Another interesting animal skin is found in the blubber of whales. Mammalian skin comprises three tissue layers—the epidermis, the dermis, and the hypodermis—and these three layers are also found in whale blubber. However, in large adult whales they can reach an overall thickness of about 20 cm. As with other skins, the whale epidermis is a keratin-based layer, which in large whales is a few millimeters thick. This keratin is typically filled with melanin granules, making the epidermis dark black in color, although some whales have melanin-free white patches as well. The dermis and the hypodermis together make up the mechanically important parts of whale blubber. The whale dermis is like the dermis layers in other mammals, being composed largely of collagen-fiber bundles, with some elastin fibers as well. However, the collagen fibers in whale blubber are organized in three dimensions, as opposed to the largely two-dimensional arrays seen in, for example, cat skin. The hypodermis in whales is quite different from that of other mammals, however. Typically in mammals, the hypodermis contains a very diffuse mesh of collagen fibers and mainly comprises fat deposits. In whales, the fat deposits are present in abundance, but the hypodermis is rich in large collagen-fiber bundles that are also organized in three dimensions. In mature whales, the epidermis makes up 2%–3% of blubber thickness, the dermis makes up about 20% of blubber thickness, and the hypodermis makes up 75%–80% of blubber thickness.

A study by Orton and Brodie (1987) provides stress-strain curves for two kinds of blubber from the fin whale: the ventral-groove blubber (VGB) that forms the throat skin of this animal, and the lateral blubber from the caudal peduncle of the whale. This throat skin expands dramatically during feeding, as shown in fig. 15.7A, where we see an outline of a blue whale that is feeding. This image was drawn from a photograph taken from an airplane flying over a blue whale that was feeding at the ocean surface. The animal has rolled onto its right side, revealing its body profile, and the expansion of the striped VGB indicates the fully distended state of this structure during feeding. The dark gray body of the whale indicates the shape of the animal's body if it were not feeding, as the VGB will have contracted to form the streamlined body shape. Orton and Brodie's data are plotted as true stress, σ_{true}, against true strain, $\varepsilon_{\text{true}}$ (solid lines), in fig. 15.7B, but this figure also shows their data plotted as engineering

A

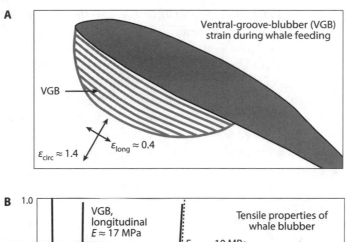

Ventral-groove-blubber (VGB)
strain during whale feeding

VGB

$\varepsilon_{circ} \approx 1.4$ $\varepsilon_{long} \approx 0.4$

B

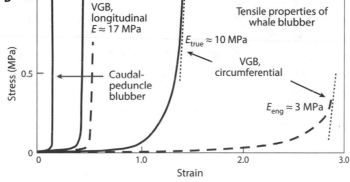

Figure 15.7. Properties of whale ventral-groove blubber. **Panel A.** Diagram of a blue whale feeding near the surface of the ocean, drawn from an aerial photograph. The animal has just finished feeding, and its ventral-groove blubber is fully distended. The *double-ended arrows* indicate the approximate strain of the blubber in the longitudinal (ε_{long}) and circumferential (ε_{circ}) directions. **Panel B.** Stress-strain behavior of blubber from a fin whale. Note that *solid lines* indicate true stress-strain behavior, and *dashed lines* indicate engineering stress-strain behavior. *E*, modulus; E_{eng}, engineering modulus; E_{true}, true modulus.

stress, σ, against engineering strain, ε, (dashed lines). The VGB was tested in both the longitudinal and circumferential directions because this tissue comprises alternating strips of a high-stiffness blubber, with properties that are likely quite similar to those of the body blubber of the whale (i.e., the blubber from the caudal peduncle), and a much softer, elastin-rich blubber. There is also a single curve for the caudal peduncle blubber because its maximal strain is about 0.14, and there is essentially no difference between the true strain and the engineering strain at this strain level.

The properties of the VGB are very different in the longitudinal and the circumferential directions. In longitudinal tests the alternating strips of stiff and soft blubber are arranged in parallel, making the tissue as a whole quite stiff (ca. 17 MPa) and the extensibility quite low, at about 50% extension ($\varepsilon \approx 0.5$). Assuming roughly equal widths of the stiff and soft blubber, the stiffer blubber is likely about twice the whole-tissue stiffness, or about 30 MPa. In circumferential tests the alternating blubber strips are arranged in series, and as a consequence the stiffness is dominated by the softer and more extensible strips of blubber. Thus, the stiffness of the whole tissue is lower and its extensibility is much greater in the circumferential direction. In this case, the stress-strain curve plotted as true stress versus true strain is very different from the curve plotted as engineering stress and engineering strain. Comparisons based on engineering stress and strain will be more useful, as they relate more directly to the physical dimensions of the animal. That is, the stiffness comparisons are not that crucial, as it is clear that stiffness rises dramatically in both directions, but the actual strain at which the stiffness rises is best revealed with engineering strain. It is very clear that the VGB has the capacity for an almost threefold expansion in the circumferential direction and only a 50% expansion in the longitudinal direction. How do these calculations match with observations of the actual expansion that occurs during normal feeding behavior? Robert Shadwick and colleagues (2013) documented the expansion that occurs in the feeding process of several species of freely swimming whales, and they calculated the typical strains that occur in the VGB in life. Interestingly, the longitudinal strain observed in feeding whales was about 0.4, which matches very nicely with the strains where stress began to rise quickly when the isolated blubber samples were tested in the longitudinal direction. However, the circumferential strain observed in these animals was only about 1.5 in fin and blue whales, which is about half the maximal strain observed in tests on the isolated tissue (fig. 15.7B). Interestingly, the authors report that the expansion seen in dead and bloated animals exhibits much greater circumferential strains that are in the range of 2.6. Thus, the potential for circumferential expansion of the structure does indeed match the behavior of the isolated tissue, but in its normal function in the living animal circumferential expansions of this level are not observed. The structural basis for this limited in vivo tissue strain is explained in this study as arising from the action of muscles that lie beneath the VGB. There are three muscle layers under the VGB, and they play an active role in controlling its extension and contraction during feeding. There is an inner layer of muscles that run parallel to the long axis of the body, followed by a pair of muscle layers that are oriented at about ±45° to the long axis of the body in the contracted state of the VGB. When the animal opens its mouth to feed, the pressure of the water causes the expansion of the VGB in both longitudinal and circumferential directions. The VGB's mechanical properties clearly limit the longitudinal expansion to about 40%, but the mechanical properties of the VGB evidently are not what limits the circumferential expansion. So, how does the animal control VGB expansion during feeding?

The answer lies in the active control of circumferential expansion by the activation of the muscles that underlie and are attached to the VGB. The geometry of the muscle system is key. At the start of VGB expansion the tissue expands in both longitudinal and circumferential directions, but longitudinal expansion is limited by the blubber to about a strain of 0.4. Circumferential expansion at this strain is not limited by the blubber, and hence it continues to rise. As a consequence, circumferential strain increases relative to longitudinal strain, and the crossed muscle layers will become increasingly oriented in the circumferential direction, where they can impose an active force that can arrest the circumferential expansion to strains that are well below the limits imposed by the blubber's mechanical properties. The logic of this system is that once the VGB is filled with water, and presumably food, the animal needs to contract the greatly expanded VGB, and it does this by the contraction of the underlying muscles. So, the stiffness of the VGB in the circumferential direction is not used to control VGB expansion because active muscles clearly provide a more versatile mechanism for controlling the complex processes associated with feeding.

Whales and other marine mammals have another skin-like structure that lies just under the blubber layer, which is called the subdermal sheath. This sheath is made primarily from collagen, and it is found with a collagen-fiber architecture that is essentially a cross-helical lattice with fiber angles in the range 48°–70°. More specifically, as described by Ann Pabst (1996) for dolphin subdermal sheath, forward-facing fibers typically have angles of about 50° and backward-facing fibers have angles of 60°–70°. This system is basically a whole-body "fabric" that would resist strain by the application of an internal pressure, but would allow bending by the reorientation of the fibers in the fabric-like structure. Alexander (1987) described the basic physics of bending in cylindrical animals with helical fibers in their skin or cuticle, and he made the interesting observation that in a cylindrical animal with a 60° cross-fiber collagen "sheath" that was attached to vertebral processes, in the midline plane the collagen fibers would not be strained. Rather, the body could bend easily. The fact that the collagen fibers in the dolphin are at fiber angles very close to 60° suggests that the subdermal sheath provides something like a whole-body tendon that encloses the entire body but does not hinder its ability to bend, a feature that the dolphin uses extensively in its very active locomotion, which involves the high-amplitude bending of the body. This kind of whole-body "tendon-like" structure is a common feature of swimming animals, and a prime example is found in sharks, as described in a study of Steve Wainwright and colleagues (1978).

The Wainwright study started with the observation that sharks employ a hydrostatic skeleton to complement their rigid cartilaginous skeleton for their locomotor activities, and they do this through the functional design of their skin. These authors observed that shark skin contains a cross-helical array of collagen fibers, and they measured the subdermal hydrostatic pressures that are generated when a shark swims. The pressures they measured were as follows: a resting shark has a tissue pressure of 7–14 kPa, and for comparison, atmospheric pressure is about 100 kPa. So resting pressure is about 10% of atmospheric pressure. In slow swimming the animal propagates relatively low-amplitude waves of bending down its body, where the body's radius of curvature reaches levels of about 40 cm, and the subdermal pressure increases about threefold to about 20–35 kPa. They calculated that this pressure would create a circumferential stress in the skin of about 0.3 MPa. In burst swimming, however, the shark's body radius of curvature is much smaller, at about 20 cm, indicating a much tighter bend, and the subdermal pressure increases about 20-fold higher than resting pressure, at about 200 kPa. This creates a circumferential stress in the skin of about 2.8 MPa.

Wainwright et al. reasoned that as the shark increased its speed, the bending curvature would increase because of the cross-helical collagen-fiber lattice that exists in shark skin. Interestingly, they found that the fiber angle of the skin collagen posterior to the caudal fin was between 59° and 62°, as shown in the diagram in fig. 15.8. That is, the collagen fibers are organized at angles where body bending should not cause large changes in volume, which would increase tissue pressure. However, as body curvature increases during a swimming stroke, fiber angles will change because of the extension or contraction of the skin in the longitudinal direction, and at higher speeds, with

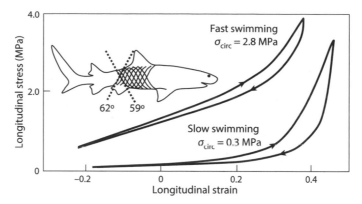

Figure 15.8. Biaxial load-cycle testing of shark skin under conditions that mimic the behavior in slow and fast swimming movements. σ_{circ}, circumferential stress.

greater body curvature, the skin strain rises because the collagen lattice is shifting well away from the 60° angle. Thus, it is likely that low-amplitude swimming, with relatively small bending of the body axis, does not apply great strains on the collagen-fiber lattice in the skin. But at increased speed the high tissue pressure would generate significant stress in the collagen lattice, which could then serve as an elastic-energy-storage device that could capture energy from the muscles that power the swimming stroke and use it to power the recovery of the locomotor system in the next bending cycle. They carried out biaxial mechanical tests on isolated shark skin samples in which they loaded one axis to a stress that corresponded to the pressure level measured in swimming animals. That is, to mimic slow swimming they held the circumferential stress to 0.3 MPa, and to mimic fast swimming they held the circumferential stress at 2.8 MPa. Then they imposed strains in the longitudinal direction to mimic the stretching and contraction of the collagen-fiber lattice. Fig. 15.8 shows the results of these biaxial tests on shark skin samples.

The stress-strain loops they generated provide an indication of the longitudinal stress that would exist at various levels of longitudinal strain in this pressurized system, and it is clear that there are significant changes in the longitudinal stress of the skin, particularly in fast swimming, where we see a very narrow stress-strain loop that suggests a high elastic resilience of this skin system. Clearly, this pressurized skin system plays an important role in the biomechanics of shark swimming. In fast swimming in particular, the stress-strain loop is very tight; that is, the area inside the loop is small, indicating a highly efficient elastic system that will provide significant stored energy to assist the return stroke of the swimming cycle.

Crossed-helical skin structures are not only seen in fish skins. Indeed, crossed-helical skin/cuticle structures are well documented in a variety of wormlike invertebrate animals, such as earthworms and nematodes. Unlike the shark, however, these animals do not have any rigid supporting structures like the shark's cartilaginous skeleton. The common earthworm provides an interesting example of a body design that takes advantage of a 45° cross-fiber collagen lattice in the cuticle of the "resting" animal. This collagen-fiber system interacts with sets of longitudinal and circumferential muscles that surround a fluid-filled body cavity. Together, these structures create a hydrostatic skeleton that allows the animal to actively shorten or extend each of its segments to create body movements that allow it to burrow effectively in soil, as described by Quillin (1999). Another interesting example of an invertebrate animal with a collagen-fiber-based cuticle is seen in the parasitic nematode *Ascaris lumbricoides*. This animal's muscle/cuticle design is similar to the shark's, in that it has only longitudinal muscles. It has a collagen-based cuticle in which the collagen fibers are organized into a crossed-helical array, with fibers that are oriented at 75° to the animal's long axis. As in the shark, the contraction of its longitudinal muscles will increase the collagen-fiber angle, and this will reduce the volume of its internal fluid cavity and hence increase the pressure within its hydrostatic skeleton. Thus, it can use its longitudinal muscles both to pressurize and to bend its cylindrical body, and this system allows it to burrow effectively in soil and in the tissues of other animals that it parasitizes, as described by Harris and Crofton (1957). These studies do not actually deal with materials design, rather they are directed at whole-body structural design, and readers who are interested in learning about these interesting animals should read the paper by Clark and Cowey (1958). It provides a very interesting and important analysis of the role of the cross-fiber collagen lattice in the structural design of invertebrate animals with hydrostatic skeletons. This completes our analysis of mechanical design of skin and skin-like pliant materials, and we will conclude our discussion of pliant composites with a analysis of the mechanical design of vertebrate arteries.

4. THE MECHANICAL DESIGN OF VERTEBRATE ARTERIES. In our analysis of microfibrillar elastomers in chapter 14 we discussed the role of the microfibril J-shaped stress-strain curve in the design of invertebrate arteries, where microfibrils are frequently found. Indeed, it is very clear from many years of analysis of artery design that a J-shaped stress-strain curve is a dominant feature of all animal arteries, as described in review articles by Shadwick (1998, 1999). The physiological function of elastic arteries in animal circulatory systems is to provide a biophysical interface between a pulsatile pump, the heart, and the requirement for reasonably steady flow of blood through vessels that take blood to the tissues of the body. The heart is a pump that is driven by the transient shortening of muscles that make up the walls of its chambers. During systole, when the muscles of the left ventricle shorten, blood is ejected into the thoracic aorta, typically at a frequency of one to three contractions per second in humans. However, between ventricular contractions, during the phase called diastole, the left ventricle is refilled with blood by the contraction of the left atrium. During this period the flow of blood into the aorta stops, and the aortic valves close to prevent blood from returning into the heart. If the heart were attached to arteries made from a rigid material, blood flow in the artery during diastole would simply come to a halt, and then in the next systolic contraction the heart would not only have to deliver energy to eject blood from the left ventricle, it would have to accelerate the blood sitting outside the heart in the aorta. This would be a very inefficient system. The soft, rubberlike elasticity of the arterial system that receives blood from the heart provides a mechanism to store some of the kinetic energy of the blood ejected during systole as strain energy in the stretching of the arterial-wall elastic tissue. In diastole, this stored energy powers the elastic recoil of the aortic wall that keeps the blood pressure in the aorta at

a reasonable pressure and maintains blood flow to the tissues while the heart prepares for its next contraction. Thus, rubberlike elasticity is a key feature of the arteries in all animal circulatory systems.

However, the form of the elastic behavior in arteries is dramatically different from the elasticity of pure rubber, which is illustrated in figs. 15.9A and 15.9B. In fig. 15.9A we see a graph of circumferential wall tension (T_{circ}) plotted against the radius of a rubber cylinder. You can think of this as a cylindrical rubber balloon. The law of Laplace defines

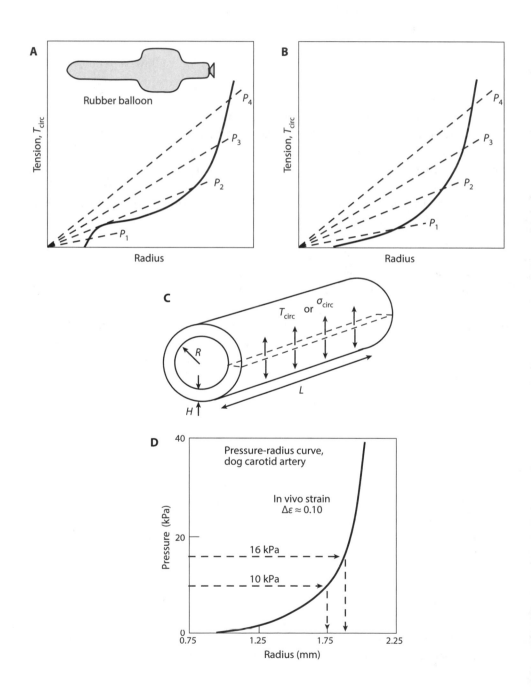

the circumferential wall tension, with units of Newtons per meter of the cylinder's length $(N\ m^{-1})$, for a pressurized cylinder at a function pressure (P) and radius (R), as:

$$T_{circ} = PR. \tag{15.2}$$

Tension is thus the circumferential force per unit length of the cylinder. The four dashed lines in fig. 15.9A represent different solutions for this equation at different pressure levels. Note that line P_1 intersects the balloon's tension/radius plot at a single point, as do lines P_3 and P_4. These lines indicate stable inflations because they cross the tension-radius-inflation curve at a single radius. Line P_2 is drawn so that it intersects the curve tangentially at low radius and then crosses the tension-radius line at a larger radius. The region between these two points indicates a region of instability in this elastic tube because the balloon can exist in two different states at the same pressure, one at low tension and low radius and one at high tension and high radius. As a consequence, there is a dramatic increase in radius between the two intersection points on line P_2, and this causes the formation of a local expansion that creates a bulge in the balloon, as drawn in fig. 15.9A. Because the second intersection with the pressure-radius line occurs at much higher radius, the wall tension in the bulge will be much higher than in the narrower parts of the balloon. In addition, the wall thickness (H) in this bulge will fall because of the increased radius and length of the balloon, and this will create a much higher circumferential wall stress, , σ_{circ}, which is defined as:

$$\sigma_{circ} = \frac{PR}{H}. \tag{15.3}$$

With continued inflation this bulge will extend until it encompasses the whole length of the balloon. In an artery, however, this kind of local expansion would be called an aneurysm, and this is clearly not a good design because the wall tension, and hence the wall stress, is dramatically higher in the "aneurysm" than in the narrower part of the artery on either side of it. Fig. 15.9B shows the type of tension-radius relationship observed in arteries, where we see that there is a continuously increasing slope, and it is

Figure 15.9. (*left*) Elastic tubes and arteries. **Panel A** shows the tension-radius behavior of a cylindrical rubber balloon, where we see that the declining stiffness of the balloon rubber at intermediate extensions allows the formation of a local expansion in the balloon. This anomaly causes the stress to be much higher in the expanded region than in other regions, creating an aneurysm-like structure. This region of instability arises from the fact that at intermediate pressures (line P_2) the pressure-radius lines intersect the inflation curve at two points, thus creating this anomaly. **Panel B.** Tension-radius plot of an artery-like structure in which the wall material shows a continuously increasing stiffness that prevents the formation of aneurysm-like structures. **Panel C.** Diagram of an arterial segment indicating the structural parameters that determine the circumferential tension and stress of an arterial segment. **Panel D.** Pressure-radius curve for a dog's carotid artery. Note that there is a continuously increasing pressure-radius relationship, which is a key characteristic of all artery wall materials. H, wall thickness; L, length; P, pressure; R, radius; T_{circ}, circumferential wall tension; $\Delta\varepsilon$, change in strain; σ_{circ}, circumferential stress.

this increasing "stiffness" of the wall material that creates a stable inflation, without the formation of aneurysms. The key question is how quickly the stiffness of the arterial-wall material needs to rise for the inflation to be stable and not lead to the creation of an aneurysm. There is, unfortunately, no simple mathematical solution to this problem, and there are a number of mathematical analyses that are much too complex for this discussion. However, it is generally agreed that the circumferential stiffness must increase dramatically with arterial expansion.

In addition, we will make another key assumption about the way that arteries function in vertebrate animals; namely, that they are almost always attached to rigid or semirigid structures in animal bodies, and that in vivo the arteries are under longitudinal tension from their attachments to these structures. As a consequence, during the cardiac cycle, in which blood pressure varies over a range of about 10–16 kPa, the diameter of arteries increases by about 10%, but there is essentially no change in the length of the artery. That is, the circumferential strain amplitude of arteries is approximately 0.1. Fig. 15.9C shows a diagram of an arterial segment that identifies the structural parameters of this cylindrical elastic system.

Now, let's look at the mechanics of artery inflation. Fig. 15.9D shows the results of an inflation test on a dog carotid artery, as measured by Dobrin and Rovick (1969). Here we see the pressure-radius relationship for the inflation of an arterial segment that was longitudinally stretched to its in vivo length in the unpressurized state. The zero-pressure resting radius is about 1 mm, but the in vivo "resting" radius for the cyclic pressure fluctuations that occur in the living animal is about 1.75 mm. That is, at diastolic pressure (ca. 10 kPa) the artery is stretched circumferentially by about 75%. Then, during the contraction of the heart the blood pressure rises to the systolic pressure, and the artery wall is stretched by an additional ≈10%. What is the structural basis for this J-shaped pressure-radius curve?

Fig. 15.10 shows the general features of arterial structure that are responsible for the mechanical properties of the artery wall. Fig. 15.10A shows the general organization of the arterial-wall connective tissue at zero pressure and at diastolic pressure (10 kPa). The artery wall's tissue is quite different in its structural organization from that of skin, because the artery wall is essentially a muscle-based system. The study by Clark and Glagov (1985) provided the first comprehensive description of the structure of the arterial media, and we can now consider the structural basis for the mechanical properties of arteries. Fig. 15.10A provides a low-magnification diagram of the tunica media, which provides the majority of the structural properties that are required for the mechanical function of arteries. Not shown, however, are two additional quite thin tissue layers that are present at the inner and outer surfaces of the artery wall. These layers are the inner, tunica intima, and the outer, tunica adventitia. The tunica intima consists largely of an endothelial layer that provides an important interface between the blood and the arterial tissue, and in addition contains a single elastin layer called the internal elastic membrane. The tunica adventitia is composed largely of wavy collagen fibers that do not contribute significantly to the mechanical properties of the artery at physiological pressures, but are recruited at pressures that exceed the normal systolic pressure of the circulatory system. The tunica media, however, is composed entirely of alternating layers of vascular smooth muscle cells, elastin, and collagen, and this structure provides the mechanical properties that dominate the functional behavior of

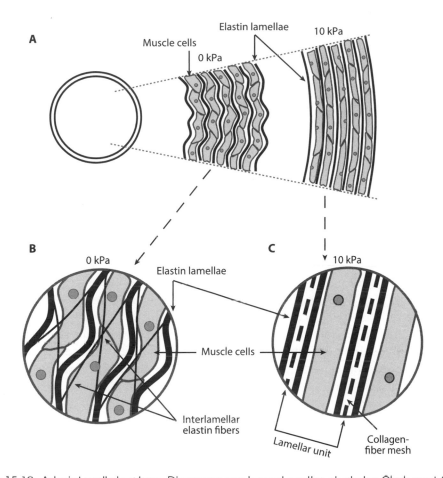

Fig 15.10. Arterial-wall structure. Diagrams are based on the study by Clark and Glagov (1985). **Panel A** shows a low-magnification view of the structural organization of the aortic tunica media as it exists at zero pressure (0 kPa) and at diastolic pressure (10 kPa). The two most prominent features of the artery structure, the elastin lamellae and the vascular smooth muscle cells, are illustrated here, where the *dark bold lines* indicate the elastin lamellae. Note that at zero pressure, a state that does not exist in the living animal, the elastin lamellae show a wavy structure, but at diastolic pressure the lamellae are straight, and as a consequence the straight lamellae are stretched during systole. Panels B and C show the lamellar structure at higher magnification, where it is possible to see two additional features of the artery-wall structure. First, in **panel B**, with the artery at 0 kPa, we can see that in addition to the thick elastin lamellae, there are thin interlamellar elastin fibers that run between the lamellae and also interact with the vascular muscle cells. It is the contraction of these fibers that causes the wavy structure of the elastin lamellae at low pressures. In **panel C**, with the artery at 10 kPa pressure, additional detail is provided for the organization of the layers previously identified as elastin lamellae. In fact, each layer is actually composed of two elastin lamellae, with a collagen-fiber-mesh layer in between. The straightening of the lamellae and the stretching of the interlamellar elastin fibers, followed by the stretching of the elastin lamellae, are responsible for the initial increase in artery-wall stiffness with inflation. However, the full increase in stiffness as pressure approaches 10 kPa requires the alignment of the collagen-fiber mesh and the transfer of stress to the collagen fibers.

the artery. Figs. 15.10B and 15.10C show magnified views of the tunica media at zero pressure and at 10 kPa. Note that in these panels the structural elements are drawn in an expanded form to make it easier to see the individual components. In the living tissue, these layers are tightly packed together. At zero pressure (fig. 15.10B) we see that the elastin lamellae and the vascular smooth muscle cells show a wavy structure. This wavy appearance is caused by the contraction of another key component of the arterial structure, the interlamellar elastin fibers. These elastin fibers run essentially parallel to the elastin lamellae, but they also interact with the vascular smooth muscle cells, and they attach between adjacent lamellae. It is the contraction of these interlamellar elastin fibers that causes the wavy appearance of the cells and lamellae at zero pressure. However, this is a state that does not exist in the living animal, because zero blood pressure is only seen in a dead animal or an isolated artery sample. Fig. 15.10C shows an expanded view of the lamellar organization at diastolic pressure (10 kPa), as it likely exists in the living animal. Here we see the straightened vascular smooth muscle cells, as well as an expanded view of the elastin lamellae. In reality, the structure of the elastin lamellae, which are shown as single black lines in figs. 15.10A and 15.10B, consists of two elastin layers, with a mesh of wavy collagen fibers in between. In fig. 15.10C, the solid black lines represent the elastin lamellae, and the dashed lines represent the collagen-fiber mesh. In the compact form of the intact tissue, the elastin lamellae are closely associated with the cells, and the collagen mesh is sandwiched between the elastin lamellae. It is this structural organization that accounts for the shape of the arterial-wall stress-strain curve.

What are the mechanical properties of the arterial tissue that underlie the pressure-radius behavior of an artery shown in fig. 15.9D? Fig. 15.11 shows several ways of presenting the mechanical properties of the artery-wall material, as determined for the carotid artery by Dobrin and Rovick (1969). In fig. 15.11A we see the inflation data converted into a stress-strain curve that shows a classic J-shape, with continuously increasing slope. Hence we can quickly conclude that the stiffness of the artery wall rises dramatically over its full range of expansion. It is important to note that this is not a curve that indicates the failure level of this material, as rupture of a healthy artery will occur at much higher pressures than exist in arteries during their normal function. Rather, this plot focuses on the material's properties over the stress and strain range that occurs during a normal heart beat. Fig. 15.11B shows a plot of the artery's elastic modulus as a function of pressure. Normal blood pressure typically oscillates between about 10 and 16 kPa, and this graph provides a direct connection between animal physiology and arterial-wall stiffness. That is, if we determine the modulus values at 10 and 16 kPa from fig. 15.11B, we see that artery-wall stiffness increases about threefold during its physiological function, from 0.25 to 0.8 MPa. We can apply these modulus values to the modulus-strain curve in fig. 15.11C, and we see again that the in vivo strain amplitude of the artery $\Delta\varepsilon \approx 0.10$.

Before we move onto other types of artery, it will be useful to think about the origins of the stiffness levels we see in the artery over its normal range of expansion. A diastolic modulus of 0.25 MPa is roughly one-quarter the stiffness of elastin, and it is likely that the elastin lamellae are fully straightened at this point, and that they occupy about one-quarter of the arterial wall's cross-sectional area. If this is correct, then the rise in stiffness to diastolic pressure should be entirely due to the recruitment and straightening

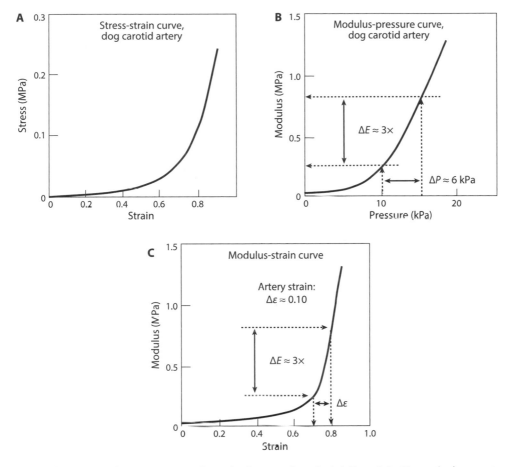

Figure 15.11. Mechanical properties of artery-wall material. **Panel A.** Stress-strain curve for a dog carotid artery. Note the rapidly rising J-shaped curve. **Panel B.** Modulus-pressure curve based on the stress-strain data in panel A. Note that the circumferential modulus of the artery wall rises about threefold, from about 0.25 to 0.8 MPa over the physiological range of blood pressure seen by the artery. ΔE, change in modulus; ΔP, change in pressure. **Panel C.** Modulus-strain curve, derived from the stress-strain data in panel A, indicating that the artery wall expands and retracts by about 10% during the cardiac cycle. $\Delta \varepsilon$, change in strain.

of collagen fibers to raise the wall stiffness threefold. This scenario is plausible and it may exist in some arteries, but there is variation in the pattern of tissue recruitment at different locations in the circulatory system, as we will soon see.

The dog carotid artery provides a good example of the properties of a muscular artery in which the muscles are under nervous control, and activation of the muscles can cause quite significant changes in arterial stiffness and diameter. The data presented here from the Dobrin and Rovick study were taken from vessels that were first treated with potassium cyanide, which inactivates the muscle cells, and thus they document the properties of the entirely passive elastin and collagen components of the vessel wall. In muscular arteries, like the carotid artery, the active control of arterial stiffness

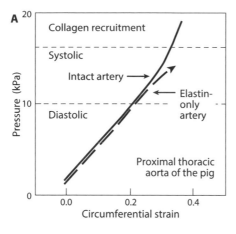
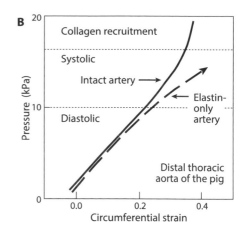

Figure 15.12. Relative contributions of collagen and elastin to the mechanical properties of the aorta. Elastin alone is responsible for the first part of aortic expansion, followed by an increasing contribution from the recruitment of collagen fibers. **Panel A.** Collagen recruitment in the proximal thoracic aorta of the pig begins at about mean arterial pressure. **Panel B.** Collagen recruitment in the distal thoracic artery occurs at pressures that are well below diastolic pressure.

and diameter is a normal functional feature. However, in the largest arteries of the mammalian circulatory system—the aortic arch and the descending thoracic and abdominal aortas—we see vessels where the muscles' major role is in the synthesis of the connective tissue, and the active contraction of muscles is considerably reduced. Margo Lillie and colleagues (2012) documented the change in the relative roles of collagen and elastin in the circumferential mechanical properties along the length of the pig thoracic aorta by comparing the inflation behavior of intact arterial segments (fig. 15.12; solid lines) with that of the purified elastin that was isolated from the arterial samples (dashed lines). Their results showed that the recruitment of collagen fibers occurred at lower extension as the distance away from the heart increased. That is, in the proximal thoracic aorta (fig. 15.12A) the recruitment of collagen occurred in the mid-physiological pressure range, at about 13 kPa, indicating that elastin alone is responsible for about the first half of aortic expansion, followed by an increasing contribution from the recruitment of collagen fibers. The distal aorta shows quite a different pattern, where collagen-fiber recruitment occurred much earlier, at somewhat less than 10 kPa, as shown in fig. 15.12B. This suggests that there is a continuously growing role for collagen-fiber reinforcement in aortic expansion as distance from the heart increases. In addition, this trend for increased collagen contribution to the artery moving down the aorta suggests that the relative role of elastin may be even greater in the largest and most central part of the aorta, the aortic arch, as we will see shortly.

To this point we have concentrated on the structure and properties of arteries from reasonably large mammals, such as dogs, pigs, and humans. Clearly, animals vary in size over a very large range, and vessel-wall tension, $T = PR$, will vary widely simply from the change in vessel radius (R). How does arterial-wall structure accommodate this change in artery size? The short answer is that if wall thickness (H) increases in

Table 15.2. The Scaling of Mammalian Arteries

Animal	Body Mass (kg)	Artery Diameter at Mean Arterial Pressure (mm)	Artery-Wall Thickness (mm)	Number of Lamellar Units	Lamellar-Unit Thickness (μm)
Mouse	0.025	1.2	0.04	5	9
Rat	0.3	2	0.1	9	11
Guinea pig	2	3	0.1	14	7
Cat	4	4.5	0.25	22	11
Rabbit	6.5	5.8	0.18	22	8
Monkey	6	5.6	0.3	22	14
Dog (small)	8	7.8	0.48	35	14
Dog (large)	24	14.5	0.8	50	16
Human	75	18	1.1	64	17
Sheep	90	21	1.1	68	16
Pig	200	23	1	70	14
Whale (aorta)	40,000	130	2	170	12
Whale (aortic arch)	40,000	380	20	820	24

The data for all animals except for the whale come from Wolinsky and Glagov (1967). The data for the fin whale come from studies by Shadwick and Gosline (1994) and Gosline and Shadwick (1996).

The average lamellar thickness for the lamellar unit of animal aortas is 12 μm, which is the same as the lamellar thickness for the whale aorta. However, the lamellar thickness for the whale aortic arch is twice that of the average aortic lamellar-unit thickness for all animals listed.

proportion to radius, the stiffness of the arterial-wall material can remain constant, and functional properties of the artery wall will remain unchanged as vessel size increases. This process is achieved by the modular nature of the artery-wall material, where the wall thickness is adjusted by changes in the number of lamellar units in the arteries in proportion to the size of the animal and its arteries. That is, the number of lamellar units, as illustrated in fig. 15.10C, is adjusted to match the size of the aorta that is required for animals of different sizes. In fact, fig. 15.10A shows an arterial media with five lamellar units, which is the number of lamellae seen in the aorta of a mouse. How, then, does the number of lamellar units scale for other animals?

Table 15.2 provides the basic information needed for analysis of the scaling of arterial dimensions in mammals. Some of these data will be provided in graphs, but at this point it will be useful to consider the final column of this table, which gives values for the thickness of the lamellar units in arteries from animals that range from a 25 g mouse to a 40,000 kg fin whale. These lamellar-thickness values were derived from measurements of artery-wall thickness at mean arterial pressure divided by the number of lamellar units seen in microscopic sections of fixed arterial samples. There is some variation in the lamellar-unit thickness of the smaller animals, but the average thickness is 12 μm. Interestingly, this is identical to the lamellar thickness in the thoracic aorta of the fin whale. So the developmental program present in smaller mammals is sufficient to control the lamellar thickness in the whale. Interestingly, however, the lamellar thickness for the fin-whale aortic arch, at 24 μm, is twice the lamellar thickness of the aorta, so there is something interesting about the lamellar design in the aortic arch.

Studies on the mechanics of the fin-whale aortic arch show that this is indeed the case. Fig. 15.13A is a diagram of the aortic arch from a fin whale in its unpressurized state. Fig. 15.13B shows the profile of this arch when inflated to pressures that are somewhat above the range of the animal's systolic blood pressure, which is assumed to be about 16 kPa. Clearly, this is a very expandable structure, likely because of its very high elastin content. The arch was attached to a water hose through the innominate artery, and the other blood vessels were tied off or were blocked with wooden plugs. The biaxial strains were measured along the long axis and around the circumference at the center of the structure with a flexible measuring tape. Inflations stopped when the 1.5 m measuring tape used to measure circumference reached its end. It is not known how far the inflation could have continued, but the trend of the pressure-volume data did not show any indication of reaching a fixed limit at pressures that are well above 16 kPa. Figs. 15.13C and 15.13D show data for the modulus of the arch material and the volume-pressure relationship for the aortic arch and descending aorta (from Shadwick and Gosline, 1994). The data for the modulus of the aortic-arch material (fig. 15.13C) are from inflation tests, as described above, and also from tensile tests on isolated aortic-arch wall tissue, in the longitudinal and circumferential directions. The results are presented as log modulus against strain, and it is clear that the arch material is about threefold stiffer in the circumferential direction than in the longitudinal direction. The stiffness for the arch material rises by about an order of magnitude over the full range of strain, from 0 to 1.0, but the rise in stiffness is not as abrupt as that seen in the dog carotid, where stiffness rises threefold over a strain of just 0.1. So, the recruitment of collagen in the whale's arch is very much less abrupt than that seen in the aorta of other animals.

Fig. 15.13D shows calculated volumes for the entire descending aorta (thoracic, abdominal, and caudal) and for the aortic arch of an 18 m fin whale, based on extensive measurements of the dimensions of the animal's aortic arch and descending aorta, including the thoracic, abdominal, and caudal portions of the aorta. These data reveal very large differences in the stiffness, and hence volume capacity, of the aortic arch and the descending aorta in the whale. Look first at the curve for the descending (thoracic) aorta, where we see that that total volume of the aorta at diastolic pressure is approximately 80 liters, but at systolic pressure the aorta's volume changes by about 1.5 liters, or less than 2%. Interestingly, at systolic pressure the arch volume is essentially identical to the volume of the whole descending aorta, but over each cardiac cycle there is about a 45 liter change in the volume of the aortic arch. As a consequence, the volume change in the arch is about 30 times greater than that in the entire descending aorta. It is also interesting that inflation of the arch to pressures above the estimated systolic pressure indicates that there remains a significant capacity for continued expansion of the arch. Is it possible that this enhanced capacity for expansion of the aortic arch beyond systolic pressure levels plays a significant role in the diving physiology of whales?

In a dive to 100 m a whale will experience a hydrostatic pressure of 10 atmospheres, or 1000 kPa, and pressures of this level are known to collapse the lungs of diving

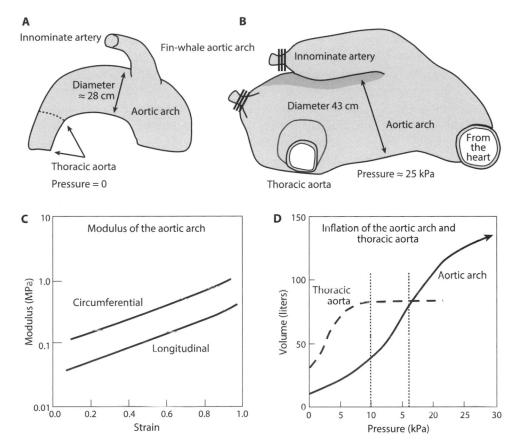

Figure 15.13. Properties of the fin-whale aorta and aortic arch. **Panel A.** Diagram of the uninflated aortic arch from an 18 m–long fin whale. **Panel B.** The whale's aortic arch inflated to a pressure of about 25 kPa. **Panel C.** Inflation of the whale's aortic arch created the expected J-shaped stress-strain behavior seen in all arteries. This panel shows the magnitude and rise of the stiffness of the aortic-arch wall in the longitudinal and circumferential directions. Note that the circumferential stiffness is about three times greater than that in the longitudinal direction. **Panel D.** Volume-pressure data obtained from the inflation of an intact aortic-arch sample, as illustrated in panels A and B, and from an isolated section of the thoracic aorta from the same whale. Note that over the physiological pressure range the thoracic aorta expands by about 1.5 liters, whereas the aortic arch expands by about 45 liters.

mammals. As a result, the thoracic cavity of these animals must be dramatically deformed. The rib cage of a whale, like that of other mammals, is supported by solid rib bones, and in whales the ribs can rotate somewhat to reduce the volume of the thoracic cavity. In addition, the muscular diaphragm that separates the thoracic from the abdominal cavity can expand forward into the thoracic cavity to further reduce this cavity's volume. However, it is not clear if these two processes are sufficient to accommodate the volume changes that must occur as the lungs collapse. It is possible that the capacity of the aortic arch to continue its expansion quite dramatically at pressures in excess of 16 kPa might help to reduce the collapse of the thorax from the rising hydro-static pressure by allowing blood to accumulate in a greatly expanded aortic arch. Note that the arch volume, as shown in fig. 15.13D, continues to rise by another 50 liters as pressure rises to 30 kPa, and perhaps the aortic arch volume can rise to considerably more than 150 liters at pressures beyond this level. If expansion of the aortic arch does indeed continue beyond the 16 kPa pressure created by the contraction of the heart, how is this expansion powered? The answer is that at 100 m depth the extreme hydro-static pressures in the water column can easily force blood from the abdominal organs and body musculature through the venous system that returns blood to the heart at an elevated pressure, and this externally generated pressure may then raise the overall pressure in the heart and the aortic arch. Thus, blood squeezed out of body tissues may accumulate in the aortic arch during a deep dive, and the increased volume of the aortic arch may limit the extent of the whale's thoracic compression. Note also the large expansion shown for the innominate artery in fig. 15.13B. This and other arteries emerging from the heart and present in the thorax also contribute to the vascular expansion that helps to maintain the volume of the thoracic cavity during a dive.

Returning to the scaling of arterial design and to table 15.2, the unusual structure of the arterial lamellae in the whale's aortic arch is likely a part of the design that allows whales to accommodate the dramatic compression of their thoracic cavity. Fig. 15.14 provides some interesting comparisons between the whale and a typical terrestrial mammal, the dog. Figs. 15.14A and 15.14B show plots of normalized aortic diameter and of aortic modulus as a function of position along the aorta, which came from the study by Gosline and Shadwick (1996). The diameter data in fig. 15.14A are normalized in a way that brings the thoracic and abdominal aorta diameters of the two animals to the same level, and this allows a direct comparison of the expansion of the whale's aortic arch relative to that of the dog's aortic arch. Clearly, the scaled diameter of the whale's arch is effectively twice that of the dog's, giving a roughly fourfold greater scaled arch volume per unit length. The modulus data in fig. 15.14B clearly demonstrate the difference in the arterial-system design, where we see that the stiffness of the aortic arch is essentially the same in both animals, but the modulus of the descending aorta in the whale is about 30 times greater than that of the dog. These comparisons clearly document the key features of connective-tissue design employed by terrestrial mammals and diving mammals.

Figs. 15.14C and 15.14D show log-log plots of the arterial diameter and the number of lamellar units as a function of body mass for the full range of animals listed in table 15.2. Fig. 15.14C shows the very strong correlation ($R^2 = 0.97$) between body mass and arterial diameter for animals ranging from the mouse to the fin whale. Interestingly, the whale's aortic arch and aorta are respectively greater than the trend and less than the trend. That is, the whale arch has a diameter that is about twice that predicted by the trend, and the aorta has a diameter that is about two-thirds of that predicted by the trend. These "deviations" reflect the need for a much larger aortic arch because virtually all of the arterial elasticity is concentrated in the arch. The volume capacity of the aortic arch, based on a twofold diameter increase, would translate to a fourfold increase in the arch volume capacity at fixed length, but the significantly lower longitudinal stiffness of the arch, relative to its circumferential stiffness (fig. 15.13C) suggests that the potential for the arch to sequester enormous volumes of blood during a dive is likely tenfold or

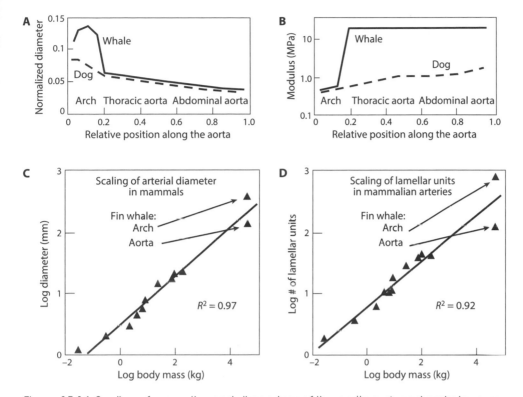

Figure 15.14. Scaling of properties and dimensions of the aortic arch and aorta in mammals. **Panel A.** Comparison of the normalized diameter of arch and aorta of a dog and a whale. Note that the normalized diameter of the whale's aortic arch is about twice that of the dog, while the scaled diameter of the whale's aorta is essentially identical to that of the dog. **Panel B.** The modulus of the aortic arches in the whale and the dog are essentially identical, but the whale's descending aorta is about 30 times stiffer than that of the dog. **Panel C.** Scaling of the arterial diameter of mammals, from the mouse to the whale. Note that the whale's arch is significantly greater than the trend, while the whale's aorta is significantly less than the trend. **Panel D.** Scaling of the number of lamellar units in the arteries of mammals, from the mouse to the whale. Note that the whale's arch has about twice the number of lamellar units predicted by the trend, and that the whale's aorta has about one-third the number of lamellae predicted by the trend.

more. As discussed above, this capacity for the arch to achieve enormous expansions must certainly play an important role in the volume compensation that occurs during the compression of the animal's thorax during a dive.

Fig. 15.14D shows another strong correlation ($R^2 = 0.92$) between the number of lamellar units and body mass. Here we see that the number of lamellar units in the arch is about 2.5 times greater than the level predicted by the trend of the data for arteries from terrestrial mammals, and the number of lamellar units in the aorta is about one-third the level predicted by the trend. The dramatically increased number of lamellar units for the arch likely is an adaptation to the very high wall tensions, as defined in eq. (15.2) as pressure times radius, that must develop during the full expansion of the aortic arch.

To maintain wall stress at reasonable levels, wall thickness must increase, and this is achieved by the very high number of lamellae observed for the arch. Clearly, it is the dominance of elastin in the properties of this tissue that gives the arch its exceptional extensibility. Thus, it is likely that the enhanced elastin content of the arch, and hence reduced collagen content, is responsible for the roughly twofold greater thickness of the lamellae in the arch, as well as the very large number of lamellae. That is, at high pressures more elastin is required to compensate for the reduced reinforcement of the aortic arch by collagen fibers.

At the opposite extreme, the number of lamellar units in the aorta is about two-thirds that predicted by the trend for the terrestrial mammals. This reduction in aortic lamellae likely reflects the dominance of much stiffer collagen fibers in the aorta and near total lack of elastin. Interestingly, polarized-light microscopy of the aorta-wall tissue by Gosline and Shadwick (1996) shows multidirectional layers of tightly packed, wavy layers of collagen fibers that look more like tendons than the more diffuse, wavy collagen fibers seen in most normal skin and arteries. Indeed, one might say that the collagen mesh in the whale aorta is the arterial analog of the ultratough skin of the rhinoceros. Thus, the unusual demands of the aortic system in whales, likely due to their function in diving, have caused evolution to create unique modifications to the general pattern of arterial design to achieve structural properties that allow air-breathing mammals to function in the depths of the ocean.

SECTION VI

CONCLUDING COMMENTS

Chapter 16

Final Thoughts

This book provides a broad overview of the mechanical design of the three major classes of structural materials in animals: tensile, rigid, and pliant materials, but it has been based on the detailed analysis of a small number of selected examples. Thus, in the section on tensile materials we considered the design of collagen fibers, as found in mammalian tendon and sea-urchin spine ligament, and the design of silk fibers, as produced by spiders. In the section on rigid materials we analyzed the structural design of mammalian bone and insect cuticle. In the section on pliant materials, we analyzed the design of sea-anemone mesoglea, slug mucus, several rubberlike animal proteins, and vertebrate skin and artery.

One of the most consistent themes in these analyses is that the basic structural organization of essentially all biological structural materials is that of fiber-reinforced composites. Indeed, in our discussion of collagen tendon, bone, and insect cuticle, we explicitly used composite-material theory to quantify the structural design of these materials. That is, we used estimates of fiber aspect ratios and volume fractions, and of the fibers' known material properties to estimate the properties of the matrix components, which are essentially impossible to quantify experimentally. This allowed us to obtain reasonable semiquantitative understandings of the composite design of these materials.

Tendon and cuticle are both made from two discrete components: short, rigid filaments and the softer matrix that surrounds them, which holds the filaments together into the composite structure of the material. In the case of tendon, the crystalline collagen fibrils are held together by a soft gel of the proteoglycan called decorin, and because the aspect ratio of the collagen fibrils is very large, of the order of 5×10^4, and the fiber volume fraction is quite high, this soft gel is sufficient to create the reasonably strong, stiff, and springlike tensile-fiber systems that we see in tendons and ligaments. Similarly, in the case of insect cuticle, we were able to take known values for the stiffness, aspect ratio, and volume fraction of chitin fibrils that have been isolated from insect cuticle, and with this information we estimated the properties of the matrix materials that exist in these composites as well.

We also attempted to apply composite-fiber theory to the stiffness of bone, and we saw that the orientation-dependent stiffness of bone did not follow the theoretical values expected for a typical fiber-reinforced composite. However, if we took account of the fact that the hydroxyapatite crystals in bone are not fibers but platelets, unlike a parallel-fiber composite, the hydroxyapatite crystals could simultaneously reinforce in two directions, at slightly different levels. With this geometry, we found that the composite of mineral-crystal platelets and a matrix formed by collagen fibers could reasonably explain the properties of bone.

One important feature of composites like these is that the structural integrity of the composite arises from the structural properties (stiffness and strength) of each component individually, as well as the strength of the interface between the two components. Indeed, the failure of composite systems is most often initiated by failure at the fiber-matrix interface. Spider silk, however, provides a different kind of composite design where there is no interface between fiber and matrix. Spider dragline silk, like many artificial polymer fibers, is made from a single kind of polymer molecule, a protein called a fibroin. When these polymers are drawn from the spinneret complex at the posterior of the spider's abdomen, the molecules become organized to create a biphasic structure that is reminiscent of the fiber-reinforced composite systems described above. The fibroins that form spider silk have amino-acid-sequence designs that establish the unique "composite" design of this material. That is, silk fibers contain polymer crystals that are interspersed in a relatively amorphous matrix, but both of these phases are created from a mixture of two silk proteins that have quite similar organizations. The sequence designs of these proteins are those of block copolymers, where there are sequence blocks that encode the crystalline structures that alternate with blocks of sequence that encode an amorphous or semicrystalline "matrix." Specifically, the crystalline structures are formed by blocks of about 10 alanine residues that form β-pleated-sheet crystals that are the reinforcing "fibers" of this composite, and the matrix component of this material is formed from longer sequence blocks that are rich in glycine and proline. The details of these sequences are given in chapter 10, but the result is that this packaging of matrix and fiber components into a single molecule allows the creation of a composite-like system at the nanometer scale that is an order of magnitude or two smaller than the composite systems seen in collagen, cuticle, and bone. The take-home message here is that this single-molecule nanocomposite structure allows spider silk to achieve mechanical properties that are truly remarkable.

The stress-strain curve for spider dragline silks (major-ampullate-gland silk) starts with a linear high-stiffness zone that extends to strains of 0.02–0.03, with initial stiffness values of about 10 GPa at low strain rates. However, at high strain rates, like those that occur during the impact of a flying insect with the spider's web, the initial stiffness approaches 20–30 GPa. Comparing these initial stiffness values for spider silk with those for collagen tendon, insect cuticle, and bone in table 16.1, we see that spider silk is 7–20 times stiffer than collagen tendon (at about 1.5 GPa) and about 2.5–7.5 times stiffer than insect cuticle (at about 4 GPa for mature locust cuticle). Bone, by virtue of its high content of rigid mineral platelets, is actually somewhat stiffer than silk that is tested at low strain rates. That is, the tensile stiffness of a bone sample loaded parallel to the long axis of a bone is about 20 GPa, which is roughly equivalent to the stiffness of spider silk at high strain rates, considering the dramatic effect of strain rate on the stiffness of silk. The presence of a high volume fraction of exceptionally stiff filler particles, the hydroxyapatite-crystal platelets ($E \approx 120$ GPa), is able to raise the composite stiffness of bone to a level that is comparable to the pure-protein silk fibers. Thus, using

Table 16.1. Summary of Material Properties of Collagen Tendon, Spider Silk, Bone, and Insect Cuticle

Material	Modulus (GPa)	Strength (MPa)	Extensibility (%)	Energy to Break (MJ m^{-3})	Energy Storage Capacity (MJ m^{-3})
Carbon fiber	300	4000	≈1	20	≈5
Carbon-fiber composite (parallel, 70% fiber)	200	2800	≈1		
Collagen tendon	1.5	100	10	7.5	7
Spider silk (dragline)	**10**	**1500**	**30**	**200**	**70**
Bone	20	100–200	2–3	5	≈0.2
Locust cuticle	4	100–200	4	≈2	≈0.2

Properties of biological materials are compared with the properties of carbon fiber, which is used in the construction of high-performance synthetic composites. Bold highlights the properties of spider silk.

tensile stiffness as a basis for comparison, spider silk and bone are at the top end of the "quality" scale, but what about strength and toughness?

In spite of the high initial stiffness of spider silk, its strength and toughness are actually the more exceptional attributes of dragline silk fibers, and it will be interesting to compare the strength and toughness of silk with the other materials in table 16.1. Let's start this comparison with an artificial carbon fiber, as an indication of the upper range of stiffness and strength possible from organic materials. Clearly, evolution has not figured out how to convert a polymer fiber, like collagen or silk, into a carbon fiber, which is the current standard for synthetic high-performance reinforcing fibers in composite materials. But carbon fibers establish a clear upper limit to what can be expected from truly high performance fibrous carbon-based tensile and rigid materials. Carbon fibers have stiffness values in the region of 300 GPa and strengths of the order of 4 GPa, and high-volume-fraction parallel composites made from carbon fibers have only slightly lower levels of strength and stiffness.

Table 16.1 provides additional information about the properties of carbon fibers, as well as information on the properties of tendon, silk, bone, and insect cuticle. None of the natural materials analyzed in this book comes close to the levels of stiffness and strength seen for carbon fibers, but the strength of spider silk is the closest, being about a third of that of carbon fibers. This strength comparison is impressive, but even more impressive is the fact that spider silk energy to break (i.e., its toughness) is an order of magnitude greater than that of carbon fibers. Spider silk is not renowned for its elastic-energy storage capacity, but this property is also tenfold greater than that of carbon fiber. This high toughness and energy storage capacity comes largely from the 30-fold greater extensibility of silk relative to that of carbon fiber.

Comparison of spider silk with the other biological materials—collagen, bone, and cuticle—reveals clearly the greatly reduced levels of strength, extensibility, energy to break, and even energy storage capacity of these other animal materials. Even collagen tendon, which is renowned as a superb tissue spring in vertebrate locomotion, has an energy storage capacity that is a factor of ten less than that of spider silk. This is interesting because spring energy capacity is not considered a particularly useful attribute for a spider web, where energy dissipation during the impact of a flying insect with the web is seen as the more useful property for the web material. Indeed, the energy storage capacity of spider silk is about one-third that of its energy to break, indicating that energy dissipation, which is key to prey capture in the web, is the crucial material property for spider draglines and the silk used to create the web's frame and radial threads.

There is another general principle that we can take from our investigation of these four materials; namely, that the properties observed for collagen, silk, bone, and cuticle are strongly influenced by the processes by which they are made in the living organism. We have seen above that spider silk is drawn from the silk-gland-spinneret complex in the abdomen of the spider, and this system provides a draw-processing mechanism that is essentially identical to the processes used by the textile-fiber industry in the production of synthetic fibers such as nylon, rayon, and polyester. These processes gradually elongate the polymer molecules in the fluid state, a process that facilitates the assembly of the crystal-forming domains of the silk molecule to form crystalline structures that establish the solid-state structure of the silk-fibroin network. Extensive draw processing in the final stages of silk formation in the terminal duct then elongates the flexible polymer domains between the crystalline cross-links, and as the silk thread emerges from the silk-gland spigot it dries, and the extended, flexible chains between the crystalline cross-links stiffen greatly to produce the material we know as spider dragline silk. As a consequence, spiders can make silk fibers for their draglines and web construction that are orders of magnitude longer than the length of their abdomen. Indeed, a spider spinning its web produces a length of silk that is at least 1000 times longer than the length of its silk-gland-spinneret complex.

This process of spider-silk synthesis is quite different from the processes involved in the formation of tendon, cuticle, and bone, whose material synthesis and assembly into the structural material occur at a size scale that in some cases is considerably smaller than the individual cells that synthesize the structural components that are transformed into these materials. For example, the chitin fibrils in insect cuticle have average lengths of about 300 nm and diameters of about 2.5 nm, giving them modest aspect ratios of about 120. Bone platelet size parallel to the long axis of a bone was estimated to be at least 110 nm but likely less than 200 nm. The platelets are 30–40 nm wide and about 3 nm thick, giving them longitudinal aspect ratios of only 40–70. The reinforcing structures of chitin and bone are indeed smaller than the cells that produce them. In the case of the collagen tendon, however, the basic structural subunit, the collagen fibril, has a considerably higher aspect ratio of about 10^4–10^5, and an overall length of the order of 5–10 mm. Thus, the linear dimensions of collagen fibrils are considerably greater than the size of the cells that synthesize the collagen molecules. However, these high-aspect-ratio fibrils are necessary because the matrix that holds the fibrils together in collagen fibers, the decorin gel, is exceptionally soft.

In each case, for the three discontinuous-composite systems, the overall material performance is limited by the aspect ratio of the reinforcing fibers, by the stiffness of the matrix, by the inherent stiffness of the reinforcing fiber, or by some combination of these attributes. This is not the case for the silk produced by spiders, where the "fiber" and "matrix" components are continuously linked through the covalent backbone of the silk-fibroin protein, and this protein composite is, as a consequence, much stronger and tougher than the discontinuous-fiber composites produced by their insect and vertebrate cousins.

This summary comparison of the tensile and rigid biological materials with high-performance synthetic materials might leave us wondering why the materials made by animals don't have more impressive properties. The answer lies in the evolutionary history of these materials. Materials design in living systems is an agelong process. As we have seen in chapter 13, the most primitive materials created during the evolution of the earliest animals, such as the sea anemone, are very soft, pliant materials that are based on collagen fibers embedded in extremely soft proteoglycan-like matrix materials. Sea anemones evolved late in the Proterozoic era, about 550 Ma, but as the more advanced animal groups evolved, more complexity developed in animal structural materials. Insects first evolved in the Carboniferous period, about 400 Ma, and the Orthoptera (grasshoppers and locusts) evolved at about 330–300 Ma. Birds first evolved in the Jurassic era, at about 160 Ma, but modern birds appeared around 75 Ma. Interestingly, insects in their early evolution were considerably larger than current insects, possibly because they did not have to compete with other animals of similar size, but most likely because atmospheric oxygen levels were considerably higher during this period, and this enabled insects, with their tracheal respiratory systems, to support the greater energy demands required for flight by larger size. However, when modern birds appeared on the scene, their competition with and predation on the insects likely drove insect size down, bringing them to the size scale that we know today.

The locust is one of the larger of the modern insects, and it is certainly much smaller than the majority of modern birds; indeed, there is very little overlap in the size range of insects and birds, so the competition between birds and insects in the modern era has been one of bird eating insect. We saw in our analysis of the scaling of the specific modulus of insect cuticle, bone, and a range of artificial composites in table 2.1 and fig. 12.13 that there is a strong correlation between the stiffness (i.e., the specific modulus) of these materials and the size of the flying "machines" made from them. Indeed, the roughly fivefold greater specific modulus of bone relative to that of insect cuticle extrapolated very nicely to the specific modulus of wood and of the synthetic materials that have been and are currently used in the construction of artificial flying machines. The implication of this fivefold greater stiffness and similar strength, as shown in table 16.1, is that skeletal stiffness based on bone has allowed birds to be much larger than insects, and this has established the predator-prey relationship that defines insect-bird competition in our modern world. Apparently, the quality of structural materials in different animal groups has helped to establish size differences between those groups, and presumably differences in locomotor performance as well. Thus, the design of structural materials has played significant roles establishing the ecological and behavioral interactions among animals.

LIST OF SYMBOLS

Symbol	Definition	Eq.	Units
A	Area	(2.2)	m^2
A_0	Unstressed area	(2.7)	m^2
A_m	Muscle cross-sectional area	(9.2)	m^2
A_p	Area created by crack propagation	(15.1)	m^2
A_t	Tendon cross-sectional area	(9.2)	m^2
AR	Aspect ratio		
a	Acceleration		$m\,s^{-2}$
a_i	Fraction of laminate thickness of a particular lamina	(8.4)	
D	Diffusion coefficient	(7.1)	$m^2\,s^{-1}$
D_0	Reference diffusion coefficient	(7.2)	$m^2\,s^{-1}$
$D(t)$	Time-dependent compliance	(3.9)	Pa^{-1}
d	Deformation	(2.1)	m
d	Fiber diameter	(8.6)	m
E	Elastic modulus	(2.3)	Pa
E_C	Modulus of a collagen molecule		Pa
E_c	Modulus of the composite	(8.1)	Pa
E_{comp}	Modulus in compression	fig. 11.8	Pa
E_{eq}	Equilibrium modulus	fig. 3.2	Pa
E_f	Modulus of the fibers	(8.1)	Pa
E_{init}	Initial modulus	fig. 10.1	Pa
E_m	Modulus of the matrix	(8.1)	Pa
$E(t)$	Time-dependent elastic modulus	(3.3)	Pa
E_{tan}	Tangent elastic modulus	(2.4)	Pa
E'	Storage modulus	(3.18)	Pa
E''	Loss modulus	(3.18)	Pa
$E^\star$	Complex elastic modulus	(3.18)	Pa
F	Force	(2.1)	N
$\mathcal{F}$	Helmholtz free energy	(6.1)	J
F_d	Drag force		N
F_f	Friction force		N
F_{max}	Breaking force	(10.1)	N
F_{net}	Net force	(4.2)	N
F_s	Force due to entropy	(6.30)	N

F_U	Force due to internal energy	(6.30)	N
f_0	Molecular friction coefficient	(7.1)	kg s^{-1}
f_n	Resonant frequency	(3.11)	Hz
f_s	Segmental friction coefficient	(7.5)	kg s^{-1}
G	Shear modulus	(2.12)	Pa
G	Strain-energy release rate	(5.9)	Pa
G_m	Matrix shear modulus	(8.8)	Pa
G'	Shear storage modulus		Pa
G''	Shear loss modulus		Pa
H	Hysteresis	(3.1)	
ΔH_{act}	Activation energy/enthalpy	(7.2)	J
h	Sample height/thickness	(2.11)	m
J	J-integral	(5.9)	Pa
K	Spring constant	(2.1)	N m^{-1}
K	Stress intensity factor	(5.10)	
k	Beam stiffness	(3.10)	N m^{-1}
k	Boltzmann constant	(6.16)	J K^{-1}
k	Permeability coefficient		m^4 N^{-1} s^{-1}
L	Length	(2.8)	m
L_0	Original length	(2.2)	m
$L_{0.5}$	Half length of fiber	(8.6)	m
L_{crit}	Critical crack length		m
L_{final}	Final length	(2.8)	m
L_{init}	Initial length	(5.4)	m
L_{tot}	Total length of fiber	(8.6)	m
ΔL	Change in length	(2.2)	m
ℓ	Length of one chain link	(6.11)	m
M	Mass	(3.10)	kg
M	Molecular weight	fig. 7.1	kg
M_c	Molecular weight between cross-links	(6.24)	kg
N	Number of microstates	(6.16)	
n	Number of links in a polymer chain	(7.5)	
n	Scaling factor	(8.8)	
P	Pressure	(6.7)	Pa
Q	Heat absorbed	(6.3)	J
R	Resilience	(3.2)	
R	Vessel radius	(15.2)	m
R_{rel}	Relative reflection of sound waves	(11.1)	
r	Distance between atoms	(4.1)	m
r	End-to-end separation	(6.10)	m
r	Fiber radius	(8.6)	m
r_0	Thickness of matrix layer	(8.7)	m
r_m	Crack-tip radius	(5.8)	m
r_{mp}	Most probable end-to-end distance	(6.14)	m
S	Surface energy	(5.5)	J
S	Entropy	(6.1)	J K^{-1}

S_{BW}	Static safety factor	(10.1)	
SF	Safety factor	(9.1)	
s	Aspect ratio	(8.6)	
T	Absolute temperature	(6.1)	K
T_{circ}	Tension in the circumferential direction	(15.2)	N
T_g	Glass-transition temperature	fig. 7.2	K
$\tan\delta$	Tangent of the phase shift, δ	(3.20)	
t_{ve}	Viscoelastic time constant	(3.4)	s
U	Internal energy	(6.1)	J
U_{net}	Bond potential energy	(4.1)	J
u	Velocity	(2.14)	m s^{-1}
V	Volume	(5.2)	m^3
V_f	Volume fraction of the fibers	(8.1)	
V_m	Volume fraction of matrix	(8.1)	
V_{min}	Volume fraction of mineral in a bone	fig. 11.4	
V_r	Volume of a spherical shell	(6.12)	m^3
W	Strain energy (work)	(5.3)	J
W_{break}	Energy to break	(10.2)	J
W_g	Gravitational energy released	(10.3)	J
x	Location on the x axis		m
y	Amplitude of oscillation	(3.12)	m
y	Location on the y axis		m
Z	Acoustic impedance	(11.1)	kg m^{-2} s^{-1}
z	Location on the z axis		m
γ	Shear strain	(2.11)	
$\dot{\gamma}$	Shear strain rate	(2.14)	s^{-1}
γ_{surf}	Energy per surface area (surface energy)	(4.6)	J m^{-2}
Δ	Logarithmic decrement	(3.12)	
δ	Phase shift	(3.17)	radians
ε	Engineering strain	(2.2)	
ε_0	Initial strain	(3.7)	
ε_C	Collagen-molecule strain		
ε_D	D-period strain		
ε_{max}	Breaking strain	(2.5)	
ε_T	Whole-tendon strain		
ε_t	True strain	(2.8)	
$\varepsilon(t)$	Time-dependent strain	(3.3)	
ε_x	Strain along the x axis	(2.9)	
ε_y	Strain along the y axis	(2.9)	
ε_z	Strain along the z axis	(2.9)	
$\Delta\varepsilon$	Change in strain	(2.4)	
η	Viscosity	(2.15)	Pa s
$\eta_{ref,\theta}$	Fiber-reinforcement efficiency at angle θ	(8.3)	Pa s
θ	Angle		°
λ	Extension ratio	(2.7)	
ν_2	Volume fraction of polymer, hydrated	(14.5)	

ν_{xy}	Poisson's ratio, y axis	(2.9)	
ν_{xz}	Poisson's ratio, z axis	(2.9)	
ρ	Mass density	(6.24)	kg m^{-3}
σ	Engineering stress	(2.2)	Pa
σ_0	Initial stress	(3.6)	Pa
σ_{circ}	Stress in the circumferential direction	(15.3)	Pa
σ_{crack}	Stress at the crack tip	(5.8)	Pa
σ_f	Tensile stress in a fiber	(8.10)	Pa
$\sigma_{in\ vivo}$	Stress in real life		Pa
σ_m	Muscle stress	(9.2)	Pa
σ_{max}	Breaking stress	(2.5)	Pa
$\sigma_{max,th}$	Theoretical tensile strength	(4.7)	Pa
σ_t	Tendon stress	(9.2)	Pa
σ_t	True stress	(2.6)	Pa
$\sigma(t)$	Time-dependent stress	(3.6)	Pa
$\Delta\sigma$	Change in stress	(2.4)	Pa
τ	Shear stress	(2.10)	Pa
τ_m	Shear stress in the matrix	(8.9)	Pa
ϕ_{break}	Breaking strain energy per volume	(2.5)	J m^{-3}
ϕ_{in}	Energy per volume put in	(3.1)	J m^{-3}
ϕ_{lost}	Energy per volume lost	(3.1)	J m^{-3}
ϕ_{out}	Energy per volume recovered	(3.2)	J m^{-3}
ϕ_{stored}	Stored energy per volume	(5.1)	J m^{-3}
ω	Natural frequency	(3.16)	radians s^{-1}
ω_n	Resonant frequency	(3.10)	radians s^{-1}

Bibliography

CHAPTERS 2 AND 3

Denny, M. 1988. *Biology and the Mechanics of the Wave-Swept Environment*. Princeton, NJ: Princeton University Press.

Gosline, J. M., and C. J. French. 1979. "Dynamic Mechanical Properties of Elastin." *Biopolymers* 18: 2091–103.

Ker, R. F. 1981. "Dynamic Tensile Properties of the Plantaris Tendon of Sheep." *J. Exp. Biol.* 93: 283–302.

Lazan, B. J. 1968. *Damping of Materials and Members in Structural Mechanics*, London: Pergamon.

Setton, L. A., Z. Wenbo, and V. C. Mow. 1993. "The Biphasic Poroviscolastic Behavior of Articular Cartilage: Role of the Surface Zone in Governing the Compressive Behavior." *J. Biomech.* 26: 581–92.

CHAPTER 4

Brenner, S. S. 1956. "Tensile Strength of Whiskers." *J. Appl. Phys.* 12: 1484–91.

Martin, J. W. 1972. *Strong Materials*. London: Wykeham.

CHAPTER 5

Anderson, T. L. 2005. *Fracture Mechanics: Fundamentals and Applications*, 3rd ed. Boca Raton, FL: CRC.

Griffiths, A. A. 1921. "The Phenomena of Rupture and Flow in Solids." *Philos. Trans. R. Soc. Lond. A* 221: 163–98.

CHAPTER 6

Treloar, L.R.G. 1975. *Physics of Rubber Elasticity*. Oxford: Clarendon.

CHAPTER 7

Bueche, F. 1962. *Physical Properties of Polymers*. New York: Interscience, John Wiley and Sons.

Bueche, F., P. J. Flory, and B. J. Kinzig. 1963. "Entanglement Effects in Concentrated Polymer Solutions from Viscosity Measurements." *J. Chem. Phys.* 39: 128–31.

Shaw, M. T., and W. J. McKnight. 2005. *Introduction to Polymer Viscoelasticity*. Hoboken, NJ: Wiley-Interscience.

CHAPTER 8

Bunsell, A. R., and J. Renard. 2005. *Fundamentals of Fibre Reinforced Composite Materials*. Bristol, UK: Institute of Physics Publishing.

Cook, J., and J. D. Gordon. 1964. "A Mechanism for the Control of Crack Propagation in All-Brittle Systems." *Proc. R. Soc. Lond. A* 262: 508–20.

Daniel, I. M., and O. Ishai. 2006. *Engineering Mechanics of Composite Materials*. Oxford: Oxford University Press.

CHAPTER 9

Bailey, A. J. 1968. "Intermediate Labile Intermolecular Crosslinks in Collagen Fibers." *Biochem. Biophys. Acta* 160: 447–53.

Bennett, M. B., R.F. Ker, M. J. Imery, and R. McN. Alexander. 1986. "Mechanical Properties of Various Mammalian Tendons." *J. Zool. Lond. A* 209: 537–48.

Bezerra, M., A. Lemos, K.D.S. Lira, P.V.C. Silveira, M.P.G. Coutinho, and S.R.A. Moraes. 2012. "Does Aerobic Exercise Training Promote Changes in Structural and Biomechanical Properties of the Tendons in Experimental Animals? A Systematic Review." *Biol. Sport* 29: 249–54.

Brinkmann, J., H. Notbohn, and P. K. Muller. 2005. *Collagen: Primer in Structure, Processing and Assembly*. Berlin: Springer.

Craig, A. S., M. J. Bertles, J. F. Conway, and D.A.D. Parry. 1989. "An Estimate of the Mean Length of Collagen Fibrils in Rat Tail Tendon as a Function of Age." *Conn. Tissue Res.* 19: 51–62.

Danielson, H. Baribault, D. F. Holmes, H. Graham, K. E. Kadler, and R. V. Iozzo. 1997. "Targeted Disruption of Decorin Leads to Abnormal Collagen Fibril Morphology and Skin Fragility." *J. Cell Biol.* 136: 729–43.

Diamant, J., A. Keller, E. Baer, M. Litt, and R.G C. Arridge. 1972. "Collagen Ultrastructure and Its Relation to Mechanical Properties as a Function of Ageing." *Proc. R. Soc. Lond. B* 180: 293–315.

Dourte, L. M., L. Pathmanathan, A. F. Jawad, R. V. Iozzo, M. J. Mienaltowski, D. E. Birk, and L. J. Soslowsky. 2012. "Influence of Decorin on the Mechanical, Compositional and Structural Properties of Mouse Patellar Tendon." *J. Biomech. Eng.* 134 (3): 031005.

Eppell, S. J., B. N. Smith, H. Kahn, and R. Ballernini. 2006. "Nano Measurements with Micro-Devices: Mechanical Properties of Hydrated Collagen Fibrils." *J. R. Soc. Interface* 3: 117–21.

Folkhard, W., E. Mosler, W. Geercken, E. Knörzer, H. Nemetschek-Gansler, and Th. Nemetschek. 1987. "Quantitative Analysis of the Molecular Sliding Mechanism in Native Tendon Collagen: Time-Resolved Dynamic Studies Using Synchrotron Radiation." *Int. J. Biol. Macromol.* 1987: 169–75.

Franchi, M., V. Ottani, R. Stagni, and A. Ruggeri. 2010. "Tendon and Ligament Fibrillar Crimps Give Rise to Left-Handed Helices of Collagen Fibrils in Both Planar and Helical Crimps." J. Anat. 216: 301–9.

Fratzl, P. 2008. *Collagen, Structure and Mechanics*. New York: Springer.

Fratzl, P., K. Misof, and I. Zizak. 1997. "Fibrillar Structure and Mechanical Properties of Collagen." *J. Struct. Biol.* 122: 119–22.

Goh, K. L., J. R. Meakin, R. M. Aspden, and D.W.L. Hukins. 2007. "Stress Transfer in Collagen Fibrils Reinforcing Connective Tissues: Effects of Collagen Fibril Slenderness and Relative Stiffness." *J. Theor. Biol.* 245: 305–11.

Gordon, J. E. 1988. *The Science of Structures and Materials*. New York: Scientific American Library.

Hidaka, M., and K. Takahashi. 1983. "Fine Structure and Mechanical Properties of the Catch Apparatus of the Sea-Urchin Spine, a Collagenous Connective Tissue with Muscle-like Holding Capacity." *J. Exp. Biol.* 103: 1–14.

Hynes, R. O. 2012. "The Evolution of Metazoan Extracellular Matrix." *J. Cell Biol.* 196: 671–79.

Kastelic, J., and E. Baer. 1980. "Deformation in Tendon Collagen." In: *The Mechanical Properties of Biological Materials*, edited by J.F.V. Vincent and J.D. Currey, 398–422. Cambridge: Cambridge University Press.

Ker, R. F., R. McN. Alexander, and M. B. Bennett. 1988. "Why Are Mammalian Tendons so Thick?" *J. Zool. (Lond.)* 216: 309–24.

Ker, R. F., X. T. Wang, and A. V. Pike. 2000. "Fatigue Quality of Mammalian Tendons." *J. Exp. Biol.* 203: 1317–27.

Koob, T. J., M. M. Koob-Emunds, and J. A. Trotter. 1999. "Cell-Derived Stiffening and Plasticizing Factors in Sea Cucumber (*Cucumaria frondosa*) Dermis." *J. Exp. Biol.* 202: 2291–2301.

Liao, J., and I. Vesely. 2007. "Skewness Angle of Inter-Fibrillar Proteoglycan Increases with Applied Load on Mitral Valve Cordae Tendineae." *J. Biomech.* 40: 390–93.

Lujan, T. J., C. J. Underwood, N. T. Jacobs, and J. A. Weiss. 2009. "Contribution of Glycosaminoglycans to Viscoelastic Tensile Behaviour of Human Ligament." *J. Appl. Physiol.* 106: 423–31.

Matson, A., N. Konow, S. Miller, P. P. Konow, and T. J. Roberts. 2012. "Tendon Material Properties Vary and Are Independent among Turkey Hindlimb Muscles." *J. Exp. Biol.* 215: 3552–58.

Nemetschek, T., R. Jonak, H. Nemetschek-Gansler, H. Riedl, and G. Rosenbaum. 1978. "On the Determination of Changes in the Large Periodic Structure of Collagen." *Z. Naturforsch.* 33: 928–36.

Patterson-Kane, J. C., E. C. Firth, A. E. Goodship, and D.A.D. Parry. 1997. "Age-Related Differences in Collagen Crimp Patterns in the Superficial Digital Flexor Tendon of Untrained Horses." *Aust. Vet. J.* 75: 39–44.

Pike, A., R. F. Ker, and R. McN. Alexander. 2000. "The Development of Fatigue Quality in High- and Low-Stressed Tendons of Sheep (*Ovis ares*)." *J. Exp. Biol.* 203: 2187–93.

Ribeiro, A. R., A. Barbaglio, C. D. Benedetto, C. C. Ribeiro, I. C. Wilkie, M.D.C. Carnivali, and M. A. Barbosa. 2011. "New Insights into Mutable Collagenous Tissue: Correlations between the Microstructure and Mechanical State of a Sea-Urchin Ligament." *PLoS One* 6: e24822.

Robinson, P. S., T. F. Huang, E. Kazam, R. V. Iozzo, D. E. Birk, and L. J. Soslowsky. 2005. "Influence of Decorin and Biglycan on Mechanical Properties of Multiple Tendons in Knockout Mice." *J. Biomech. Eng.* 127: 181–85.

Screen, H.R.C., and S. L. Evans. 2009. "Measuring Strain Distributions in the Tendon Using Confocal Microscopy and Finite Elements." *J. Strain Anal. Eng. Des.* 44 (5): 327–35.

Screen, H.R.C., D. E. Berk, K. E. Kadler, F. Ramiriez, and M. F. Young. 2015. "Tendon Functional Extracellular Matrix." *J. Orthop. Res.* 33 (6): 793–99.

Screen, H.R.C., D. A. Lee, D. L. Bader, and J. C. Shelton. 2004. "An Investigation into the Effects of the Hierarchical Structure of Tendon Fascicles on Micromechanical Properties." *Proc. Inst. Mech. Eng. H* 218: 109–19.

Shadwick, R. E. 1990. "Elastic Energy Storage in Tendons: Mechanical Differences Related to Function and Age." *J. Appl. Physiol.* 68: 1033–40.

Shen, Z. L., M. R. Dodge, H. Kahn, R. Ballarini, and S. J. Eppell. 2008. "Stress-Strain Experiments on Individual Collagen Fibrils." *Biophys. J.* 95: 3956–63.

Smith, D. S., S. A. Wainwright, J. Baker, and M. L. Cayer. 1981. "Structural Features Associated with Movement and 'Catch' of Sea-Urchin Spines." *Tissue Cell* 2: 299–320.

Svensson, R. B., P. Hansen, T. Hassenkam, B. T. Haraldsson, P. Aaggaard, V. Kovanen, M. Krogsgaard, et al. 2012. "Mechanical Properties of Human Patellar Tendon at the Hierarchical Levels of Tendon and Fibril." *J. Appl. Physiol.* 112: 419–26.

Svensson, R. B., T. Hassenkam, C. A. Grant, and S. P. Magnusson. 2010. "Tensile Properties of Human Collagen Fibrils and Fascicles Are Insensitive to Environmental Salts." *Biophys. J.* 99: 4020–27.

Trotter, J., and T. Koob. 1989. "Collagen and Proteoglycan in a Sea Urchin Ligament with Mutable Mechanical Properties." *Cell Tissue Res.* 285: 527–39.

Trotter, J. A., and C. Wofsy. 1989. "The Length of Collagen Fibers in Tendon." *Trans. Orthop. Res. Soc.* 14: 180.

Wang, X. T., and R. F. Ker. 1995. "Creep Rupture of Wallaby Tail Tendons." *J. Exp. Biol.* 198: 831–45.

Wang, X. T., R. F. Ker, and R. M. Alexander. 1995. "Fatigue Rupture of Wallaby Tail Tendons." *J. Exp. Biol.* 198: 847–52.

Woo, S. L.-Y., M. A. Ritter, D. Amiel, T. M. Sanders, M. A. Gomez, S. C. Kuei, S. R. Garfin, et al. 1980. "The Biomechanical and Biochemical Properties of Swine Tendons: Long-Term Effects of Exercise on the Digital Extensors." *Connect. Tissue Res.* 7: 177–83.

Woo, S. L., M. A. Gomez, D. Amiel, M. A. Ritter, R. H. Gelberman, and W. H. Akeson. 1981. "The Effects of Exercise on the Biomechanical and Biochemical Properties of Swine Digital Flexor Tendons." *J. Biomech. Eng.* 103: 51–56.

Wright, D. M., V. C. Duance, T. J. Wess, C. M. Kielty, and P. P. Purslow. 1999. "The Supramolecular Organisation of Fibrillin-Rich Microfibrils Determines the Mechanical Properties of Bovine Zonular Filaments." *J. Exp. Biol.* 202: 3011–20.

Yamada, A., M. Tamori, T. Iketani, K. Oiwa, and T. Motokawa. 2010. "A Novel Stiffening Factor Inducing the Stiffest State of Holothurian Catch Connective Tissue." *J. Exp. Biol.* 213: 3416–22.

CHAPTER 10

Agnarsson, I., M. Kuntner, and T. A. Blackledge. 2010. "Bioprospecting Finds the Toughest Biological Material." *PLoS One* 5 (9): e11234.

Askarieh, G., M. Hedhammar, K. Nordling, A. Saenz, C. Casals, A. Rising, J. Johansson, et al. 2010. "Self-Assembly of Spider Silk Proteins Is Controlled by a pH-Sensitive Relay." *Nature* 465: 236–38.

Blackledge, T. A., and C. Y. Hayashi. 2006. "Silken Toolkits: Biomechanics of Silk Fibers Spun by the Orb Web Spider, *Argiope argentata*." *J. Exp. Biol.* 209: 2452–61.

Denny, M. W. 1976. "The Physical Properties of Spiders' Silk and Their Role in the Design of Orb-Webs." *J. Exp. Biol.* 65: 483–506.

Eisoldt, L., C. Thamm, T. Scheibel. 2012. "Role of Terminal Domains during Storage and Assembly of Spider Silk Proteins." *Biopolymers* 97: 355–61.

Fudge, D. S., K. H. Gardner, V. T. Forsyth, C. Riekel, and J. M. Gosline. 2003. "The Mechanical Properties of Hydrated Intermediate Filaments: Insights from Hagfish Slime Threads." *Biophys. J.* 85: 2015–27.

Garrido, M. A., M. Elices, C. Viney, and J. Pérez-Rigueiro. 2002. "Active Control of Spider Silk Strength: Comparison of Dragline Spun on Vertical and Horizontal Surfaces." *Polymer* 43: 1537–40.

Gatesy, J., C. Hayashi, D. Motriuk, J. Woods, and R. Lewis. 2001. "Extreme Diversity, Conservation and Convergence of Spider Fibroin Sequences." *Science* 291: 2603–5.

Gosline, J. M., P. A. Guerette, C. S. Ortlepp, and K. N. Savage. 1999. "The Mechanical Design of Spider Silks: From Fibroin Sequence to Mechanical Function." *J. Exp. Biol.* 202: 3295–303.

Guinea, G. V., M. Cerdiera, G. R. Plaza, M. Elices, and J. Pérez-Rigueiro. 2010. "Recovery in Viscid Line Fibers." *Biomacromolecules* 11: 1174–79.

Hagn, F., L. Eisoldt, J. G. Hardy, C. Vendrely, M. Coles, T. Scheibel, and H. Kessler. 2010. "A Conserved Spider Silk Domain Acts as a Molecular Switch that Controls Fiber Assembly. *Nature* 465: 239–42.

Harrington, M. J., and H. Waite. 2007. "Holdfast Heroics: Comparing the Molecular and Mechanical Properties of *Mytilus californianus* Byssal Threads." *J. Exp. Biol.* 210: 4307–18.

Hayashi, C. Y., T. A. Blackledge, and R. V. Lewis. 2004. "Molecular and Mechanical Characterization of Aciniform Silk: Uniformity of Iterated Sequence Modules in a Novel Member of the Spider Silk Fibroin Gene Family." *Mol. Biol. Evol.* 21: 1950–59.

Hermann, H., H. Bär, L. Kreplak, S. V. Strelkov, and U. Aebi. 2007. "Intermediate Filaments: From Cell Architecture to Nanomechanics." *Nat. Rev. Mol. Cell Biol.* 8: 562–73.

Knight, D. P., and F. Vollrath. 1999. "Liquid Crystals and Flow Elongation in a Spider's Silk Production Line." *Proc. R. Soc. Lond. B* 266: 519–23.

Liu, Y., A. Spopnner, D. Porter, and F. Vollrath. 2008. "Proline and Processing of Spider Silks." *Biomacromolecules* 9: 116–21.

Ortlepp, C., and J. Gosline. 2004. "Consequences of Forced Silking." *Biomacromolecules* 5: 727–31.

———. 2008. "The Scaling of Safety Factor in Spider Draglines." *J. Exp. Biol.* 211: 2832–40.

Perry, D. J., D. Bittencourt, J. Siltberg-Liberles, E. L. Rech, and R. V. Lewis. 2010. "Piriform Silk Sequences Reveal Unique Repetitive Elements." *Biomacromolecules* 11: 3000–3006.

Savage, K. N., and J. M. Gosline. 2008. "The Effect of Proline on the Network Structure of Major Ampullate Silks as Inferred from Their Mechanical and Optical Properties." *J. Exp. Biol.* 211: 1937–47.

———. 2008. "The Role of Proline in the Elastic Mechanism of Hydrated Spider Silks". *J. Exp. Biol.* 211: 1948–57.

Tian, M., and R. V. Lewis. 2006. "Tubuliform Silk Protein: A Protein with Unique Molecular Characteristics and Mechanical Properties in the Spider Silk Fibroin Family." *Appl. Phys. A* 82: 265–73.

Vincent, J. 2012. *Structural Biomaterials*, 3rd ed. Princeton, NJ: Princeton University Press.

Vollrath, F., and D. Knight. 2001. "Liquid Crystalline Spinning of Spider Silk." *Nature* 410: 541–45.

Vollrath, F., J. Fairbrother, R.J.P. Williams, E. K. Tillinghast, D. T. Bernstein, K. S. Gallagher, and M. A. Townley. 1990. "Compounds in the Droplets of the Orb Spider's Viscid Spiral." *Nature* 345: 526–28.

Wilson, R. 1969. "Control of Drag-Line Spinning in Certain Spiders." *Am. Zool.* 9: 102–11.

Yang, Y., X. Chin, Z. Shao, P. Zhou, D. Porter, D. P. Knight, and F. Vollrath. 2005. "Toughness of Spider Silk at High and Low Temperatures." *Adv. Mater.* 17: 84–88.

Zhao, A., et al. 2005. "Unique Molecular Architecture of Egg Case Silk Protein in a Spider, *Nephila clavata*." *J. Biochem.* 138: 593–604.

CHAPTER 11

Adharapurapu, R. R., F. Jiang, and K. S. Vecchio. 2006. "Dynamic Fracture of Bovine Bone." *Mater. Sci. Eng.* 26: 1325–32.

Currey, J. D. 1975. "Effects of Strain Rate, Reconstruction and Mineral Content on Some Mechanical Properties of Bone." *J. Biomech.* 8: 18–86.

———. 2002. *Bones, Structure and Mechanics.* Princeton, NJ: Princeton University Press.

———. 2010. "Mechanical Properties and Adaptations of Some Less Familiar Bony Tissues." *J. Mech. Behav. Biomed. Mater.* 3: 357–72.

Currey, J. D., K. Brear, and P. Zioupos. 1994. "Dependence of Mechanical Properties on Fiber Angle in Narwhal Tusk, a Highly Oriented Biological Composite." *J. Biomech.* 27: 885–97.

Currey, J. D., T. Landete-Castillejos, J. Estevez, F. Ceacero, A. Olguin, A. Garcia, L. Gallego. 2009. "The Mechanical Properties of Red Deer Antler Bone when Used in Fighting." *J. Exp. Biol.* 212: 3985–93.

Erickson, G. M., J. Catanese III, and T. M. Keaveny. 2002. "Evolution of the Biomechanical Material Properties of the Femur." *Anat. Rec.* 268: 115–24.

Fratzl, P., N. Fratzl-Zelman, K. Klaushofer, G. Vogl, and K. Koller. 1991. "Nucleation and Growth of Mineral Crystals in Bone Studied by Small-Angle X-ray Scattering." *Calcif. Tissue Int.* 48: 407–13.

Gupta, H. S., P. Fratzl, M. Kerschnitzki, G. Benecke, W. Wagemaier, and H.O.K. Kirchner. 2007. "Evidence for an Elementary Process in Bone Plasticity with an Activation Enthalpy of 1 eVm." *J. R. Soc. Interface* 4: 277–82.

Gupta, H. S., J. Seto, W. Wagemaier, P. Zaslansky, P. Boesecke, and P. Fratzl. 2006. "Cooperative Deformation of Mineral and Collagen in Bone at the Nanoscale." *Proc. Natl. Acad. Sci. U.S.A.* 103: 17741–46.

Hansen, U., P. Zioupos, R. Simpson, J. D. Currey, and D. Hynd. 2008. "The Effect of Strain Rate on the Mechanical Properties of Human Cortical Bone." *J. Biomech. Eng.* 130: 011011.

Hoo, R. P., P. Fratzl, J. E. Daniels, J.W.P. Dunlop, V. Honkimaki, and M. Hoffman. 2011. "Comparison of Length Scales and Orientations in the Deformation of Bovine Bone." *Acta Biomater.* 7: 2943–51.

Kalamajski, S., A. Aspberg, K. Lindblom, D. Heinegård, and Å. Oldberg. 2009. "Asporin Competes with Decorin for Collagen Binding, Binds Calcium and Promotes Osteoblast Collagen Mineralization." *Biochem. J.* 423: 53–59.

Kim, J. H., M. Niinomi, T. Akahori, and H. Toda. 2007. "Fatigue Properties of Bovine Compact Bones That Have Different Microstructures." *Int. J. Fatigue* 29: 1039–50.

Koester, K. J., J. W. Ager III, and R. O. Ritchie. 2008. "The True Toughness of Human Cortical Bone Measured with Realistically Short Cracks." *Nat. Mater.* 7: 672–77.

Landis, W. J., and F. H. Silver. 2009. "Mineral Deposition in the Extracellular Matrices of Vertebrate Tissues: Identification of Possible Apatite Nucleation Sites on Type I Collagen." *Cells Tissues Organs* 189: 20–24.

Landis, W. J., M. J. Song, A. Leith, L. McEwen, and B. F. McEwen. 1993. "Mineral and Organic Matrix Interaction in Normally Calcifying Tendon Visualized in Three Dimensions by High-Voltage Electron Microscopic Tomography and Graphic Image Reconstruction." *J. Struct. Biol.* 119: 39–54.

McElhaney, J. 1966. "Dynamic Response of Bone and Muscle Tissue." *J. Appl. Phys.* 21: 1231–36.

Montalbano, T., and G. Feng. 2011. "Nanoindentation Characterization of the Cement Lines in Ovine and Bovine Femurs." *J. Mater. Res.* 26: 1036–41.

Ruben, J. A., and A. A. Bennett. 1987. "The Evolution of Bone." *Evolution* 41: 1187–97.

Skedros, J., J. Holmes, E. Vajda, and R. Bloebaum. 2005. "Cement Lines of Secondary Osteons in Human Bone Are Not Mineral-Deficient: New Data in a Historical Perspective." *Anat. Rec. A* 286: 781–803.

Wang, R., and H. S. Gupta. 2011. "Deformation and Fracture Mechanisms of Bone and Nacre." *Ann. Rev. Mater. Res.* 41: 41–73.

Yang, L., K. O. van der Werf, C.F.C. Fitié, M. L. Bennink, P. J. Dijkstra, and J. Feijen. 2008. "Mechanical Properties of Native and Crosslinked Type-I Collagen Fibrils." *Biophys. J.* 94: 2204–11.

Zioupos, P., J. D. Currey, and A. Casinos. 2001. "Tensile Fatigue in Bone: Are Cycles-, or Time to Failure, or Both, Important?" *J. Theor. Biol.* 210: 389–99.

CHAPTER 12

Andersen, S. O., P. Højrup, and P. Roepstorff. 1986. "Characterization of Cuticular Proteins from the Migratory Locust." *Insect Biochem.* 16: 441–47.

———. 2001. "Matrix Proteins from Insect Pliable Cuticles: Are They Less Flexible and Easily Deformed?" *Insect Biochem. Mol. Biol.* 31: 445–52.

———. 2010 "Insect Cuticular Sclerotization: A Review." *Insect Biochem. Mol. Biol.* 40: 166–78.

———. 2011. "Are Structural Proteins in Insect Cuticles Intrinsically Disordered Regions?" *Insect Biochem. Mol. Biol.* 41: 620–27.

Atkins, E. 1985. "Conformations in Polysaccharides and Complex Carbohydrates." *J. Biosci.* 8: 375–87.

Barbakadze, N., S. Enders, S. Gorb, and E. Arzt. 2006. "Local Properties of the Head Articulation Cuticle in the Beetle." *J. Exp. Biol.* 209: 722–30.

Bennet Clark, H. C. 1975. "The Energetics of the Jump of the Locust *Schistocerca gregaria.*" *J. Exp. Biol.* 63: 53–83.

Chen, P., Y. Nishiyama, J.-L. Puteaux, and K. Mazeau. 2014. "Diversity of Potential Hydrogen Bonds in Cellulose I Revealed by Molecular Dynamics Simulation." *Cellulose* 21: 897–908.

Cribb, B. W., C.-L. Lin, L. Rintoul, R. Rasch, J. Hasenpusch, and H. Huang. 2010. "Hardness in Arthropod Exoskeletons in the Absence of Transition Metals." *Acta Biomater.* 6: 3152–56.

Currey, J. D. 2002. *Bones, Structure and Mechanics.* Princeton, NJ: Princeton University Press.

Diani, J., B. Fayolle, and P. Gilormini. 2009. "A Review on the Mullins Effect." *Eur. Polymer J.* 45: 601–12.

Dirks, J.-H., and D. Taylor. 2012. "Fracture Toughness of Locust Cuticle." *J. Exp. Biol.* 215: 1502–8.

Dirks, J.-H., E. Parle, and D. Taylor. 2013. "Fatigue of Insect Cuticle." *J. Exp. Biol.* 216: 1924–27.

Fraenkel, G., and K. M. Rudall. 1947. "The Structure of Insect Cuticles." *Proc. R. Soc. Lond. B.* 134: 111–43.

Gotte, L., M. Mammi, and G. Pezzin. 1968. "Some Structural Aspects of Elastin Revealed by X-ray Diffraction and Other Physical Methods." In *Symposium on Fibrous Proteins*, edited by W. G. Crewther, 236–45. New York: Plenum.

Hepburn, H. R., and H. D. Chandler. 1976. "Material Properties of Arthropod Cuticles: The Arthrodial Membranes." *J. Comp. Physiol.* 109: 177–98.

Hepburn, H. R., and I. Joffe. 1974. "Locust Solid Cuticle: A Time Sequence of Mechanical Properties." *J. Insect Physiol.* 20: 497–526.

Hillerton, J. E. 1980. "Electron Microscopy of Fibril-Matrix Interaction in a Natural Composite, Insect Cuticle." *J. Mater. Sci.* 15: 3109–12.

Hillerton, J. E., S. Reynolds, and J. Vincent. 1982. "On the Indentation Hardness of Insect Cuticle." *J. Exp. Biol.* 96: 45–52.

Hillerton, J. E., and J. Vincent. 1982. "The Specific Location of Zinc in Insect Mandibles." *J. Exp. Biol.* 101: 333–36.

———. 1983. "Consideration of the Importance of Hydrophobic Interactions in Stabilizing Insect Cuticle." *Int. J. Macromol.* 5: 163–66.

Jensen, M., and T. Weis-Fogh. 1962. "Biology and Physics of Locust Flight: V. Strength and Elasticity of Locust Cuticle." *Philos. Trans. R. Soc. Lond. B* 245: 137–69.

Katz, S., and J. M. Gosline. 1992. "Ontogenetic Scaling and Mechanical Behaviour of the Tibae of the African Desert Locust." *J. Exp. Biol.* 168: 125–50.

———. 1994. "Scaling Modulus as a Degree for Freedom in the Design of Locust Legs." *J. Exp. Biol.* 187: 207–23.

Klocke, D., and H. Schmitz. 2011. "Water as a Major Modulator of the Mechanical Properties of Insect Cuticle." *Acta Biomater.* 7: 2935–42.

Lillie, M. A., and J. M. Gosline. 1990. "The Effects of Hydration on the Dynamic Mechanical Properties of Elastin." *Biopolymers* 29: 1147–60.

Lin, H-T., A. L. Dorfmann, and B. A. Trimmer. 2009. "Soft-Cuticle Biomechanics: A Constitutive Model of Anisotropy for Caterpillar Integument." *J. Theor. Biol.* 256: 447–57.

Meyers, M. A., P.-Y. Chen, A. Y.-M. Lin, and Y. Seki. 2008. "Biological Materials: Structure and Properties." *Prog. Mater. Sci.* 53: 1–206.

Muller, M., M. Olek, M. Gierseg, and H. Schmitz. 2008. "Micromechanical Properties of Consecutive Layers in Specialized Insect Cuticle: The Gula of *Pachnoda marginata* (Coleoptera, Scarabaeidae) and the Infrared Sensilla of *Melanophila acuminata* (Coleoptera, Buprestidae)." *J. Exp. Biol.* 211: 2576–83.

Nation, J. L. 2008. *Insect Physiology and Biochemistry*. Boca Raton, FL: CRC.

Neville, A. C. 1970. "Cuticle Ultrastructure in Relation to the Whole Insect." In *Insect Ultrastructure*, 17–39. Oxford: Blackwell Science.

Neville, A. C., and B. M. Luke. 1969. "A Two-System Model for Chitin-Protein Complexes in Insect Cuticles." *Tissue Cell* 1: 689–707.

Neville, A. C., D. A. Parry, and J. Woodhead-Galloway. 1976. "The Chitin Crystallite in Arthropod Cuticle." *J. Cell Sci.* 21: 73–82.

Nikolov, S., M. Petrov, L. Lymperakis, M. Friák, C. Sachs, H. O. Frabritius, and D. Raabe. 2010. "Revealing the Design Principles of High-Performance Biological Composites Using Ab Initio and Multiscale Simulations: The Example of Lobster Cuticle." *Adv. Mater.* 22: 519–26.

Petrov, M., L. Lymperakis, M. Friák, J. Neugebauer. 2013. "Ab Initio–Based Conformational Study of Crystalline α-Chitin." *Biopolymers* 99: 22–34.

Rebers, J. E., and J. H. Willis. 2001. "A Conserved Domain in Insect Cuticular Proteins Binds Chitin." *Insect Biochem. Mol. Biol.* 40: 1083–93.

Reynolds, S. 1975a. "Mechanical Properties of the Abdominal Cuticle of *Rhodnius* Larvae." *J. Exp. Biol.* 62: 69–80.

———. 1975b. "The Mechanism of Plasticization of the Abdominal Cuticle in *Rhodnius*." *J. Exp. Biol.* 62: 81–98.

Rudall, K. M. 1963. "The Chitin/Protein Complexes of Insect Cuticles." *Adv. Insect Physiol.* 1: 257–313.

Stephens, R., and J. Vincent. 1976. "Infrared Spectroscopy of the Locust Extensible Intersegmental Membrane Cuticle and its Components." *J. Insect Physiol.* 22: 601–5.

Tychsen, P., and J. Vincent. 1976. "Correlated Changes in Mechanical Properties of Intersegmental Membrane and Bonding between Proteins in the Female Adult Locust." *J. Insect Physiol.* 22: 115–25.

Vincent, J. 1975. "Locust Oviposition: Stress Softening of the Extensible Intersegmental Membranes." *Proc. R. Soc. Lond. A* 188: 189–201.

———. 2012. *Structural Biomaterials*, 3rd ed. Princeton, NJ: Princeton University Press.

Vincent, J., and S. Ablett. 1987. "Hydration and Tanning in Insect Cuticle." *J. Insect Physiol.* 33: 973–79.

Vincent, J., and J. Hillerton. 1979. "The Tanning of Insect Cuticle: A Critical Review and Revised Mechanism." *J. Insect Physiol.* 25: 653–58.

Vincent, J., and U. Wegst. 2004. "Design and Mechanical Properties of Insect Cuticle." *Arthropod Struct. Dev.* 33: 187–99.

Wainwright, S. A., W. D. Biggs, J. D. Currey, and J. M. Gosline. 1976. *Mechanical Design in Organisms*. Princeton, NJ: Princeton University Press.

Weis-Fogh, T. 1961. "Molecular Interpretation of the Elasticity of Resilin, a Rubber-like Protein." *J. Mol. Biol.* 3: 648–667.

CHAPTER 13

Alexander, R. McN. 1959. "Visco-elastic Properties of the Body-Wall of Sea Anemones." *J. Exp. Biol.* 39: 373–86.

Batham, E. J., and C. Pantin. 1950. "Muscular Activity in the Sea Anemone *Metridium senile*." *J. Exp. Biol.* 27: 264–89, 377–98.

Gosline, J. M. 1971. "Connective Tissue Mechanics of *Metridium senile*." *J. Exp. Biol.* 55: 763–74, 775–95.

Grimstone, A. V., R. W. Horne, C.F.A. Pantin, and A. Robson. 1958. "Fine Structure of Mesenteries of Sea Anemone *Metridium senile*." *Q. J. Microsc. Sci.* 99: 523–40.

Katzman, R. L., and R. W. Jeanloz. 1970. "Carbohydrate Chemistry of Invertebrate Connective Tissue." *Chem. Mol. Biol. Intercell. Matrix* 1: 217–27.

Koehl, M. 1977a. "Effects of Sea Anemones on the Flow Forces They Encounter." *J. Exp. Biol.* 69: 87–105.

Koehl, M. 1977b. "Mechanical Diversity of Connective Tissue of the Body Wall of Sea Anemones." *J. Exp. Biol.* 69: 107–25.

Koehl, M. 1977c. "Mechanical Organization of Cantileverlike Sessile Organisms: Sea Anemones." *J. Exp. Biol.* 69: 127–42.

Nordwig, A., and U. Hayduk. 1969. "Invertebrate Collagens: Isolation, Characterization and Phylogenetic Aspects." *J. Mol. Biol.* 44: 161–72.

Orgel, J.P.R.O., T. C. Irving, A. Miller, and T. J. Wess. 2006. "Microfibril Structure of Type I Collagen." *Proc. Natl. Acad. Sci. U.S.A.* 103: 9000–9005.

Wess, T. J., A. P. Hammersley, L. Wess, and A. Miller. 1998. "A Consensus Model for Molecular Packing of Type I Collagen." *J. Struct. Biol.* 122: 92–100.

CHAPTER 14

Aaron, B., and J. M. Gosline. 1981. "Elastin as a Random Network Elastomer." *Biopolymers* 20: 1247–60.

Andersen, S. O. 1967. "Isolation of a New Type Of Cross Link from the Hinge Ligament Protein of Molluscs." *Nature* 216: 1029–30.

———. 2010. "Studies on Resilin-like Gene Products in Insects." *Insect Biochem. Mol. Biol.* 40: 541–51.

Bowie, M., J. D. Layes, and M. E. Demont. 1993. "Damping in the Hinge of the Scallop." *J. Exp. Biol.* 175: 311–15.

Cao, Q., Y. Wang, and H. Bayley. 1997. "Sequence of Abductin, the Molluscan Rubber Protein." *Curr. Biol.* 7: R677–78.

Davis, E. C. 1993. "Stability of Elastin in the Developing Mouse Aorta." *Histochemistry* 100: 17–26.

DeMont, M. E., and J. M. Gosline. 1988. "Mechanics of Jet Propulsion in the Jellyfish, *Polyorchis*: Mechanical Properties of the Locomotor Structure." *J. Exp. Biol.* 134: 313–32.

Denny, M. E., and L. Miller. 2006. "Jet Propulsion in the Cold: Mechanics of Swimming in the Antarctic Scallop." *J. Exp. Biol.* 209: 4503–14.

Glab, J., and T. Wess. 2008. "Changes in Molecular Packing of Fibrillin Microfibrils during Extension Indicate Intrafibrillar and Interfibrillar Reorganization in the Elastic Response." *J. Mol. Biol.* 383: 1171–80.

Gosline, J. M., and C. J. French. 1979. "Dynamic Mechanical Properties of Elastin." *Biopolymers* 18: 2091–103.

Jensen, M., and T. Weis-Fogh. 1962. "Biology and Physics of Locust Flight: V. Strength and Elasticity of Locust Cuticle." *Philos. Trans. R. Soc. Lond. B* 245: 137–69.

Jensen, S. A., I. B. Robertson, and P. A. Handford. 2012. "Dissecting the Fibrillin Microfibril: Structural Insights into Organization and Function." *Structure* 20: 215–25.

Jensen, S. A., D. Yadin, I. B. Robertson, and P. A. Handford. 2013. "Evolutionary Insights into Fibrillin Structure and Function." In *Evolution of Extracellular Matrix*, edited by F. W. Keeley and R. P. Mecham, 121–62. Berlin: Springer.

Lillie, M. A., and J. M. Gosline. 2002. "Viscoelastic Basis for the Tensile Strength of Elastin." *Intl. J. Biol. Macromol.* 30: 119–27.

McConnell, C. J., G. M. Wright, and M. E. DeMont. 1996. "Modulus of Elasticity of Lobster Aorta Microfibrils." *Experientia* 52: 918–21.

Megill, W. M., J. M. Gosline, and R. Blake. 2005. "Modulus of Fibrillin-Containing Elastic Fibers in the Mesoglea of a Hydromedusa." *J. Exp. Biol.* 208: 3819–34.

Muiznieks, L. D., A. Weiss, and F. W. Keeley. 2010. "Structural Disorder and Dynamics of Elastin." *Biochem. Cell Biol.* 88: 239–50.

Rauscher, S., S. Baud, M. Miao, F. W. Keeley, and R. Pomès. 2006. "Proline and Glycine Control Protein Self-Organization in Elastomeric or Amyloid Fibrils." *Structure* 14: 1667–76.

Reber-Müller, S., T. Spissinger, P. Schuchert, J. Spring, and V. Schmid. 1995. "An Extracellular Matrix Protein of Jellyfish Homologous to Mammalian Fibrillins Forms Different Fibrils Depending on the Life Stage of the Animal." *Dev. Biol.* 169: 662–72.

Shadwick, R. E., and J. M. Gosline. 1981. "The Mechanics of the Octopus Aorta." *Science* 213: 759–61.

Shapiro, S. D., S. K. Endicott, M. A. Province, J. A. Pierce, and E. J. Campbell. 1991. "Marked Longevity of Human Lung Parenchymal Elastic Fibers Deduced from Prevalence of D-Aspartate and Nuclear Weapons-Related Radiocarbon." *J. Clin. Invest.* 87: 1828–34.

Sherratt, M. J., C. Baldock, J. L. Haston, D. F. Holmes, C.J.P. Holmes, C. A. Shuttleworth, T. J. Wess, et al. 2003. "Fibrillin Microfibrils Are Stiff Reinforcing Fibres in Compliant Tissues." *J. Mol. Biol.* 332: 183–93.

Vanacore, R., A.-J. L. Ham, M. Voehler, C. R. Sanders, T. P. Conrads, T. D. Veenstra, K. B. Sharpless, et al. 2009. "Sulfamine Bond Identified in Collagen IV." *Science* 325: 1230–34.

Weis-Fogh, T. 1961. "Molecular Interpretation of the Elasticity of Resilin, a Rubber-like Protein." *J. Mol. Biol.* 3: 648–67.

Wright, D. M., V. C. Duance, T. J. Wess, C. M. Kielty, and P. P. Purslow. 1999. "The Supramolecular Organisation of Fibrillin-Rich Microfibrils Determines the Mechanical Properties of Bovine Zonular Filaments." *J. Exp. Biol.* 202: 3011–20.

Wu, Y. J., D. P. La Pierre, J. Wu, A. J. Yee, and B. B. Yang. 2005. "The Interaction of Versican with its Binding Partners." *Cell Res.* 15: 483–94.

CHAPTER 15

Aaron, B. B., and J. M. Gosline. 1981. "Elastin as a Random Network Elastomer." *Biopolymers* 20: 1247–60.

Alexander, R. McN. 1959. "Visco-elastic Properties of the Body-Wall of Sea Anemones." *J. Exp. Biol.* 39: 373–86.

———. 1966. "Rubber-like Properties of the Inner Hinge Ligament of Pectinidinae." *J. Exp. Biol.* 44: 119–30.

———. 1987. "Bending of Cylindrical Animals with Helical Fibers in Their Skin or Cuticle." *J. Theor. Biol.* 124: 97–110.

Andersen, S. O. 1967. "Isolation of a New Type of Cross Link from the Hinge Ligament Protein of Molluscs." *Nature* 216: 1029–30.

———. 2010. "Studies on Resilin-like Gene Products in Insects." *Insect Biochem. Mol. Biol.* 40: 541–51.

Ateshian, G. A., and C. T. Hung. 2003. "Functional Properties of Native Articular Cartilage." In *Functional Tissue Engineering*, edited by F. Guilak, D. L. Butler, S. A. Goldstein, and D. Mooney, 46–68. New York: Springer.

Bansil, R., and B. Turner. 2006. "Mucin Structure, Aggregation, Physiological Function and Biomedical Applications." *Curr. Opin. Colloid Interface Sci.* 11: 164–70.

Batham, E. J., and C. Pantin. 1950. "Muscular Activity in the Sea Anemone *Metridium senile*." *J. Exp. Biol.* 27: 264–89, 377–98.

Bowie, M., J. D. Layes, and M. E. Demont. 1993. "Damping in the Hinge of the Scallop." *J. Exp. Biol.* 175: 311–15.

Cao, Q., Y. Wang, and H. Bayley. 1997. "Sequence of Abductin, the Molluscan Rubber Protein." *Curr. Biol.* 7: R677–78.

Carmichael, S. W. 2014. "The Tangled Web of Langer's Lines." *Clin. Anat.* 27: 162–68.

Celli, J., B. S. Turner, N. H. Afdal, R. H. Ewoldt, G. H. McKinley, R. Bansil, and S. Erramilli. 2007. "Rheology of Gastric Mucin Exhibits pH-Dependent Sol-Gel Transition." *Biomacromolecules* 8: 1580–86.

Chung, M., M. Miao, R. J. Stahl, E. Chan, J. Parkinson, and F. W. Keeley. 2006. "Sequences and Domain Structures of Mammalian, Avian, Amphibian and Teleost Tropoelastins: Clues to the Evolutionary History of Elastins." *Matrix Biol.* 25: 492–504.

Clark, J. M., and S. Glagov. 1985. "Transmural Organization of the Arterial Media." *Atheroscleorsis* 5: 19–34.

Clark, R. B., and J. B. Cowey. 1958. "Factors Controlling the Change of Shape of Certain Nemertean and Turbellarian Worms." *J. Exp. Biol.* 35: 731–48.

Davis, E. C. 1993. "Stability of Elastin in the Developing Mouse Aorta." *Histochemistry* 100: 17–26.

DeMont, M. E., and J. M. Gosline. 1988. "Mechanics of Jet Propulsion in the Jellyfish, *Polyorchis*: Mechanical Properties of the Locomotor Structure." *J. Exp. Biol.* 134: 313–32.

Denny, M., and J. M. Gosline, J. 1980. "The Physical Properties of the Pedal Mucus of the Terrestrial Slug, *Ariolimax columbianus*." *J. Exp. Biol.* 88: 375–94.

Denny, M., and L. Miller. 2006. "Jet Propulsion in the Cold: Mechanics of Swimming in the Antarctic Scallop." *J. Exp. Biol.* 209: 4503–14.

Dobrin, P., and A. Rovick. 1969. "Influence of Vascular Smooth Muscle on Contractile Mechanics and Elasticity of Arteries." *Am. J. Physiol.* 6: 1644–51.

Glab, J., and T. Wess. 2008. "Changes in Molecular Packing of Fibrillin Microfibrils during Extension Indicate Intrafibrillar and Interfibrillar Reorganization in the Elastic Response." *J. Mol. Biol.* 383: 1171–80.

Gosline, J. M. 1971. "Connective Tissue Mechanics of *Metridium senile*." *J. Exp. Biol.* 55: 763–74, 775–95.

———. 1978. "Temperature-Dependent Swelling of Elastin." *Biopolymers* 17: 697–707.

Gosline, J. M., and C. J. French. 1979. "Dynamic Mechanical Properties of Elastin." *Biopolymers* 18: 2091–103.

Gosline, J. M., and R. E. Shadwick. 1996. "Mechanical Properties of Fin Whale Arteries Are Explained by Novel Connective Tissue Designs." *J. Exp. Biol.* 199: 985–97.

Grimstone, A. V., R. W. Horne, C.F.A. Pantin, and A. Robson. 1958. "Fine Structure of Mesenteries of Sea Anemone *Metridium senile*." *Q. J. Microsc. Sci.* 99: 523–40.

Harris, J. E., and H. D. Crofton. 1957. "Structure and Function in the Nematodes: Internal Pressure and Cuticular Structure in Ascaris." *J. Exp. Biol.* 34: 116–30.

Huang, C., A. Stankiewicz, G. A. Ateshian, and V. C. Mow. 2005. "Anisotropy, Inhomogeneity and Tension-Compression Nonlinearity of Human Cartilage." *J. Biomech.* 38: 799–809.

Jeffery, A. K., G. W. Blunn, C. W. Archer, and G. Bentley. 1991. "3-D Collagen Architecture of Articular Cartilage." *J. Bone Joint Surg.* 73: 795–801.

Jensen, M., and T. Weis-Fogh. 1962. "Biology and Physics of Locust Flight: V. Strength and Elasticity of Locust Cuticle." *Philos. Trans. R. Soc. Lond. B* 245: 137–69.

Jensen, S. A., I. B. Robertson, and P. Handford. 2012. "Dissecting the Fibrillin Microfibril: Structural Insights into Organization and Function." *Structure* 20: 215–25.

Jensen, S. A., D. Yadin, I. B. Robertson, and P. A. Handford. 2013. "Evolutionary Insights into Fibrillin Structure and Function." In *Evolution of Extracellular Matrix*, edited by F. W. Keeley and R. P. Mecham, 121–62. Berlin: Springer.

Katzman, R. L., and R. W. Jeanloz. 1970. "Carbohydrate Chemistry of Invertebrate Connective Tissue." *Chem. Mol. Biol. Intercell. Matrix* 1: 217–27.

Kiani, C., L. Chen, Y. J. Wu, A. J. Yee, and B. B. Yang. 2002. "Structure and Function of Aggrecan." *Cell Res.* 12: 12–32.

Koehl, M. 1977a. "Effects of Sea Anemones on the Flow Forces They Encounter." *J. Exp. Biol.* 69: 87–105.

Koehl, M. 1977b. "Mechanical Diversity of Connective Tissue of the Body Wall of Sea Anemones." *J. Exp. Biol.* 69: 107–25.

Koehl, M. 1977c. "Mechanical Organization of Cantileverlike Sessile Organisms: Sea Anemones." *J. Exp. Biol.* 69: 127–42.

Koenders, M., L. Yang, R. G. Wismans, K. O. van der Werf, D. P. Reinhardt, W. Daamen, M. L. Bennink, et al. 2009. "Microscale Mechanical Properties of Single Elastic Fibers: The Role of Fibrillin Microfibrils." *Biomaterials* 30: 2425–32.

Laasanen, M. S., J. Toyräs, R. K. Korhonen, J. Rieppo, S. Saarakkala, M. T. Nieminen, J. Hirvone, et al. 2003. "Biomechanical Properties of Knee Articular Cartilage." *Biorheology* 40: 133–40.

Lillie, M. A., T. E. Armstrong, S. G. Gérard, R. E. Shadwick, and J. M. Gosline. 2012. "Contribution of Elastin and Collagen to the Inflation Response of the Pig Thoracic Aorta." *J. Biomech.* 45: 2133–41.

Lillie, M. A., and J. M. Gosline. 2002. "Viscoelastic Basis for the Tensile Strength of Elastin." *Intl. J. Biol. Macromol.* 30: 119–27.

Lu, X. L., C. Miller, F. H. Chen, X. E. Guo, and V. C. Mow. 2007. "The Generalized Triphasic Correspondence Principle for Simultaneous Determination of the Mechanical Properties and Proteoglycan Content of Articular Cartilage by Indentation." *J. Biomech.* 40: 2434–41.

McConnell, C. J., G. M. Wright, and M. E. DeMont. 1996. "Modulus of Elasticity of Lobster Aorta Microfibrils." *Experientia* 52: 918–21.

Megill, W. M. , J. M. Gosline, and R. Blake. 2005. "Modulus of Fibrillin-Containing Elastic Fibers in the Mesoglea of a Hydromedusa." *J. Exp. Biol.* 208: 3819–34.

Nordwig, A., and U. Hayduk. 1969. "Invertebrate Collagens: Isolation, Characterization and Phylogenetic Aspects." *J. Mol. Biol.* 44: 161–72.

Orgel, J.P.R.O., T. C. Irving, A. Miller, and T. J. Wess. 2006. "Microfibril Structure of Type I Collagen." *Proc. Natl. Acad. Sci. U.S.A.* 103: 9000–9005.

Orton, L., and P. Brodie. 1987. "Engulfing Mechanics of Fin Whales." *Can. J. Zool.* 65: 2898–907.

Pabst, A. D. 1996. "Morphology of the Subdermal Connective Tissue Sheath of Dolphins." *J. Zool. (Lond.)* 238: 35–52.

Purslow, P. P. 1983. "Measurement of the Fracture Toughness of Extensible Connective Tissues." *J. Mater. Sci.* 18: 3591–98.

Quillin, K. 1999. "Kinematic Scaling of Locomotion by Hydrostatic Animals: Ontogeny of Peristaltic Crawling by the Earthworm *Lumbricus terrestris*." *J. Exp. Biol.* 202: 661–74.

Rauscher, S., S. Baud, M. Miao, F. W. Keeley, and R. Pomès. 2006. "Proline and Glycine Control Protein Self-Organization in Elastomeric or Amloid Fibrils." *Structure* 14: 1667 76.

Reber-Müller, S., T. Spissinger, P. Schuchert, J. Spring, and V. Schmid. 1995. "An Extracellular Matrix Protein of Jellyfish Homologous to Mammalian Fibrillins Forms Different Fibrils Depending on the Life Stage of the Animal." *Dev. Biol.* 169: 662–72.

Ridge, M., and V. Right. 1966. "Directional Effects of Skin." *J. Invest. Dermatol.* 46: 341–46.

Setton, L. A., Z. Wenbo, and V. C. Mow. 1993. "The Biphasic Poroviscolastic Behavior of Articular Cartilage: Role of the Surface Zone in Governing the Compressive Behavior." *J. Biomech.* 26: 581–92.

Shadwick, R. E. 1998. "Elasticity in Arteries." *Am. Sci.* 86: 535–41.

———. 1999. "Mechanical Design in Arteries." *J. Exp. Biol.* 202: 3305–13.

———. 2008. "Foundations of Animal Hydraulics: Geodesic Fibres Control the Shape of Soft-Bodied Animals." *J. Exp. Biol.* 211: 289–91.

Shadwick, R. E., J. A. Goldbogen, J. Potvin, N. D. Pyenson, and A. W. Vogl. 2013. "Novel Muscle and Connective Tissue Design Enables High Extensibility and Controls Engulfment Volume in Lunge-Feeding Rorqual Whales." *J. Exp. Biol.* 216: 2691–701.

Shadwick, R. E., and J. M. Gosline. 1981. "The Mechanics of the Octopus Aorta." *Science* 213: 759–61.

———. 1994. "Arterial Mechanics in the Fin Whale Suggest a Unique Hydrodynamic Design." *Am. J. Physiol.* 267: R805–18.

Shadwick, R. E., A. P. Russell, and R. F. Lauff. 1992. "The Structure and Mechanical Design of Rhinoceros Dermal Armour." *Philos. Trans. R. Soc. Lond. B* 337: 419–28.

Shapiro, S. D., S. K. Endicott, M. A. Province, J. A. Pierce, and E. J. Campbell. 1991. "Marked Longevity of Human Lung Parenchymal Elastic Fibers Deduced from Prevalence of D-Aspartate and Nuclear Weapons-Related Radiocarbon." *J. Clin. Invest.* 87: 1828–34.

Sherratt, M. J., C. Baldock, J. L. Haston, D. F. Holmes, C.J.P. Holmes, C. A. Shuttleworth, T. J. Wess, et al. 2003. "Fibrillin Microfibrils Are Stiff Reinforcing Fibres in Compliant Tissues." *J. Mol. Biol.* 332: 183–93.

Taylor, C., A. Allen, P. W. Dettmar, and J. P. Parsons. 2003. "The Gel Matrix of Gastric Mucus Is Maintained by a Complex Interplay of Transient and Nontransient Associations." *Biomacromolecules* 4: 922–27.

Vanacore, R., A.-J. L. Ham, M. Voehler, C. R. Sanders, T. P. Conrads, T. D. Veenstra, K. B. Sharpless, et al. 2009. "Sulfamine Bond Identified in Collagen IV." *Science* 325: 1230–34.

Veronda, D., and R. Westman. 1970. "Mechanical Characterization of Skin." *J. Biomech.* 3: 111–24.

Wainwright, S. A., F. Vosburgh, and J. H. Hebrank. 1978. "Shark Skin: Function in Locomotion." *Science* 202: 747–49.

Weis-Fogh, T. 1961. Molecular Interpretation of the Elasticity of Resilin, a Rubber-like Protein." *J. Mol. Biol.* 3: 648–67.

Wess, T. J., A. P. Hammersley, L. Wess, and A. Miller. 1998. "A Consensus Model for Molecular Packing of Type I Collagen." *J. Struct. Biol.* 122: 92–100.

Wolinsky, H. and S. Glagov. 1967. "A Lamellar Unit of Aortic Medial Structure and Function in Mammals." *Circ. Res.* 20: 99.

Wright, D. M., V. C. Duance, T. J. Wess, C. M. Kielty, and P. P. Purslow. 1999. "The Supramolecular Organisation of Fibrillin-Rich Microfibrils Determines the Mechanical Properties of Bovine Zonular Filaments." *J. Exp. Biol.* 202: 3011–20.

Wu, Y. J., D. P. La Pierre, J. Wu, A. J. Yee, and B. B. Yang. 2005. "The Interaction of Versican with Its Binding Partners." *Cell Res.* 15: 483–94.

Index

Aaron, B., 305–6
abductin: amino acid composition of, 298–301; amino acid sequence and cross-linking of, 301, 302–3; dynamic mechanical properties of, 310–12; network structure and mechanical properties of, 307–8
aciniform gland, silk produced by, 156, 161, 162, 164
acoustic impedance, 228
activation energy: to form hole in fluid, 74–75, 80; glass transition and, 80; for postyield strain in bone, 212
Adharapurapu, R. R., 220
affine-network assumption, 67, 68
agar, 288
aggrecan, 321–22, 324
aggregate glands, 152, 156
alanine: in insect cuticle matrix, 244, 245, 246, 247, 249, 256; in rubberlike proteins, 298, 302, 303, 311; in silks, 156–58, 160, 170, 354
Alexander, R. McN., 281–82, 335
allysine, 115, 116
α chitin, 234–35, 236, 238, 239
α helices: in female locust intersegmental membrane, 271; opening and unfolding of, in hair keratins, 171; in rubberlike proteins, 302; in terminal domains of spidroin proteins, 176, 177, 178
amorphous polymers, 60–61. See also random-coil molecules; viscoelastic behavior
Andersen, S. O., 247, 248, 249, 303, 306
annelids: earthworm cross-fiber collagen lattice, 336–37; fibrillin microfibrils in, 296
Anthopleura xanthogrammica, 288–89
antler, 225–27
apodeme (cuticular tendon), 238–39, 240, 253–56, 258
apolysis, 232
Araneae, 158
Araneus, 158
Araneus diadematus: dragline function in, 183–85; FL silk from, 152–54; MaA silks from, 152–54, 164, 165, 170, 172, 173, 183–84
Araneus sericatus, 166, 180
Argiope, 158, 160, 166
Argiope argentata, 162, 164–65, 167
Argiope trifasciata, 160, 161, 162–63
Ariolimax columbianus, pedal mucus of, 316–18

arteries, 337–50; blood flow to tissues and, 337–38; circumferential wall stress and, 338–40; collagen fibers in, 340, 342, 343, 344, 346, 350; elastin in, 37, 82, 297, 312–13, 340–44, 346, 350 (see also elastin); fin-whale aorta and aortic arch, 345–50; invertebrate, fibrillin microfibrils in, 151, 297; J-shaped pressure-radius curve of, 340; J-shaped stress–strain curves of, 297, 337, 342; mechanical properties of, 337–40, 342–44; scaling for different mammals, 345–46, 348–50; wall structure of, 340–42, 344–46
arthropods: body plan of, 231–33; fibrillin microfibrils in, 296–97; range of material stiffness in, 271. See also insect cuticle; spider silks
Ascaris lumbricoides, 337
Askarieh, G., 176
aspect ratio: of cellulose microfibrils, 238, 273; of chitin microfibrils, 233, 236–37, 239, 254, 257, 271, 273, 353, 356; of collagen fibrils in mesoglea, 285; of collagen fibrils in sea urchin catch ligament, 147–48, 149; of collagen fibrils in tendon, 119–20, 127, 140, 142, 143, 353; defined, 90–91; in estimating matrix properties, 353; of fibrillin microfibrils, 290; of inclusions in major ampullate silk gland, 175; material performance and, 356; of mineral platelets in bone, 204, 207, 209, 210, 356; of mineral platelets in dentine, 205; reinforcement efficiency and, 91–92; stiffness of discontinuous-fiber composite and, 92–93; of tendons in mechanical testing, 129–30
asporin, 193
assassin bug (Rhodnius prolixus), 266–267
Atkins, E., 237
atomic-force microscopy (AFM), collagen fibril stiffness and, 140, 208–9
avian tendons, calcified, 131, 192–93, 210, 228

Baer, E., 125
Bailey, A. J., 116
banana slug, pedal mucus of, 316–18
Bansil, R., 315
Barbakadze, N., 262
Batham, E. J., 280
Bennet-Clark, H. C., 253
Bennett, A. A., 191

β-sheet crystals: in *Bombyx* silk, 156; in FL spidroins, 160; in major ampullate silks, 170, 176, 178; in *Salticus* dragline silk, 186; structure of, 156–58

β-sheets, in insect cuticle proteins, 247

Bezerra, M., 131

biglycan, 121, 124, 142

bird bones: mechanical properties of, 228–29; specific modulus of, 272–73, 357. *See also* avian tendons, calcified

birds as predators on insects, 357

Blackledge, T. A., 164–65, 167

Bombyx mori (silkworm): cuticle properties of, 265; silk of, 156, 157–58

bond-energy elasticity: basic theory of, 43–45; of crystalline polymers, 60; of hydrated spider silks, 172–73; soft elasticity and, 63, 71–73; strength of materials and, 46–49

bone, 191–230; of birds, 228–29, 272–73, 357 (*see also* avian tendons, calcified); cells of, 200–201; collagen fibers in, 191, 195, 197, 198, 201–3, 205; collagen fibrils in, 192–95, 206–13; comparison to other materials, 354–55; critical flaw length of, 57; differences between Haversian and fibrolamellar, 222; diverse adaptations of, 225–30; evolution of, 191–92, 231; fatigue behavior of, 218–19, 225; fracture toughness of, 212–14, 216–17; mechanical properties of, 214–25; mechanical properties of materials in, 201–6; mineral volume fraction of, 203–4, 225–29; modeling of, 222–23; nanoscale composite models for, 206–14; Poisson's ratios for, 20; reinforcement efficiency of, 209–10; remodeling of, 197, 200, 201, 222, 223; shear deformation in, 207–8; strain-rate dependence of properties, 220–22; stress and strain in normal activities, 220, 223–25; structural hierarchy of, 192–200; in structural system of animal, 6–7, 8; summary of composite structure, 353–54; viscoelastic behavior of, 212, 217, 220; yield behavior of, 203, 212–13. *See also* fibrolamellar bone; Haversian (secondary lamellar) bone; lamellar bone; woven bone

Bowie, M., 311

breaking stress. *See* strength

Brinkmann, J., 107, 115

Brodie, P., 332–33

Bueche, F., 76

Bunsell, A. R., 88, 89

byssal threads, 186

Caerostris darwini, 166, 170

calcium carbonate, in invertebrates, 191–92, 229–30, 273

calcium phosphate: acid resistance and, 191–92; amorphous, 191, 192; as hydroxyapatite, 191–93, 203, 205

canaliculi, 201

cancellous bone, 195

Cao, Q., 303

carbon fibers, properties of, 273, 355

Carmichael, S. W., 329

cartilage, 8, 319–27; collagen fibers in, 319–22, 324, 325–26; creep test of, 29–30, 322; fixed charge density in, 326; proteoglycans in, 8, 142, 319, 320, 321–22, 324–26; triphasic theory of, 326; viscoelastic behavior of, 29–30, 142

catch ligament of sea urchin, 144–51, 296

Celli, J., 316

cellulose: crystal structure of, 233–34, 236; in flying machines, 272–73; in plant cell walls, 59, 186, 234, 236, 237–38, 273

cellulose I, 234, 236

cellulose microfibrils, 233, 234, 236; dimensions of, 237–38; stiffness and strength of, 238–39, 273

cement layer around secondary osteons, 200, 201, 213, 217

ceramics, mineralized, 273

Chandler, H. D., 264

Chen, P., 236

chitin, 186; α chitin, 234–35, 236, 238, 239; crystal structure of, 233–36; evolutionary origin of, 231. *See also* insect cuticle

chitin microfibrils: cellulose microfibrils compared to, 273; crystalline organization of chains in, 233–36; dimensions of, 236–37; in discontinuous-fiber composite, 236, 254, 257; estimating matrix properties and, 353; material performance and, 356; matrix protein binding to, 246–47, 302; organization in cuticle composite, 239–43; perpendicular to beetle gula surface, 262; with resilin in locust ligament, 308, 309, 310; stiffness and strength of, 238–39, 273; structure of, 236–37; transverse in female locust, 268, 270, 271. *See also* insect cuticle

choanoflagellates, 107

chondroitin sulfate: of aggrecan, 322, 326; of decorin, 121–22; structure of, 121–22; of versican, 327

ciliary ligament, 292. *See also* zonular filaments of vertebrate eye

cis-isoprene. *See* natural rubber

Clark, J. M., 340

Clark, R. B., 337

cnidarians, 107

coelenterates, 229, 231, 273, 277. *See also* corals; sea anemones

collagen: amino acid sequence of, 108–10; evolution of, 107, 231, 357; insects that have lost gene for, 271–72; as main tensile fiber in animals, 107; as primary fiber-forming polymer in animals, 6–8; resilience of, 22; of tendons (*see* tendon collagen); types of, 107. *See also* tropocollagen

collagen crimp pattern, 124–25; absent in echinoderms, 146, 151; age of tendon and, 126–27; functional requirements of particular tendon and, 144

collagen fibers: absent in mesoglea, 279; of artery, 340, 342, 343, 344, 346, 350; of bone, 191, 195, 197, 198, 201–3, 205; calculated stiffness of, 140–42; of cartilage, 319–22, 324, 325–26; cross-helical lattice in skin or cuticle, 335–37; in echinoderms (*see* sea-urchin spine ligament); fascicles of, 124, 127, 141, 143–44; proteoglycans in, 121–24 (*see also* decorin);

of skin, 13, 123–24, 327, 329–32, 335–37; structural organization in tendons and ligaments, 118–25

collagen fibrils: age of animal and, 116, 118, 127; assembly of, 109, 110, 112–18; in bone, 192–95, 206–13; cross-links within and between, 115–18, 127; in echinoderms, 140, 147–49, 151; evolution of, 277–80; length of, in tendons, 119–20, 127; in mesoglea, 277–80, 283–87, 292, 331; organization into fibers, 118–25; stiffness of, 140, 201, 203; strain of, in tendons, 139–40; tapered, 148

collagen microfibrils, 279–80, 290

collagen molecule. See tropocollagen

compact bone, 195; tensile mechanical properties of, 206

compliance, time-dependent, 28; of mesoglea, 282

composite materials. See fiber-reinforced composites

compression modulus of elasticity, 12, 18–19; of insect cuticle, 262

compression strength, 19

continuous dynamic analysis, of spider silk, 167

Cook, J., 95–97

Cook-Gordon effect, 213

corals: mineralized skeletons of, 229, 231, 273; polyps of, 277

cork, 18

cost-benefit ratios, 8–9

Cowey, J. B., 337

crack growth. See fracture

crack-stopping mechanisms, 95–98

Craig, A. S., 119

creep behavior of tendons, in static-fatigue testing, 136–37, 142–43

creep tests, 24–27, 28–30; of assassin bug cuticle, 266, 267; of cartilage, 29–30, 322; of mesoglea, 281–82, 286, 287

Cribb, B. W., 263

critical flaw length, 56–58; for insect cuticle, 258–59; for selected materials, 258

Crofton, H. D., 337

cross-helical collagen-fiber lattice, 335–37

cross-laminate composite, 88; bone compared to, 204

cross-links: of collagen fibrils, 115–18, 127; of insect cuticle proteins, 246, 247–49, 251, 256, 260, 261, 265, 271, 308; in mesoglea matrix, 285; in rubberlike proteins, 298, 299, 300, 301–3, 304–8; in sea-urchin ligament, 149, 150; spidroins' rubberlike elasticity and, 160

crystalline polymers, 59–60, 103–6. See also β-sheet crystals; cellulose; chitin

Currey, J. D., 191, 192, 195, 204–5, 214, 220, 225, 226, 228, 229, 272

cuticle: arthropod, 231–32 (see also insect cuticle); earthworm, 336–37; nematode, 327, 337

cylindrical (tubuliform) gland, 156; silk produced by, 160–61, 162

Cyrtophora citricola, 173

damping coefficient (loss tangent), 35, 37; glass transition and, 80, 81

Daniel, I. M., 89

Danielson, H., 123

Davis, E. C., 313

decorin, 121, 122–24, 142, 143, 356; in bone tissues, 193, 211, 212; mesogleal collagen compared to, 287–88; pliant composites and, 314

DeMont, M. E., 291

Denny, M., 18, 166, 180, 311, 316

dentine in narwhal tusk, 204–5

dermatan sulfate, 121–22, 124

design, definitions of, 3–4

design quality: basic measures of, 8–9; energy transformations and, 22; time-dependent elasticity and, 24; true stress-strain analysis and, 17

desmosine, 300, 302

Diadema, 146. See also sea-urchin spine ligament (catch ligament)

Diamant, J., 126

Diani, J., 271

diffusion, 74–75

diffusion coefficient, 74–75

Dirks, J., 253, 257, 259

discontinuous-fiber composites, 89–93; bone as, 204, 209–11; cellulose as, 238; chitin in, 236, 254; collagen as, 119, 127, 140–44, 353; locust cuticle as, 254–57; mesoglea as, 285, 287; in sea-urchin catch ligaments, 149–50; spider silk compared to, 357; synthetic, 93

dityrosine cross-links, 302, 303

Dobrin, P., 340, 342, 343

Dolomedes, 158

dolphin, subdermal sheath of, 335

Dourte, L. M., 124

Drosophila, 271, 302, 306

D-spacing in collagen fibril, 113, 115, 116

dynamic fatigue testing, of wallaby tail tendons, 136, 138

dynamic mechanical testing, 33–39; continuous, of MaA silk, 167; forced-oscillation methods in, 33, 35–39; resonant methods in, 33–35

earthworm, 336–37

ecdysis (molting), 232–33, 236, 242–43, 249–52, 271–72

echinoderms, 140, 144–51; mineralized, 229, 231, 273

effectiveness ratios, 9; specific properties used as, 15

efficiency: defined, 8–9; of energy transformations, 22–24

Eisoldt, L., 176

elastic-energy storage capacity: of apodeme, 254; of bone, 223–25; of spider silk, 355–56; of tendon collagen, 132–33, 355–56

elasticity: ideal, 11–12, 43–45; time-dependent, 24–25. See also modulus of elasticity (stiffness); network elasticity; rubberlike elasticity

elastin: amino acid composition of, 298–301; amino acid sequence of, 301–2; in arteries, 37, 82, 297, 312–13, 340–44, 346, 350; cartilage structures containing, 319; constant-strain-rate tests of, 38, 82–83;

elastin (*continued*)
    cross-linking of, 302, 304–5, 307–8; cuticle matrix
        proteins compared to, 254, 256; dynamic mechani-
        cal properties of, 308–10, 312–13; fatigue behavior
        of, 312–13; fibrillin microfibrils in assembly of,
        297; glass transition of, 82–83; hydrophobicity of,
        244–46, 300, 302, 305; network structure of, 304–6;
        in skin, 8, 13, 327, 329–31; viscoelastic behavior of,
        35–38, 81–83; in whale blubber, 332, 333
endocuticle, 232, 233, 242–43
endosteum, 200
energy budgets, 4; design quality and, 8
energy dissipation: by spider silk, 23–24, 166–67,
    356; by viscous processes, 21, 22–24, 98. *See also*
    resilience
energy storage capacity. *See* elastic-energy storage
    capacity
energy to break, 13–15; bond energy and, 46–49; for
    true stress-strain data, 17. *See also* toughness
engineering strain, 16–18
engineering stress, 16–18
entanglement of random-coil molecules, 76–77, 81
entropy and elasticity, 61–63, 65–66; of fibrillin, 290,
    293; of rubber network, 67, 70–73
environmental scanning electron microscope (ESEM),
    217
epicuticle, 232, 242; of insect wing membranes, 259
Eppel, S. J., 140
equivalent random links, 66, 69–70, 75
Erickson, G. M., 228–29
Eugarus, 158
exocuticle, 232, 233, 242–43
extensibility, defined, 13
extension ratio, 16

fascicles, of collagen fibers, 124, 127, 141, 143–44
fatigue behavior: of bone, 217–19, 225; of elastin,
    312–13; of Haversian bone, 201; of locust wing and
    leg cuticles, 257; of tendons, 136–38, 142–43, 219;
    testing of, 98–99
Feng, G., 201
fiberglass. *See* glass fibers
fiber-reinforced composites, 5, 84–99; direction of
    loading and, 84–86; in engineered structures, 89,
    93; failure at fiber-matrix interface, 99, 354; fiber
    angle and, 86–89; fiber fraction and, 84–86; short-
    fiber, 89–91, 271–73; spider silk as single-molecule
    version of, 354; strength of, 93–95; summary
    comparison of, 353–54; with 3-D fiber orientations,
    89; toughness of, 84, 95–98; in tubular and shell-like
    structures, 89. *See also* bone; collagen; discontin-
    uous-fiber composites; insect cuticle; laminated
    composite structure; reinforcement efficiency
fibrillin microfibrils, 151, 290–97; amino acid compo-
    sition of, 299, 300; in elastin-based tissues, 297; in
    invertebrate arteries, 151, 297; in jellyfish *Polyorchis*,
    151, 291–92; in sea-urchin catch ligaments, 151,
    296; in vertebrate eye, 151, 292–96, 297
fibroblasts: collagen fibril length and, 119; collagen
    synthesis in, 108, 119

fibroin, defined, 158
fibrolamellar bone: architecture of, 197–99; me-
    chanical properties of, 214, 215, 216, 218–20, 222;
    orientation of materials in, 205–6; remodeling of,
    200; stress-strain curves for, 203
filled rubbers, stress softening in, 271
flagelliform glands, 152
FL (flagelliform) silk: amino acid sequence motifs
    in, 160, 162; compared to other materials, 153;
    functions of, 152; mechanical properties of, 152–54;
    protein size of, 164
flying machines, specific modulus of, 272–73, 357
forced-oscillation methods, 33, 35–39
fracture: of composites, 95–99; release of strain energy
    and, 50–56, 58
fracture-mechanics properties, materials compared, 258
fracture toughness, 56–58; of antler, 226–27; of bone,
    213–14, 216–17; of fiber-reinforced composites, 98;
    of insect cuticle, 257, 258; of rat skin, 331–32; of
    rhinoceros and wild-ass skin, 331; of sea anemone
    mesoglea, 331. *See also* toughness
Fraenkel, G., 247
Franchi, M., 125
Fratzl, P., 107, 115, 139, 140, 193, 206, 212
free energy, 61–63
French, C. J., 35–36
Fudge, D. S., 186

Garrido, M. A., 180
Gatesy, J., 158
Gaussian behavior, of rubber networks, 63–65, 68–69,
    293
gelatin, 288
Glab, J., 294
Glagov, S., 340
glass, stress-strain curve for, 12–13, 14
glass fibers: composites made with, 84–85, 93, 95;
    critical flaw length, 57; tensile strength of, 48, 54, 58
glass transition, 32–33; of elastin, 36, 38, 82–83; of
    natural rubber, 61; of PMMA, 61, 79–81
glass-transition temperature, 33, 61; measurement
    of, 79
glycine: in collagen, 109–12; in insect cuticle matrix,
    244, 245, 246, 247, 249, 256; in rubberlike proteins,
    298, 300, 301–3, 311–12; in silks, 156, 157, 158–60,
    161
glycosaminoglycans (GAGs), 121–24; of aggrecan,
    322, 324, 326; in pliant composites, 314
Gordon, J. D., 95–97
Gordon, J. E., 132
Gosline, J. M., 35–36, 168, 172, 179, 181, 183, 186,
    245, 256, 259, 283, 285, 288, 291, 297, 305, 312, 316,
    346, 348, 350
Gotte, L., 246
Griffiths, A. A., 50, 53, 54, 55, 56, 57
Grimstone, A. V., 279
Gupta, H. S., 191, 206–7, 211, 212, 213, 229–30

hagfish, tensile fibers produced by, 186–87
Hagn, F., 176

hair: keratin of, 13, 273; stress-strain curve for, 13; yield behavior of, 13, 58

Hansen, U., 220, 222

Harrington, M. J., 186

Harris, J. E., 337

Haversian (secondary lamellar) bone: cement layer in, 200, 201, 213, 217; formation of, 200; function of, 222–23; lacunae in, 200; mechanical properties of, 214–25; orientation of collagen and mineral platelets in, 205

Hayashi, C. Y., 165, 167

Hayduk, U., 277

helicoidal insect cuticle, 241, 242, 250, 251, 254, 257, 258, 265, 266

Helmholtz free energy, 61–63

Hepburn, H. R., 249–52, 253, 258, 264

Hermann, H., 187

Hidaka, M., 146

hierarchical complexity, 5; composite failure modes and, 99; self-assembly and, 106

Hillerton, J. E., 237, 256, 260, 263

HLKNL (hydroxylysino-ketonorleucine), 116, 118

ΔHLNL (dehydro-hydroxylysino-norleucine), 116, 118

Hoo, R. P., 206–7, 209, 211

hoof, keratin in, 187, 273; stiffness of, 258; toughness of, 57, 58, 217, 331, 332

Hooke's Law, 10, 44

horn, 187, 273

horse: skin of wild ass similar to, 330–31; structural materials of, 6–8; tendons of, 6, 119, 124, 128, 130, 144

Huang, C., 325

hyaluronic acid: aggrecan bound to, 322; versican bound to, 327

hydromedusa Polyorchis, 151, 291–92

hydrostatic skeleton, 337

hydroxyallysine, 115, 116

hydroxyapatite, 191–93, 203, 205

hydroxylysine, 109, 115–16

hydroxyproline, 109–10, 112

Hynes, R. O., 107

hypodermal cells under insect cuticle, 242

hysteresis, 22–23

ideal elasticity, 11–12; bond energy and, 43–45

insect cuticle, 231–73; abdominal, of assassin bug, 266–67; of caterpillars, 265; evolution of, 231–32; functional consequences of short-fiber composites in, 271–73; hardness testing of, 260–63; hydration of, 233, 254, 256, 257–59, 260, 261, 262–63, 265–66, 268, 271; intersegmental, of locust, 264, 267–71; of locust prealar arm ligament, 239–40, 308, 309, 310; of mandibles with or without minerals, 263; matrix of (see insect cuticle protein matrix); mechanical properties of rigid cuticle, 249–63; scaling of modulus from juvenile to adult, 259–60; soft type of, 244, 257, 263–71; stiffened by dehydration, 246, 249, 256, 257, 265; structure of, in locust's hind leg, 242–43; summary comparison of, 353–57; tendon composed of (apodeme), 238–39, 240, 253–56, 258;

three general classes of, 244. See also chitin; chitin microfibrils; ecdysis (molting)

insect cuticle protein matrix, 243–49; amino acid composition of, 244–47, 256; of exocuticle vs. endocuticle, 251; of female locust intersegmental membrane, 271; with incorporated metal ions, 244, 263; protein modifications in, 246–49, 256; shear modulus and tensile modulus of, 254; water content and, 256, 261, 266. See also resilin

insects: competition between birds and, 357; evolution of, 357

insect tendons: apodeme (cuticular tendon), 238–39, 240, 253–56, 258; chitin-free dragonfly tendon, 306–7, 308, 309–10

intermediate filaments, 186–87; of keratin, 187, 273

Ishai, O., 89

cis-isoprene. See natural rubber

Jeanloz, R. W., 288

Jeffery, A. K., 319

jellyfish Polyorchis, 151, 291–92

Jensen, M., 252, 253

Jensen, S. A., 291, 296

J-integral, 56; antler and, 226; for bone, 217

Joffe, I., 249–52, 253, 258

jumping spiders, 186

Kalamajski, S., 193

Kastelic, J., 125

Katz, S., 256, 259

Katzman, R. L., 288

Ker, R. F., 38, 133, 135, 136, 138, 142, 143

keratan sulfate, 322

keratin: of epidermis, 327; fracture-mechanics properties of, 258; intermediate filaments in, 187, 273; materials formed from, 273; of whale epidermis, 332; yield behavior of hair and, 13. See also hoof, keratin in

Kevlar: MaA spider silk compared to, 153, 170, 183; structure of, 106

Kevlar-epoxy composite, 86–88, 90, 92–93

Kiani, C., 322

Kim, J. H., 218

Klocke, D., 262

Knight, D. P., 174–75, 178

Koehl, M., 280, 282, 283, 288–89

Koester, K. J., 216–17

Koob, T., 147–48

Laasanen, M. S., 323

lacunae, osteocyte, 200–201

lamellar bone, 195–201. See also Haversian (secondary lamellar) bone; primary lamellar bone

laminar bone. See fibrolamellar bone

laminated composite structure, 86–89; bone mechanical properties and, 204–5; weakness at interfaces in, 99. See also quasi-isotropic composites

Landis, W. J., 193, 210

Langer's lines, 329

Laplace's law, 338–39

Lazan, B. J., 38

Liao, J., 123

ligaments: collagen of, 6; elastin in, 36, 305, 308; function of, 6, 144; hinge ligament of scallop, 307, 310–12; locust resilin in, 239–40, 308, 309, 310; mechanical properties of, 124, 125, 129; sea-urchin spine ligament (catch ligament), 144–51, 296; similarity to tendons, 144

Lillie, M. A., 244, 312, 344

Lin, H-T., 265

linear elasticity, 11–12; bond energy and, 43–45

linear polymers, 59–61. *See also* amorphous polymers; crystalline polymers

Liu, Y., 173–74

lobster, microfibrils in arteries of, 297

locomotor energetics, 4

logarithmic decrement, 34

loss modulus, 35, 36; of pig gastric mucus, 318; of rubberlike proteins, 308, 309

loss tangent (damping coefficient), 35, 37; glass transition and, 80, 81

Lu, X. L., 326

Lujan, T. J., 124

Luke, B. M., 241

lysyl oxidase, 116

MaA (major ampullate) silk: amino acid sequence motifs in, 158–60, 162, 164; compared to other materials, 153, 354–57; formation in gland/spinneret complex, 155, 174–80; functions of, 152; 3_1 helices in, 171, 173; mechanical properties of, 152–54, 165–70, 172–74; network structure of, 170–74; N-terminal and C-terminal domains of, 176–78; supercontraction of, 172–73; tension applied by spider and, 178–80; viscoelastic behavior of, 166–70. *See also* spider draglines; spider dragline silks

major ampullate glands, 152, 155–56; structure of spinneret and, 174, 179

Manduca sexta larval cuticle, 265

Marfan syndrome, 291

marrow cavity, 197

master curve, 32–33; for elastin, 36

matrix materials of composites: collagen molecules in bone as, 208, 210, 354; crack growth and, 95–98; interfacial weakness in laminates and, 99; of mesoglea, 287–89, 292. *See also* insect cuticle protein matrix; proteoglycans

Matson, A., 131

Maxwell model, 27–28, 30–32

McConnell, C. J., 296–97

McElhaney, J., 220, 222

Megill, W. M., 292

MEMS (microelectromechanical systems), and stiffness of collagen fibrils, 140

mesoglea: collagen fibrils in, 277–80, 283–87, 292, 331; composite model of, 284–87; fibrillin microfibrils in, 291–92; fracture toughness of, 331; matrix chemistry in, 287–89; matrix in jellyfish *Polyorchis*, 292; viscoelastic properties of, 280, 281–83, 287, 289

Metridium, 280, 282–84, 288–89

Metridium dianthus, 288

Metridium senile, 279, 331

Meyers, M. A., 273

microfibrillar elastomers. *See* fibrillin microfibrils

microfibrils. *See* cellulose microfibrils; chitin microfibrils; collagen microfibrils; fibrillin microfibrils

Miller, L., 311

minor ampullate gland, 156; silk produced by, 160, 162, 164

mitral valve chordae tendineae, 123

modulus of elasticity (stiffness): complex, 35–36; defined, 11–12; fracture toughness and, 58; of selected materials, 15–16; specific, 15, 272–73, 357; spring constant and, 10; stress-strain curves and, 12–13; tangential, 13, 18; time-dependent, 24–25, 27, 28, 30, 32–33; units of, 12. *See also* loss modulus; shear modulus of elasticity; storage modulus

molecular friction coefficient, 74–75

mollusks: abductin in, 298, 307, 308, 310–12 (*see also* abductin); byssal threads of, 186; calcium carbonate in shells of, 229–30, 273; evolution of, 231; fibrillin microfibrils in, 296–97; mucus of, 316–19; nacre of, 229–30

molting. *See* ecdysis (molting)

Montalbano, T., 201

mother-of-pearl, 229–30

Mow, V. C., 326

mucin, 314–15, 318–19

mucus, 314–19

Mullins effect, 271

mutable connective tissues, 144–51

N-acetylgalactosamine, 315

N-acetylglucosamine: of aggrecan, 322; of chitin, 233, 235, 237; of mucin, 315

nacre, 229–30

nanoindentation studies, of insect cuticle, 261–63

narwhal dentine, 204–5

Nation, J. L., 233

natural rubber: Poisson's ratio for, 18; as random-coil molecule, 61; stress-strain curves, engineering vs. true, 16–17; structure of, 61. *See also* rubberlike elasticity

nematodes, 232, 327, 336

Nemetschek, T., 139

Nephila, 158

Nephila clavipes, 160, 162, 170, 172

Nephila edulis, 167–68, 178

network elasticity, 66–70; of protein rubbers, 304–8

Neville, A. C., 239, 241, 242

Newtonian viscosity, 21

Nikolov, S., 238

non-Gaussian behavior: microfibrils and, 293, 295, 297; of rubber networks, 69–70, 293, 306, 307

nonlinear elasticity, 13

Nordwig, A., 277

nylon, 103–5, 106

octopus, fibrillin microfibrils in, 297

Ortlepp, C., 179, 181, 183–84, 186

Orton, L., 332–33
osteoblasts, 200
osteoclasts, 201
osteocytes, 200–201
osteons, 198–200; of fibrolamellar bone, 198, 199, 200, 205–6; of Haversian bone, 199, 200, 201, 213, 217

Pabst, A. D., 335
Pachynoda arthrodial membranes, 264–65
Pantin, C., 280
parallel-composite structure, 85–86; bone model and, 207; in insect cuticle, 241, 242, 251, 253, 257, 258
patellar tendon, 124, 140
PBS (phosphate-buffered saline): cartilage in, 326; collagen fibrils in, 206, 209
pennate muscle-fiber design, 134–35
periosteum, 200; antler velvet as, 226
periotic bone, 227, 228
Petrov, M., 235
Plectreurys, 158
Plexiglas. *See* polymethyl methacrylate (PMMA)
pliant materials: evolutionary origins of, 107, 277–89, 357; functions in animals, 6, 8; Poisson's ratios for, 20, stress-strain curves for, 13; true stress and strain for, 17. *See also* arteries; cartilage; mucus; skin
Poisson's ratios, 18, 20; for female locust's abdominal cuticle, 270
polyethylene, 60, 66, 103, 106
polymer glass. *See* glass transition
polymers, 4–5, 59–61. *See also* crystalline polymers; random coil molecules
polymethyl methacrylate (PMMA), 60–61; visco-elasticity in, 61, 75–81
Polyorchis, 151, 291–92
polyproline-II helix, 110–12
polysaccharides, 59. *See also* cellulose; chitin; glycos-aminoglycans (GAGs)
pore channels in insect cuticle, 242
preprocollagen, 108–9
primary lamellar bone, 197–98, 199–200; lacunae in, 200–201; orientation of collagen and mineral platelets in, 205; stress-strain curves for, 203; weak interfaces in, 213
primary osteons, 198, 199, 200, 205–6
procollagen, 109; hydroxylation of lysines in, 115
proline: in collagen, 109–10; hydrophobic structure of, 244; in rubberlike proteins, 298, 300, 301, 302; in spidroins, 160, 170, 173
proteins, 59; equivalent random links for, 66. *See also* collagen; insect cuticle protein matrix; proteogly-cans; rubberlike proteins; spider silks
proteoglycans: in bone, 211, 212; in cartilage, 8, 142, 319, 320, 321–22, 324–26; in collagen, 121–24, 127 (*see also* cartilage; decorin); in jellyfish mesogleal matrix, 292; in mammalian skin, 327; in mucus, 314–15; in pliant composites as matrix, 314; in sea-urchin ligament, 150
Pryor, M., 247
pseudo-orthogonal structure, 242

Purslow, P. P., 331–32
pyriform glands, 155; silk produced by, 161, 163, 164

quasi-isotropic composites, 88; bone compared to, 204. *See also* helicoidal insect cuticle
Quillin, K., 337
quinone tanning, 247–49, 261

random-coil molecules: entanglement of, 76–77, 81; in female locust intersegmental membrane, 271; linear force-extension behavior of, 65–66; mechanism of elasticity of, 61–66; networks of, 66–70, 290; random-walk model for, 63–66; rubberlike protein structures and, 301, 303, 306; thermodynamics of, 61–63, 65. *See also* rubberlike proteins
random-fiber composite, 329
random walk, 63–66
Rauscher, S., 298, 300
Reber-Muller, S., 291
Rebers, J. E., 247
Rebers and Riddiford consensus blocks, 246–47, 302
reinforcement efficiency: of collagen fibers, 140; of discontinuous-fiber composites, 91–92; of insect cuticle, 254, 257; of laminate structures, 88; of nanoscale bone models, 209–10; of random-fiber composite, 329; of rubberlike proteins, 302
Renard, J., 88, 89
resilience, 22–24; calculated from loss tangent, 37–38; in resonant system, 34–35; of rubberlike proteins, 308–12; of spider silk, 23–24, 166–67; of tendons, 22, 132
resilin: amino acid composition of, 298–301; amino acid sequence and cross-linking of, 301–2; in chitin-free dragonfly tendon, 306–7, 308, 309–10; dynamic mechanical properties of, 308–10; in locust prealar arm ligament, 239–40, 308–9, 310; network structure and mechanical properties of, 306–8
resonant oscillation methods, 33–35
reversible deformation, 10–12
Reynolds, S., 260, 266
Rhodnius prolixus (assassin bug), 266–67
Ribeiro, A. R., 149
Ridge, M., 329
Right, V., 329
rigid materials, 5–6, 7–8, 273; origins of elasticity and strength in, 43–49. *See also* bone; insect cuticle
Robinson, P. S., 124
Rovick, A., 340, 342, 343
R-R consensus elements, 246–47, 302
rubberlike elasticity: cross-linked networks and, 66–70; molecular basis of, 61–66; temperature and, 70–73. *See also* natural rubber
rubberlike proteins: amino acid composition of, 298–301; amino acid sequences of, 301–3; cross-links in, 298, 299, 300, 301–3, 304–8; dynamic mechanical properties and fatigue of, 308–13; network structure and mechanical properties of, 304–8. *See also* abductin; elastin; fibrillin microfibrils; random-coil molecules; resilin

rubber plateau, 77, 81
Ruben, J. A., 191
Rudall, K. M., 247

safety factor: of jumping spider dragline, 186; of orb-weaving spider dragline, 181–84, 185, 186; of tendon within locomotor limb, 133–34, 135–36; of wallaby plantaris tendon, 219; of wallaby tail tendon, 136, 138, 143, 219
Salticus scenicus, 186
Savage, K. N., 172
scallop hinge ligament, 307, 310–12
Schmitz, H., 262
sclerotization, of cuticle matrix proteins, 248–49, 250, 251, 256, 257
β sclerotization, 248
Screen, H. R. C., 143, 144
sea anemones: anatomy and function of, 280–81; collagen-based composite of, 277; evolution of, 357; intertidal, 288–89. *See also* mesoglea
sea cucumbers: control of body-wall stiffness in, 148; stiffness of collagen fibrils from, 140
sea-urchin spine ligament (catch ligament), 144–51, 296
secondary lamellar bone. *See* Haversian (secondary lamellar) bone
secondary osteons, 200, 201, 213, 217
self-assembly, 106–7; of chitin microfibrils, 241; of collagen fibrils, 109
series-composite structure, 85–86; bone model and, 207, 209; in insect cuticle, 262
Setton, L. A., 322, 326
Shadwick, R. E., 128, 297, 330–31, 334, 337, 346, 348, 350
Shapiro, S. D., 313
shark skin, 335–36
shear deformation, 19–20
shear modulus of elasticity: of collagen fibers, 141–42, 149–50; of cuticle matrix, 254, 256, 257, 273; defined, 20; of mesoglea matrix, 285–86; of overlapping tropocollagen molecules, 210–11; of rubber network, 68–69, 304–7; in sea-urchin catch ligament, 150; of slug mucus, 318
shear strain, 20
shear strain rate, 21; stiffness and, 24
shear stress, 19–20, 21
shear stress, fiber-matrix, 90, 91; in bone, 209–10
Sherratt, M. J., 290, 294
sialic acid, 315
silks: β-sheet crystals in, 156–58. *See also* spider silks
silkworm *Bombyx mori*: cuticle properties of, 265; silk of, 156, 157–58
Silver, F. H., 193
siphonoglyph, 280–81, 287
Skedros, J., 201
skin: of cat, 328–31; collagen fibers in, 13, 123–24, 327, 329–32, 335–37; cross-helical collagen-fiber lattice in, 335–37; elastin in, 8, 13, 327, 329–31; of horse, 8; invertebrate, 327, 336–37; keratin in, 187, 273, 327, 332; mechanical anisotropy of, 329;

of rhinoceros, 330–31; stress-strain curve for, 13; whale blubber, 332–35; whale subdermal sheath, 335; of wild ass, 330–31
slug, pedal mucus of, 316–18
Smith, D. S., 151
soft elasticity: polymer structures and, 59–61. *See also* rubberlike elasticity
soft tissues. *See* pliant materials
solvent effects on viscoelastic properties, 24, 37, 81–82
specific modulus, 15; of flying machines, 272–73, 357
specific properties, 15–16
specific-volume method, 79
Spectra fiber, 60, 106, 153
spider draglines: functional design of, 180–87; variation with body mass, 183–84
spider dragline silks: attachment discs for, 155; comparison to other materials, 354–55; evolution of amino acid sequence motifs in, 158–60; formation and properties of, 82–83; as strongest of natural fibers, 106. *See also* MaA (major ampullate) silk
spider silks, 152–87; amino acid sequence evolution, 158–60; amino acid sequence general patterns, 156–58; amino acid sequence motifs in all orb-weaver spidroins, 160–64; chitin stiffness compared to, 238; compared to other natural and synthetic materials, 153; draw-processing in industry compared to, 356; forcibly drawn vs. naturally spun, 165–66, 179–80; formation in gland/spinneret complex, 82–83, 174–80; impact testing of, 168–70; mechanical properties of, 164–70; resilience of, 23–24, 166–67; size of proteins constituting, 161, 164; structures for producing, 155–56, 174–76; summary of structure and properties, 354–56; toughest known, from *Caerostris*, 166; true stress-strain analysis of, 17–18, 152–53; types of, 152, 155–56, 160–64; viscoelastic behavior of, 166–70. *See also* FL (flagelliform) silk; MaA (major ampullate) silk; spider draglines
spidroin, defined, 158
spongin, 107
spring constant, 10
spring tendons, 132–33, 134–36, 138, 254
standard linear solid (SLS) model, 28; of mesoglea, 282
static fatigue testing, 136–37, 138, 142
Stephens, R., 271
stiffness. *See* modulus of elasticity (stiffness)
stiparin, 148
storage modulus, 35, 36; of insect cuticle, 260; of pig gastric mucus, 318; polymer viscoelasticity and, 77, 80, 82; of resilin and elastin, 308–9; spider silk viscoelasticity and, 167; water content of elastin and, 245
strain: defined, 10; shear strain, 20; true strain, 16–18; units of, 12
strain-energy release, 50–51; in composites, 97–98
strain-energy release rate, 56; for bone, 217; for insect cuticle, 258; for skin, 331
strain-energy storage capacity: of apodeme, 254. *See also* elastic-energy storage capacity

strain rate: bone properties as function of, 220–22; shear strain rate, 21, 24; stiffness and, 24

strength: of composite materials, 93–99; in compression, 19; defined, 13; density and, 15; of selected materials, 15–16, 355; specific, 15; theoretical, 46–49. *See also* toughness

stress: defined, 10; true stress, 16–18; units of, 12

stress concentrations at flaws, 50, 54–55; in bone, 200–201, 213; in composites, 95–98

stress fractures, 225

stress intensity factor, 57–58; for bone, 217; for insect cuticle, 258

stress relaxation, 27–28, 30–32; of assassin bug cuticle, 266; of collagen fibril strain, 139; of mesoglea, 283, 285–86; of tendon fascicle strain, 143

stress softening, 271

stress-strain curves, 11–13; for compression, 18–19; for shear deformation, 20

stress trajectories, 50

sulfilimine cross-link, 303

surface energy, 47, 55–56

Svensson, R. B., 140

Takahashi, K., 146

tangential modulus of elasticity, 13; true stress-strain analysis and, 18

Taylor, C., 318

Taylor, D., 253, 257

telopeptides, 109; cross-links formed at, 115–16; stretching of, 139

temperature: fatigue lifetime of tendons and, 142; hydroxyproline of ectotherms vs. endotherms and, 112; random-coil molecules and, 65; rubber elasticity and, 70–73, 77; spider silk properties and, 167–68; static-fatigue behavior and, 137, 138. *See also* time-temperature superposition principle

tendon collagen: comparison to other materials, 354–56; crimp pattern in, 124–25, 126–27, 144; discontinuous fiber composite structure of, 119, 127, 140–44, 353; elastic-energy storage capacity of, 132–33, 355–56; fibril length in, 119–20, 127; proteoglycans in, 121, 123–24 (*see also* decorin); strain of fibrils in, 139–40; stress-strain plot for, 106; structural organization of fibers, 118–25. *See also* collagen

tendons: accuracy issues in testing of, 129–31; age dependence of rat tail tendon, 116, 125–28, 139; avian, calcified, 131, 192–93, 210, 228; collagen in (*see* tendon collagen); energy storage in locomotion and, 132–33; exercise training and, 131–32; fatigue behavior of, 136–38, 142–43, 219; flexor vs. extensor tendons, 128–29, 131–32; function of, 6; horse tendons, 6, 119, 124, 128, 130, 144; mechanical properties of, 125–38; nanomechanics of, 138–44; nonspring tendons, 135–36; pig limb tendons, 128–29, 131, 132, 134, 136; rat limb tendons, 119–20, 128, 132; rat tail tendon, 116, 119, 124, 125–28, 129, 139, 143, 144; resilience of, 22, 132; spring tendons, 132–33, 134–36, 138; stiffness value used for, 140,

141; structural design of, 133–36; summary of structure, 353; turkey limb tendons, 131, 193, 210; typical values for mechanical properties, 132–33; wallaby plantaris tendon, 219; wallaby tail tendon, 136–38, 143, 219. *See also* insect tendons

tensile materials, 6, 186–87. *See also* cellulose; chitin; collagen; spider silks; tendons

tensile modulus of elasticity, 12. *See also* modulus of elasticity (stiffness)

tensile strength, 13–15. *See also* strength

tensilin, 148

thermoelastic experiments, 71–73; on hydrated spider silks, 172

time-dependent compliance, 28; of mesoglea, 282

time-dependent modulus of elasticity, 24–25, 27, 28, 30, 32–33

time-temperature superposition principle, 31–33; fatigue lifetime of tendons and, 142–43; spider silk properties and, 167–68

toe zone of stress-strain curve: for rhinoceros skin, 330; for tendon, 125, 126, 127, 128, 132, 144, 201

toughness: defined as energy to break, 13–15; of nacre, 229–30; of spider silk, 166, 170, 355. *See also* fracture toughness

trabeculae of bone, 195

Treloar, L. R. G., 70

trityrosine cross-links, 302

tropocollagen, 107–12; in bone, 207, 208; cross-links within and between α chains of, 115–18; decorin binding to, 121; D-spacing in collagen fibril and, 115, 116, 139; in mesoglea, 277–80; sliding relative to one another, 139; stiffness of, 140; strain of, in tendons, 139; stretched within fibril crystal lattice, 139. *See also* collagen

Trotter, J., 119–20, 147–48

trouser-tear test, 331

true strain, 16–18

true stress, 16–18

tubuliform (cylindrical) gland, 156; silk produced by, 160–61, 162

Turner, B., 315

Tychsen, P., 268

Vanacore, R., 303

Veronda, D., 327–28

versican, 327, 330

Vesely, I., 123

Vincent, J., 186, 233, 237, 238, 249, 256, 260, 263, 268, 269–70, 271, 273

viscid silk. *See* FL (flagelliform) silk

viscoelastic behavior: of amorphous polymers, 60; of bone, 212, 217, 220; of cartilage, 29–30, 142; creep testing of, 24–27, 28–30; dynamic mechanical testing of, 33–39; of elastin, 35–38, 81–83; energy dissipated as heat in, 21; hysteresis and resilience of, 22–24; linear, 38–39; of mesoglea, 280, 281–83, 287, 289; models of, 25–33; molecular origins of, 74–83; of pig gastric mucus, 318; of PMMA (Plexiglas), 61, 75–81, 150; proteoglycans contributing to, 121; of

viscoelastic behavior (*continued*)
sea-urchin spine ligament, 147, 150; of soft insect cuticle, 264, 265, 266, 270; solvent effects on, 24, 37, 81–82; of spider silks, 166–70; stress relaxation of, 27–28, 30–32; temperature dependence of, 24; tendon fatigue and, 142–43; time-dependent elasticity and, 24–25; time-temperature superposition principle and, 31–33, 142–43, 167–68; of versican gel in skin, 327
viscoelastic time constant, 26, 27, 28
viscosity: activation energy related to, 75; defined, 21; polymer molecular weight and, 76–79; of slug mucus, 316–18; temperature dependence of, 24, 31–32
Voigt model, 25–27, 29–30, 31; mesoglea creep and, 282
Vollrath, F., 174–75, 178
volume change in deformation, 16, 18, 20

Wainwright, S. A., 335–36
Waite, H., 186
Wang, R., 191, 213, 229–30
Weis-Fogh, T., 252, 253, 306–7
Wess, T., 294
Westman, R., 327–28

whales: blubber of, 332–35; diving physiology of, 346–50; fin-whale aorta and aortic arch, 345–50; subdermal sheath of, 335; tympanic bulla bone of, 225, 227–28
whiskers, 48. *See also* glass fibers
Willis, J. H., 247
Wilson, R., 179
wing membranes of insects, 231, 259
WLF (Williams, Landel, and Ferry) equation, 80
Wofsy, C., 119–20
Woo, S. L.-Y., 131
woven bone: in fibrolamellar bone, 197, 199, 206; lacunae in, 200, 201; properties of, 195; transition to primary lamellar bone, 197
Wright, D. M., 151, 292–93, 296
Wu, Y. J., 327

Yamada, A., 148
Yang, L., 208–9
Yang, Y., 167
yield behavior, 13, 58; of bone, 203, 212–13; of insect cuticle, 260

Zioupos, P., 218
zonular filaments of vertebrate eye, 151, 292–96, 297